Managing Project Risks

Managing Project Risks

Peter J Edwards
RMIT University
Australia

Paulo Vaz Serra
University of Melbourne
Australia

Michael Edwards

Registered Offices
John Wiley & Sons, Inc., 111 River Street, Hoboken, NJ 07030, USA
John Wiley & Sons Ltd, The Atrium, Southern Gate, Chichester, West Sussex, PO19 8SQ, UK

Editorial Office
9600 Garsington Road, Oxford, OX4 2DQ, UK

For details of our global editorial offices, customer services, and more information about Wiley products visit us at www.wiley.com.

Wiley also publishes its books in a variety of electronic formats and by print-on-demand. Some content that appears in standard print versions of this book may not be available in other formats.

Library of Congress Cataloging-in-Publication Data
Names: Edwards, Peter J. (Peter John), 1940- | Vaz Serra, Paulo, 1966-
   author. | Edwards, Michael, 1969- author.
Title: Managing project risks / Peter J. Edwards, Paulo Vaz Serra, Michael
   Edwards.
Description: Hoboken, NJ : Wiley-Blackwell, 2020. | Includes bibliographical
   references and index. |
Identifiers: LCCN 2019010837 (print) | LCCN 2019016405 (ebook) | ISBN
   9781119489764 (Adobe PDF) | ISBN 9781119489733 (ePub) | ISBN 9781119489757
   (hardcover)
Subjects: LCSH: Risk management. | Project management.
Classification: LCC HD61 (ebook) | LCC HD61 .E3744 2019 (print) | DDC
   658.4/04--dc23
LC record available at https://lccn.loc.gov/2019010837

Cover Design: Wiley
Cover Image: © Vijay Patel / iStockphoto

Set in 10/12pt WarnockPro by SPi Global, Chennai, India
Printed and bound in Singapore by Markono Print Media Pte Ltd

10 9 8 7 6 5 4 3 2 1

Contents

List of Tables

List of Figures

Preface

If 'project' is part of your daily vocabulary, then this book is aimed at you. It is intended to appeal to practitioners of project management across a wide range of industries and professions; to people working in the private and public sectors, and those in the arts and entertainment; as well as to business organisations, service providers, and manufacturers. Students are very much included in our target readership as they pursue their academic journeys on the way to entering hopefully satisfying and rewarding careers.

An overview of the content is provided in Chapter 1. Besides offering a systematic approach to project risk management that we hope is easy to follow and understand, we have introduced topics generally not found in other books on this subject but which have an important bearing on how risks are managed, particularly those associated with today's projects. The additional matters we have dealt with include risk knowledge management, cultural risk-shaping, project complexity, and political risks. Strategic risk management is also considered. These topics are based upon our own project experiences, and reflections on how they might influence project risk management practice. Six project case studies (located as Appendices) are used to exemplify many of the points we make, together with many examples within the chapters.

We have adopted generic and multi-stakeholder perspectives of projects. This means that, whatever the types of projects in which you are involved, and whatever role you play in them, you should be able to apply the principles and processes of systematic and effective risk management in your work without constantly having to recontextualise them.

If you are a practitioner, as either a project manager or someone who specialises in risk management, we concede that you probably just want to get on with managing your projects and the risks associated with them. The inevitable time constraints for all projects will almost certainly already impact severely on the opportunities you have for reading. If this is so, then the arrangement of topics should help you. While they are predominantly sequential (in a flow process sense), the topics are distinguished as separate chapters, easily enabling you to dip in and out of them in a convenient way. The contents should meet several needs: as a refresher for your current risk management processes; as a guide to benchmarking them; or as a framework for replacing informal, reactive, and intuitive ways of dealing with project risks with a more formal, systematic, and proactive approach.

If you are a student, whatever your academic discipline, you will almost certainly be expected to take a project-oriented approach to your studies, and will also experience the pressure of time. You have to read and investigate so much about so many matters, and to demonstrate your knowledge acquisition through examinations and assignments, that what looks to be a 'quick fix' solution to learning about project risk management may look very attractive. You are right – it is! The 'dipping' topic arrangement should also suit you, but we suggest only after you have read and reflected upon the basics of risk management in the early chapters. Risk and risk management *are* big learning topics (as is project management), so dealing with them is never going to be just a quick process. We offer no easy solutions, but rather a systematic and comprehensive approach to project risk management that will serve you well in study and eventually in practice. Our book will not only provide you with a fundamental grasp of the principles and processes of project risk management, but should also help you to maximise the value of the experiential learning you gain from your own projects, now and in the future.

Instructors will find the structure of the book useful for preparing programmed reading guides for their students.

It is fashionable these days to argue that the internet will completely replace the need for books. While the Web is a huge and useful resource, it does come with its own risks. In Chapter 18 (Computer Applications), we note the vast number of hits following the entry of a risk-related term into an internet search engine. Not only would this result impose a huge task in sifting what is relevant from what is not, but there is also a substantial risk of finding information that is simply incorrect – the Web offers no certain guarantees for accuracy, reliability, and authenticity. We hope our book satisfies all three criteria.

Our aim is to provide an introduction to, and comprehensive treatment of, project risk management that will guide and assist people and organisations tasked with dealing with those risks.

As authors, our objectives are to:

- Effectively communicate a conceptual and philosophical understanding of risk.
- Establish the nature of projects and the stakeholders involved in them.
- Present a systematic and logically progressive approach to the processes of project risk management.
- Discover the drivers of project risks and the factors which shape them.
- Emphasise the importance of capturing and exploiting project risk knowledge.
- Provide guidance about implementing and building (or improving) project risk management systems in organisations.

We are friends, colleagues, and family, coming from different generations and different backgrounds and professions. We think those differences contribute much to the strength of the book.

Peter J Edwards trained originally as a quantity surveyor in the construction industry in the United Kingdom, and holds a Master of Science degree from the University of Natal and a PhD awarded by the University of Cape Town. Although now in retirement, he is currently an Adjunct Professor at Royal Melbourne Institute of Technology (RMIT) University in Melbourne, Australia and continues to be active in research and writing. He has authored and co-authored more than 170 peer-reviewed journal and conference papers, two books, and five book chapters. Many of these publications relate to project management and risk management. He has worked in the United Kingdom, South Africa, the USA, Australia, and South East Asia, and has taught undergraduate and postgraduate project risk management courses at universities in several of those countries. He has also undertaken consultancy work in project risk management.

Paulo Vaz Serra is a civil engineer with over 20 years of experience in construction management, including operational, research, and development responsibilities in construction companies in Portugal and Spain. He holds a Master of Science degree in Construction and a PhD in Civil Engineering with a focus on knowledge management. Paulo is currently a Senior Lecturer in the Faculty of Architecture, Building and Planning at the University of Melbourne, where he coordinates the courses Risk, Means and Methods and Procurement Methods in Construction within a Master of Construction Management degree programme. Paulo is a Senior Member of the Order of Portuguese Engineers, and a Chartered Member of the Institution of Civil Engineers (MICE) of the United Kingdom.

Michael Edwards has a Bachelor of Science degree, majoring in Mathematics, awarded by Monash University in Melbourne. Over more than 20 years, his work in a large department of the Australian federal government has involved initiating and managing projects for services, and service improvements, implemented not only within the department but also offered on a tender or fee-for-service basis to other government departments and to private sector organisations. He is thus experienced in stakeholder management.

Peter J Edwards was a co-author in an earlier book about project risk management (Edwards and Bowen 2005) which is no longer in print. While some of the material of that book has been included in this one, sufficient new material (and thinking about project risk management) has emerged over the past decade to justify describing this as a new book (with a new publisher) rather than a revised edition of the old one.

We hope this book meets with your expectations, and that it will provide a solid foundation and guidance for your practice in project risk management.

Reference

Edwards, P.J. and Bowen, P.A. (2005). *Risk Management in Project Organisations*. Sydney, NSW: University of New South Wales Press. ISBN: 0868405744.

Acknowledgements

We offer sincere thanks to the many people who have helped us with this book. Their contributions have enriched the content in ways that always exceeded our expectations.

In particular we thank Mike King, Andy Kwek, Peter Lawther, Sean McGoohan, and Gary Ullmann, not only for their willingness to contribute their knowledge and wisdom, but also for their time and patience in doing so. The information contributed by postgraduate students Moses Chiropa, Donald Matjuda, Lisalokuhle Mbobo, and Dube Ndabezinhle is gratefully acknowledged.

Our gratitude is also due to Rozanne Edwards for her comprehensive text editing work and design suggestions. It was a huge support for us.

Of course, none of our writing effort would have been possible without the encouragement and forbearance of our beloved families.

Glossary

Term	Amplification	Explanation
AI	Artificial intelligence	A process whereby knowledge is generated automatically through learning algorithms incorporated into a computer-based application.
AS/NZS	Australian Standard/ New Zealand Standard	Previous joint publishers of standards for Australia and New Zealand (see their replacement, SA/SNZ).
BOO; BOOT	Build-own-operate; Build-own-operate-transfer	Building procurement system alternatives which define larger and longer project roles for the construction contractor.
CAD	Computer-aided design	Computer application with graphic design interface capability.
CPN	Critical path network	An analytic project scheduling technique.
DB	Design-build	A procurement system for construction projects (see also D & C) in which the contractor has responsibility for both design and construction.
DBFO	Design-build-finance-operate	See DB and D & C. A procurement system whereby the contractor not only has responsibility for project design and construction, but also has an equity share in the investment and will operate the completed facility.
D & C	Design and construct	A procurement system for construction projects (see also DB).
DCF	Discounted cash flow	A mathematical technique for modelling the effects of time on the cash flows occurring over the life cycle of an investment.
DTA	Decision Tree Analysis	A quantitative decision support tool.

Term	Amplification	Explanation
ECP	Elemental cost planning	A technique, based upon quantitative measures of the discrete design elements, used by professional quantity surveyors to estimate the probable tender price for a proposed construction project or to achieve a balanced distribution of element costs by comparing them to historic projects.
EMV	Expected monetary value	A quantitative financial decision support tool.
EOI	Expression of interest	Issued as an invitation to participate in a project bidding process.
EPM	Enterprise project management	Total in-house responsibility for managing the delivery of projects in an organisation (see also PMO).
ETA	Event Tree Analysis	A quantitative decision support tool.
FM	Facilities management	The ongoing management of activities relating to maintenance, repair, component replacement, and energy efficiency during the operational phase of a facility.
FMECA	Failure Mode and Events Criticality Analysis	An engineering technique used in manufacturing to analyse the causes and seriousness of component failure.
HAZOPS	Hazard and Operability Study	An engineering technique, using predetermined conditional statements, to explore operational cause and effect situations during the project design stage.
HSE	Health and Safety Executive	Quasi-government authority in the United Kingdom responsible for establishing and administering national health and safety compliance requirements.
IP	Intellectual property	Rights to the legal ownership of ideas.
IRR	Internal rate of return	A form of DCF modelling which finds the percentage rate that will discount all cash flows occurring over the life cycle of an investment to a zero net present value for the whole investment (also known as the 'yield rate').
ISO	International Standards Organisation	Publisher of worldwide standards.
IT; ICT	Information technology; information and computer technology	Technologies (usually computer-based) that deal with the processing of data and information.
KMS	Knowledge management system	The arrangement of explicit knowledge in an organisation in order to facilitate inputs and access.

Term	Amplification	Explanation
NLP	Natural language programming	The use of computers to understand and process natural language (text or speech) in order to carry out required functions.
OHS	Occupational health and safety	A term used to typify situations pertaining to the workplace health and safety of people.
OR	Opportunity risk	Uncertainty with beneficial effect upon project objectives.
ORR	Organisational risk register	An interactive collection of risk information and knowledge at the organisation level.
PM	Project manager	The person given responsibility for managing all activities and processes required to bring a project from inception to completion.
PMI	Project Management Institute	US-based organisation for professional project managers.
PMO	Project Management Office	A unit within an organisation that is made responsible for managing the delivery of its projects (see also EPO).
PPP	Public-Private-Partnership	A procurement system for integrating the delivery and operation of public infrastructure and services projects.
PRM	Project risk management	Activities at the project level pertaining to the management of project risks.
PRMS	Project risk management system	A structured, organised, and documented system established by an organisation for the purpose of dealing with project risks (see also RMS).
PRR	Project risk register	An interactive collection of risk information and plans for risk management activity at the project level.
RFID	Radio frequency identification device	A wireless-enabled electronic identification tag or marker.
RFT	Request for tender	Issued as an invitation to participate in a project bidding process.
RKMS	Risk knowledge management system	A knowledge management system separately dedicated to project risks and not incorporated with an organisation's general knowledge management systems (see also ORR).
RM	Risk manager	The person responsible for ensuring that the risks an organisation faces are managed proactively as far as possible.
RMS	Risk management system	A structured, organised, and documented system for dealing with risks (see also PRMS).

Term	Amplification	Explanation
SA/SNZ	Standards Australia/ Standards New Zealand	Joint publishers of standards for Australia and New Zealand (see also the earlier AS/NZS).
SGBB	Singapore Gardens by the Bay	Botanical gardens project in Singapore.
SHA	Safety hazard analysis	A prescribed format for analysing and recording potential threats to work safety and the responses proposed to avoid or mitigate them.
TR	Threat risk	Uncertainty with adverse effect upon project objectives.
VCE	Virtual constructed environment	A dynamic computerised graphical simulation, usually three-dimensional, of a building design or construction process.
VE *or* VM	Value engineering *or* Value management	A management technique used in the project design stage and based upon identifying required functions for project components and then speculating about alternatives that could deliver the same function at lower cost, better function at the same cost, or better function at lower cost. Value is defined as a measure of worth calculated from the delivered function and the cost to achieve it.
WBS	Work Breakdown Schedule	A project planning technique which analyses a project by the activities required to undertake and complete it.

1

Introduction

1.1 Introduction

In this introductory chapter, we describe the project and project stakeholder perspectives that we have adopted to frame this book. Since we cover a range of topics and readers will have different levels of knowledge and experience about projects, project risks, and their management, we also provide a brief overview of the contents of the book. The chapter synopses will guide you in choosing the actual sequence you wish to follow for individual reading, but we recommend that you do follow the order for Chapters 4–10, as these chapters embrace the sequential and systematic application of project risk management processes.

1.2 The Project Perspective

We live in a world that is highly focused on 'development' and has become increasingly 'project-driven'. This is largely because projects are seen to be more 'containable' than other methods of achieving development goals. Projects are perceived as having clearly identifiable beginnings and finite endings (although sometimes these are hard to pinpoint precisely). The fulfilment of sought-for objectives is intended to deliver desirable (and hopefully measurable) outcomes. It is thus assumed that the project approach is more manageable than other ways of doing things, although that assumption may not always translate easily or fully into reality.

Projects are endeavours usually surrounded by uncertainty and often cloaked in risk. While we tend to regard them as exclusively human undertakings, projects do occur in the natural world. Beavers build dams across watercourses; termites construct elaborate edifices to shelter themselves from harsh extremes of weather; birds build nests to accommodate their young. These creatures also face risks as they go about their 'project' work.

Managing risks is an important part of managing projects, as much for human society as for natural fauna. This book describes a comprehensive and systematic approach to the management of project risks. Whilst we have no plans for further references to animals and insects, their potential contribution to risk management should not be ignored. Biomimicry has become an important source of innovation for contemporary society in many fields, and there is every reason to suppose that it could also contribute to risk management.

Project management, as an art and a science (hence its vulnerability to many interpretations), stems largely from the construction industry, which has been project-based since human beings first attempted to create shelter for themselves. Since then, we have become increasingly aware of the need to organise the ways in which our building activities are planned, resourced, and carried out in order to satisfy our need to develop our physical environment. Traditionally, therefore, project management has been associated with building projects, and many books (including those on risk management) retain that perspective exclusively.

In this book, however, we have tried to embrace the project-driven nature of contemporary society more fully and have deliberately adopted a generic project perspective.

All projects are exposed to risks. While particular risks will be different for varying projects and project environments, we intend to demonstrate that it is possible to adopt a systematically uniform approach in order to deal with those risks. Thus, while many of the examples presented in this book are taken from projects in the construction industry, we have sought to include others from a range of different fields. The actual risks will not be identical (although many will be similar), but the risk management principles remain the same.

1.3 The Project Stakeholder Perspective

All projects involve *stakeholders*: those people or entities that have capacity to influence the decision making associated with projects. We explore this concept in greater depth later in this book. Suffice it to say here that every project involves multiple stakeholders (or at least more than one). I may decide to embark on a renovation project on my house. While it is 'my' project, it is likely that other family members will be involved, that tradesmen will be engaged and external suppliers sourced. I may have to approach consultants for advice or even apply for permits from local authorities. To a greater or lesser extent, each and all of these will influence the decision making that inevitably surrounds the project. Anyone with that capacity has to be regarded as a stakeholder. How much influence they have will determine the nature, level, and treatment of the risks involved.

Similarly, you may propose a project to write a book as a sole author. However, if you want others to read it and if you want to earn royalties from its publication, other people will become involved in and help to make decisions about the publication process. The same scenario applies to artistic and creative works. While the intellectual inputs may be entirely individual on the part of the artist, if the project outcomes are intended to become available to others, or even to just a single end-user or purchaser, then we might argue that the follow-up process is also part of the project and thus susceptible to decision making beyond that of the original artist. Few artists can afford to ignore their 'market'.

A single project stakeholder perspective is thus only tenable if the project outcomes were never meant to be available to anyone other than the project originator.

However, while all may be involved in bringing a project to fruition, each stakeholder is likely to have at least some objectives that are different to those of other stakeholders. By definition, as we shall see in Chapter 3, this means that each stakeholder will be exposed to different risks, albeit possibly of a similar type but of varying uncertainty in terms of likelihood and consequence. Each stakeholder may have to manage its risks in ways that may be subtly different to those of other project stakeholders.

Logically, therefore, whatever the *organisational* arrangement of stakeholders in a project, any attempt to insist upon a common risk management system for that project is neither practical nor advisable, particularly where the stakeholders are autonomous entities. Even where projects are undertaken 'in-house' by an organisation (e.g. under project management office [PMO] or enterprise project management [EPM] arrangements), there will still be other stakeholders involved, including other departments within the host organisation and external stakeholders supplying goods or services to the project.

In this book, we adopt a stakeholder perspective that assumes that each stakeholder implements its own risk management system for each of the projects in which it is involved. Ideally, each stakeholder will employ an overarching approach that, while dealing individually with all of its risks on each of the projects in which it is involved, will apply common principles of risk management throughout, and will capture risk knowledge from each project to the benefit of the whole stakeholder organisation.

The project and project stakeholder perspectives outlined here provide the essential context for the whole of this book.

1.4   Overview of Contents

As noted in Section 1.1, the chapter synopses in this section should help you to determine the topic sequence you wish to follow. For those who are involved in teaching project risk management, the synopses may help you to formulate a useful reading programme for students.

In Chapter 2, attention is given to understanding risk itself. Definitions of risk are explored, and common risk terms set out. Positive and negative concepts of risk (threat risk and opportunity risk) are presented. The psychology of risk is considered, together with risk awareness. Risk and uncertainty are distinguished, and their association is clarified. The dynamic nature of risk is discussed. Approaches to classifying risks are considered. The important topic of risk communication is introduced here, but it is treated more comprehensively in Chapter 19 (Communicating Risk).

Chapter 3 is all about projects, further consolidating the essential platform upon which the processes of managing project risks can be presented. The nature of projects is considered, in terms of their life cycles and processes. Additional thought is given to project stakeholders and their influence. Project decision making is considered, and the chapter concludes with some wisdom about what may constitute a risky project.

National and international risk management standards are described in Chapter 4, which then presents a systematic approach to project risk management in the form of an experiential learning cycle. This provides an essential precursor for the more detailed presentation of the stages of the risk management process in subsequent chapters.

In Chapter 5, the important preliminary task of establishing the internal and external contexts for a project is presented, together with the importance of considering the risk drivers operating in those contexts.

For risks to be managed, they must first be identified. This process is dealt with in Chapters 6 and 7. Approaches to identifying project risks are first considered and then followed by a presentation of several risk identification tools.

Following identification, risks should be analysed and assessed in terms of their individual and comparative magnitudes or levels. Chapter 8 presents simple ways of doing this so as to provide an informed basis for subsequently deciding what should be done about the risks. The emphasis in this chapter is upon qualitative risk assessment.

The response options and types of proposed treatment actions available for project risks are presented in Chapter 9. At this point, the risk management process usually moves from exploration and planning to the active reality of implementing the project. Most risks are now 'closer' in time. Chapter 10 therefore deals with activities related to monitoring and controlling risks during the project delivery process.

It is said that 'if we do not remember history, we are doomed to repeat it' (George Santayana, 1863–1952: https://en.wikipedia.org/wiki/George_Santayana). In Chapter 11, the importance of project risk learning is considered, specifically through risk knowledge management. Knowledge about risks, captured from individual projects, is systematically recorded by the stakeholder organisation as a means of gaining important wisdom about risk that can be exploited for future projects. Much of this chapter content is unique in the risk management literature.

While Chapter 11 concludes coverage of the essential processes of systematic project risk management, we believe that our book would be incomplete without some attention to other topics closely associated with risk.

Relatively new to the risk management literature is consideration of the way in which risks are culturally shaped. Chapter 12 explores this concept from the perspectives of society in general and from the organisational characteristics of project stakeholders.

Modern projects are often described as complex, especially when they fall into the category known as 'mega-projects'. Complexity and its implications for project risk management are discussed in Chapter 13.

In addition to complexity, many projects (regardless of nature or scope) are beset by political influences that affect how they are conceived and delivered. This has impacts upon the risk management activities of the stakeholders. Political risks are discussed in Chapter 14.

In Chapter 15, opportunity risk is considered as a desirable obverse of the two-sided coin of risk. Differences in the management of threat and opportunity risks are considered.

Strategic risk management, as a responsibility of senior management that is distinct from the everyday processes of systematic risk management but still highly relevant to them, is discussed in Chapter 16.

Chapter 17 provides guidance about the process of building and maturing a risk management system in a project-based organisation, and Chapter 18 briefly considers computer-based risk management software applications. Other information technology (IT)-based tools are described.

The important topic of communicating risk is expanded in Chapter 19. Starting with a theoretical foundation, this chapter presents a model of communication and its components. The chapter then considers the implications of all this for communicating risk information within a project stakeholder organisation, externally to other project stakeholders, and beyond that to the public. The placing of this chapter late in the book is deliberate, as it allows us to draw on all the various aspects of project risk management and reconsider them from a communication perspective.

Chapter 20 presents conclusions about project risk management and offers some views about its future.

Appendices then present six case study projects:

A. A correctional facility project
B. A rail improvement project
C. An aid-funded project and project consultant
D. A train mock-up project
E. A hot-rod car project
F. An aquatic theme park project.

These studies exemplify topics discussed in earlier chapters, and references to them are made in appropriate places throughout the book. They provide comparisons and contrasts in terms of project risk management principles and processes.

Without wanting to be too radical, may we suggest that you consider reading the case studies first! They will give you a greater awareness of the different contexts for different projects. Then, as you read the rest of the book, you will better appreciate the references we make to them. We are also certain that you will find other instances where no case study references are made but which are highly relevant to your own project risk management experience – that creates a 'win-win' outcome for you and for us!

The case studies, together with the many other examples included in the book, should provide a rich menu of topics for discussion and tutorial groups.

1.5 Limitations Caveat

Our book provides a comprehensive treatment of systematic risk management for projects that is lacking in only two aspects. The synopsis for Chapter 8 does not include comprehensive or sophisticated mathematical techniques for analysing and assessing risks, nor does it provide detailed information about specialised computer applications associated with such decision support analysis. The two omissions are deliberate.

A sound understanding of the concepts of risk and the principles and processes of risk management is an essential prerequisite to all mathematical risk modelling. If risks are not understood conceptually, then no amount of mathematical treatment and analysis will resolve that deficiency, and the data inputs and outputs associated with sophisticated computerised modelling tools are likely to be spurious – a truly 'black box' situation that presents a comprehension dilemma. Obtaining adequate and reliable input data to service such models may itself be a difficult and expensive, if not impossible, task.

We do not claim that mathematical modelling has no place in project risk management, but rather that it is not a critical requirement for *every* project. Where such modelling is needed, the necessary expertise can be acquired or hired separately, as long as the project stakeholder fully appreciates the need for such effort, is willing to commit the resources required for it, and also understands the value, implications, and limitations of what the modelling will deliver.

Highly mathematical approaches in risk management are beyond the purview and objectives for this book, and really warrant a separate treatment.

In practice, most project stakeholders rarely need to undertake complex mathematical risk analysis. What they most want is to identify the risks they face and assess them in order to determine or prioritise resource requirements, explore treatment options, decide upon appropriate responses, and then successfully monitor and control the treated (or untreated) risks as the project proceeds from inception through the procurement process and beyond. The content of this book is therefore based upon these premises and focuses upon the more practical requirements of project risk management.

Although some discussion of computerised systems is presented in Chapter 17, detailed information about currently available risk management software applications is not provided. The frequent upgrading of such applications, and the rapid pace of development in modern IT and artificial intelligence, would almost certainly impact negatively on the 'shelf life' of the book.

We hope you will find this book useful in your project work, and that you will enjoy reading it.

2

An Overview of Risk

2.1 Introduction

If project risk management is to be effective, the people directly involved with it need to share a common understanding of what risk actually means for that project. This may be easier to say than to achieve, since risk is a 'social construct'. It is perceived and acted upon (or ignored) by people. Given the vagaries of human nature, it is understandable that perceptions of risk will differ, not only between individuals but also between groups (e.g. organisations and professions) and even among societies.

In this chapter, we explore definitions of risk and how it is perceived. The 'two-sided coin' of risk (i.e. threat and opportunity) is advocated. The association between risk and uncertainty is explored, and we consider the dynamic nature of risk. Theoretical precepts that constitute the psychology of risk – thus making it a 'psychosocial construct' – are discussed, together with risk awareness.

The classification of risk is important, as familiarity with different types of risk facilitates shared understanding about them. Achieving shared understanding requires effective communication of risk and risk management.

2.2 Risk Definitions

Risk is sociologically grounded. We derive our general understanding about risks, and our attitudes towards them, largely from the society in which we live and work.

For example, a community that knew nothing about money, as a means of payment for goods and services, would not appreciate financial risk per se, but instead might have a strong awareness of the risks associated with bartering goods and services (i.e. exchange risk). Similarly, a society that had no knowledge of the practice of human surgery could have no understanding of surgical risk, but might have developed a prescient knowledge of the efficacies (and dangers) of natural remedies in the treatment of illness and disease. A religious sect with particular values and beliefs might even understand the risk of dying through an outbreak of food poisoning by ascribing it to the divine disposition of some higher being or force.

For most of our encounters with risk, the concept of risk as a social construct makes little difference to the way in which we respond but, if you are working in a multicultural environment, the various ways in which risk is understood cannot be ignored entirely.

Managing Project Risks, First Edition. Peter J Edwards, Paulo Vaz Serra and Michael Edwards.
© 2020 John Wiley & Sons Ltd. Published 2020 by John Wiley & Sons Ltd.

For example, the approach to occupational health and safety risks on a construction project in a country such as Singapore, with its different ethnic cultures and its extensive use of foreign labour drawn from many countries, needs to be carefully considered, especially in terms of inculcating safe working practices among on-site workers. Increasing globalisation in many aspects of human endeavour means that achieving mutual understanding of risk, particularly in a project context, is also increasing in importance.

Beyond the social construct of risk, there exists a *psychological* perspective. This is discussed (albeit briefly) in Section 2.6. For now, there is sufficient context for us to consider how risk may be defined.

According to the Oxford Dictionary of English (2005), the noun 'risk' is defined as: 'a situation involving exposure to danger or loss'. Additional offerings include: 'the possibility that something unwelcome may happen', 'a person or thing regarded as a likely source of danger', 'a possibility of harm or damage against which something is insured', and 'the possibility of financial loss'. For the term 'risky', the same dictionary suggests: 'full of the possibility of danger, failure or loss'. Used as a verb, 'risk' denotes engagement (consciously or unconsciously) with a risky situation or risk-laden activity; hence: 'she risks everything by marrying such a person from that family'.

In the United Kingdom, a comprehensive Royal Society (1991) report about risk, drawing upon the work of more than 50 authors, proposed that: 'Risk is the probability that an adverse event occurs during a stated period of time'.

An early Australian standard, dealing with risk analysis of technological systems, defined risk as: 'the combination of the frequency, or probability, of occurrence and the consequence of a specified hazardous event' (AS/NZS 3931 1998).

Each of these sources frames risk negatively (danger, harm, damage, loss, adverse, hazardous), in terms of the outcomes or consequences of risk events. The latter two emphasise that probability (chance) is attached to risk events in terms of their likelihood of occurrence. All three sources imply (but do not make explicit) that the negative outcomes, should the event occur, are known and assessable.

So far, we may summarise that a risk means that something bad may happen. The inference is not unreasonable, given that the term originates from the seventeenth-century French 'risqué', itself derived from the Italian 'risco' (danger). However, just as 'risqué' now has a different contemporary connotation, so too might other definitional sources lead us to reconsider this negative 'threat' view of risk.

The AS/NZS 4360 (1999) standard for risk management, a forerunner in this field, states that risk is: 'the chance of something happening that will have an impact upon objectives. It is measured in terms of consequences and likelihood'. Here, the concept of purpose ('objectives') is introduced and a neutral frame is adopted.

ISO 31000 (2018) is an international standard for risk management, largely drawn from the earlier AS/NZS 4360. This standard trims the definition to: 'Risk is the effect of uncertainty upon objectives'. While the neutral frame and the concept of objectives are maintained, uncertainty is introduced (or rather made explicit instead of being implied), and an additional inference is that the uncertainty may be associated with the risk event, the consequences, or both.

The Project Management Institute (PMI) offers yet another definition: 'risk is an uncertain event or condition that, if it occurs, has an effect on at least one project objective' (PMI 2013). This is really just a slight variation on ISO 31000, but uncertainty has been confined to the risk trigger – the event or condition.

Perhaps of less interest to us, given our focus on managing project risks, is a contemporary publishing development whereby specialist dictionaries define risk in different ways for different disciplines such as economics, finance, management, environmental science, statistics, and computer science. However, we have deliberately adopted a generic project perspective for risk management, and projects may be found in any of those disciplines. Therefore, we cannot ignore this development entirely, as it will strongly affect how risks are classified.

Readers, and risk managers, must choose the definition they prefer. In this book, we generally follow that offered by ISO 31000, although with some reservations, as we will explain in Section 2.3.

Whichever definition is preferred, in dealing with project risk two questions arise:

- How likely is it that something will happen that could affect this project?
- Are the event and its consequences sufficiently important to us that we should consider doing something about either or both of them?

In exploring answers to those questions, four aspects will have to be considered:

- Each event and its nature
- The consequences of each event
- The period of exposure to each event (and to its consequences, if that is also relevant)
- The uncertainties associated with any of these (event, consequences, and timing).

Chapters 6–8, and parts of Chapters 9 and 10, explore these aspects in greater detail for project risk management, but first we still have some way to go in our conceptual understanding of risk.

In the opening to this chapter, we referred to the 'two-sided coin' of risk, implying a 'heads you win; tails you lose' framing for risk. This warrants further discussion.

2.3   Threat and Opportunity

In Section 2.2, we detected a shift in the contemporary definitional framing of risk, from a mostly negative to a neutral connotation. With some trepidation, we refer to yet another definition of risk: 'the potential for gaining or losing something of value'. This is found through an online source: https://en.wikipedia.org/wiki/Risk. The same reference notes that risk involves an intentional interaction with uncertainty, and we expand upon that in Section 2.4. From this definition, we justify our earlier metaphor of the two-sided coin of risk.

However, if we set that analogy against the neutrally framed contemporary definitions of risk (ISO 31000 and AS/NZS 4360), the latter imply that the coin, if tossed in the air, will tend to land on its edge! Either the analogy is wrong, or the neutral framing is inappropriate. Hence our reservations with the definition of risk offered by ISO 31000.

The dual view of risk – loss or gain – is attractive, especially when seen in a context such as gambling (i.e. wagering with a coin) or financial investment. From the latter, we also find mention of 'upside risk' and 'downside risk', but these terms lack precision unless you already understand them as representing positive and negative outcomes. For someone unfamiliar with finance, 'upside' might be inferred as something happening before 'downside', much like 'upstream' precedes 'downstream'. Commentators also use 'north' and 'south' with respect to the positive or negative direction of contemporary trends and outcomes (e.g. 'the company's dividends are heading south'), but precision is also lost with these terms, since your interpretation might be influenced by the world hemisphere in which you live!

For the purposes of project risk management, we prefer to use 'threat' and 'opportunity' as our negative and positive frames for risk. The consequences of some risk events may threaten the achievement of project objectives. Alternatively, the events may, if exploited as opportunities, result in project objectives being exceeded. The corollary to this is that, if the event outcomes neither threaten project objectives nor provide an opportunity for exploitation (thus reflecting a state representing the edge of the two-sided coin), they cannot be risks for that project.

Figure 2.1 attempts to show this graphically, although we admit that in some ways it falls short of the intended message. While it does reflect the old adage that 'one person's loss may be another's opportunity', as well as an 'upside' and 'downside' risk picture, not every risk has potential for both threat and opportunity outcomes. Nor is a real-life situation akin to that of a playground see-saw with a possible balancing point depending upon the 'weight' sitting upon each end. Finally, our purpose in this book is to help you manage the risks you face on the projects with which you are involved. We do not intend to deal with the threat or opportunity risks of other parties.

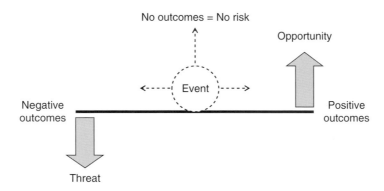

Figure 2.1 Threat and opportunity risk.

A separate chapter (Chapter 15) offers some guidance for managing opportunity risk, but, apart from occasional references to positive risk, the remainder of this book generally adopts a negative, threat risk perspective. We are not being mean here. While managing opportunity risks can be creatively (and financially) rewarding and should not be completely ignored, society generally takes the view that risk management should give more attention to negative, rather than positive, outcomes. Although proponents of, and advertisements for, gambling promote the thrill and positive chance of winning, society in general warns against the far greater probability of losing and the social problems that may arise from gambling addiction.

It is understandable, therefore, that few texts about risk and risk management present examples and case studies of risk analysis that deal with purely positive opportunity risk outcomes, as a balance to those depicting negative consequences. This book is no different. Our focus is deliberately on threat risks because this reflects contemporary project risk management practice. Project management, through its array of tools and techniques, tends to concentrate upon protecting the efficient and effective achievement of project objectives. It rarely suggests expending effort and resources on exceeding them: so too with risk management. In most instances, this approach aligns closely with the aims and objectives of the organisation itself.

Given our qualified preference for the ISO 31000 definition of risk, we now explore uncertainty and risk more fully.

2.4   Risk and Uncertainty

Knight (1921) surmised that uncertainty has never been properly separated from risk. Almost a century later, we might say the same.

Frank Knight's view, based upon the prevailing paradigms and definitions of his time, was that risk is measurable, whereas uncertainty is not. Today we tend to believe that both risk and uncertainty, if not precisely measurable, can at least be assessed in some degree, so the early twentieth-century view no longer holds as strongly as it did.

Instead, we propose that you cannot have risk without uncertainty, while you may encounter uncertainty without risk. If you are uncertain about the weather prospects for tomorrow but have no outdoor activities planned, you face no weather risk. Similarly, you may be unsure if a distant bird that you have observed in the sky is a sparrow or a hawk, but, if you have no keen interest in ornithology, no intention to enter the bird's environment in a fragile balloon, nor any plan to walk about with your head unprotected in the bird's immediate vicinity, then you face no risk arising from your uncertain identification.

Society tends to demonstrate a view of 'certainty' that is somewhat less than its dictionary meaning of 'beyond all doubt; unquestionably'. Counsel in a legal case will strive to upset the certainty claimed for the evidence of a witness, whether for the prosecution or for the defence, in order to elicit acknowledgement of a degree of doubt about its veracity or accuracy.

Similarly, in most legal jurisdictions, a jury in a criminal case is expected to deliver a verdict: guilty or not guilty. A 'guilty' decision is required by law to be 'beyond all reasonable doubt', but note how 'reasonable' has now entered the earlier definition of certainty with respect to doubt and qualifies it in a way that retreats from its former absolute connotation.

Three further observations can be made regarding the attitude of the law with respect to certainty. A 'not guilty' verdict is made with no explicit qualifications about doubt, but that does not thereby denote a certain and unqualified innocence. If a 'not proven' verdict is possible, then this does infer that there is some degree of doubt about guilt or innocence. Finally, in civil, as distinct from criminal, proceedings, a finding of 'guilty on the balance of probabilities' may be handed down, suggesting that a much lower threshold of certainty is acceptable to society in some circumstances.

We need to exercise care, therefore, over what we believe and state to be certain – even for the old adage about the only two certainties in life being death and taxes! While one may be true, the other may be doubtful – one cannot evade eventual death, but you might be able to avoid payment of taxes not only during your lifetime but also beyond it!

Uncertainty, on the other hand, represents a state that is somewhere short of certainty. Our dilemma is that uncertainty occupies a far greater intellectual space than certainty – we can be uncertain about a great many things, but certain about only a few.

Compounding the dilemma is the variability in human nature. I may disagree not only with the things that you hold as certain, but also with the extent of uncertainty for the things about which we share doubts.

Certainty implies completeness (or at least a sufficiency) of information about the matter of interest. We know, or readily access, enough information about the matter to be sufficiently certain about it in terms of veracity, accuracy, and reliability. How we know what we know is a field of philosophy (epistemology) that lies beyond the scope of this book, but we can assume here that uncertainty represents a lack of information and knowledge. How we know what we don't know is an even more abstruse area, but we may need to observe care in deciding what is always unknowable and what might become knowable if we were able to devote sufficient effort and resources in discovery.

For the purpose of managing project risks, uncertainty as a lack of complete information is a useful concept to build upon. Figure 2.2 juxtaposes the level of project information and the level of project uncertainty as a symmetry against time.

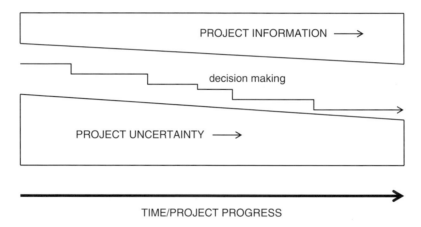

Figure 2.2 Project information/uncertainty symmetry.

As the project progresses, project information increases and uncertainty therefore diminishes. The link between the two is project decision making, the process that stimulates activity and generates information. As each project decision is made, more may become known about the project and more certainty is introduced.

The perfect symmetry conceptualised here (and used again in later figures) is impossible in reality, of course. While it is not totally asymmetrical, the relationship between project decision making, project information, and project uncertainty over time is only roughly symmetrical. Project decisions are never made at equidistant intervals, nor with equal levels of profundity or instantaneous effect. Also, external events or circumstances may add much or little to project information or may retroactively affect the outcomes of decisions already made, thus adding further uncertainty to the project. Over time, therefore, the amount of uncertainty surrounding a project is inevitably itself an uncertain quantity.

Although uncertainty might not be measurable in terms of objectively and quantitatively determining a size or quantity in specific units, there are more subjective, qualitative assessment approaches that may be suitable and sufficient. Linguistic assessment is one such method, and this is described more fully in Chapter 7 (Project Risk Identification Tools). Before conducting a qualitative assessment of project risks, however, it should be possible to explore the *nature* of the uncertainty associated with each identified risk.

As we noted in Section 2.2, project risks may embrace multiple uncertainties. These could be displayed in matrix form, as in Table 2.1. A table such as this, albeit more specifically formatted, can help to clarify possible risks for projects and even guide response options for them.

Table 2.1 A certainty/uncertainty matrix for project risk management.

		Status	
Risk element	**Constituent**	**Certain?**	**Uncertain?**
Event	Type		
	Occurrence		
	Period of exposure		
Consequence(s)	Type		
	Magnitude		
	Period of exposure		

If there is certainty about *every* aspect of an event and its consequences, then by definition no risk arises in terms of our involvement in the project. The situation may require resolution, but it is not a risk. Further, while uncertainties may be present, it was noted in this section that, if event outcomes have no effect on project objectives, they are not risks for that project. Our interest here is therefore with any uncertainties relating to risk (trigger) events and to uncertainties pertaining to the consequences of those events in terms of how they may affect project objectives.

Chapters 6–10 explore items in this matrix in more specific detail for the purposes of project risk management, but we will briefly consider the matrix constituents here from the perspective of the general nature of risks and try to provide contextual examples.

2.4.1 Uncertainties in the Type of Risk Trigger Events

Events that have the capacity to trigger risk consequences may involve encounters with uncertainty in terms of their type and nature, their likelihood of occurrence, and the period during which exposure to the event is possible.

For example, a risk event might comprise a breach of safety codes or regulations pertaining to a particular project activity (or series of activities), such as scaffolding for a construction project. However, while the breach constitutes the type of event, its nature may give rise to several possibilities. The breach may be accidental or deliberate. It may happen during the erection of the scaffold (an installation breach), through accidental or deliberate misuse (an operational breach), as a failure of proper maintenance during the building work (a maintenance breach), or during the eventual removal process (a dis-assembly breach).

Similarly, for an information technology (IT) software customer application project for a commercial bank, the threat risk of a 'glitch' event may comprise several different uncertainties. Is the 'glitch' on the bank side or the customer side, or on both sides? Is it due to algorithm logic error, coding error, or processing error? Is the error accidental or deliberate?

2.4.2   Uncertainties in the Occurrence of Risk Events

The likelihood of occurrence for the breach in the scaffolding example might be represented by a probability (e.g. 0.005). For this level of accuracy about the uncertainty, however, sufficient historic evidence (data) would have to be collected and analysed – perhaps from as many as 1000 projects – in order to make a reliable assignment of probability; but few construction companies would have access to sufficient data about safety breaches generally, let alone for each of the different possibilities for the nature of the breach. Furthermore, what would such a probability indicate? What does it relate to? Would it mean that a safety breach (whatever its nature) is likely to occur on five out of every 1000 projects undertaken? Or might it be based upon some other criterion of scope or size – on five occasions for every 1000 hours worked or per $1000\,m^2$ of building floor area? Should we distinguish between types of project, such as scaffolding safety breaches on school building projects or on high-rise commercial projects? Is it necessary to separate accidental and deliberate safety breaches? Should we moderate the probabilities in terms of the influence of worker and management experience and any quality performance improvement? Compounding risk likelihood uncertainty in this way may complicate risk management to an extent that could render it impracticable. Risk managers should always question whether or not such levels of analysis (i.e. more precise expressions of probability) are essential for a project.

2.4.3   Uncertainties in the Period of Exposure to Risk Events

The period of exposure to a risk is often incorporated into an expression of the probability of occurrence for the event. Thus, for example, you might want to know the odds of a particular cause (event) leading to consequential death in any year (see http://www .nsc.org/learn/safety-knowledge/Pages/injury-facts-chart.aspx). Such statistics will be heavily influenced by the number of remaining years in your lifetime, your lifestyle and the type of activities that you engage in, and even the country where you live. For example, the probability of dying from a gunshot wound in any year is likely to be higher in the United States than in a country such as Australia where an aggressive 'gun culture' is far less prevalent.

From a project perspective, the period of exposure to a risk event is most often reflected in the stage(s) or phase(s) of the project during which it can happen. Can it occur during the design/development period (but not thereafter), during the execution phase, or at any time between commencing and completing the project?

A further consideration about risk trigger events is whether they occur as single events, as multiple events, or as multiple events in a distinct sequence. A weather cyclone may hit a coastal town once and pass on, or it may move away but then intensify and return through a slight wind shift. Earthquakes are known to be followed by a series of aftershocks.

What we have discussed in this section are the uncertainties that may be associated with single, stand-alone events, but uncertainties multiply exponentially for multiple and sequentially multiple events.

2.4.4 Uncertainty in the Type of Consequences of Risk Events

The consequences of project risk events are most often experienced in terms of the time, cost, and quality aspects of a project. Something happens that causes the project to be delayed, to cost more, or to be of a lower standard or less functional than planned. Combinations of two or all three of these types of impact are quite commonly encountered, but may not be known with certainty beforehand. Similarly, other consequences may be experienced: for example, an event occurs which leads to a time delay for the project, in turn leading to impaired relationships with other project stakeholders. Cheaper materials may be purchased for the project, in lieu of those originally specified, but with adverse consequences for worker (and even project user) health and safety. For example, a cheaper paint finish may contain solvents that permit faster application, but the fumes given off in the process may seriously damage the painters' respiratory systems and even cause headaches among people occupying the newly finished rooms. Again, none of these outcomes may be known with certainty beforehand.

Exploration of the consequences of risk events is a vital part of the risk identification process and is dealt with more fully in Chapters 7 and 8.

2.4.5 Uncertainty in the Magnitude of Risk Consequences

Occasionally, it is possible to determine beforehand not only the type of consequence for a risk event, but also the magnitude of that consequence. Thus, if storm damage should affect a partially completed building, we may know with almost complete certainty that insurers will condemn the newly laid but now damaged carpets and cover the cost of replacement. Similarly, we may have established that there is a 15% chance of a mechanical defect causing a bulldozer to break down during the bulk excavation operations for the earthworks stage of a bridge construction project. We may also know that the minimum downtime for any such breakdown is four hours. That information is quantitative and objective, also with a high level of certainty. However, we cannot be *fully* certain in either case, and the amount of impact uncertainty for each situation will be influenced by their individual circumstances. More often than not, project risk management has to deal with high levels of uncertainty in terms of the consequences of risk events.

An extreme example may be found in the case of the cladding material, whose fire-resistant performance was not fully known or certified, used as the external façade replacement for a refurbishment project on a high-rise apartment block in London. No one could know with certainty that a small fire occurring internally in a room on one level of the refurbished building would quickly spread to the outside and destroy the entire façade, affect the availability of emergency exit routes, and lead to the deaths of more than 70 people trapped inside. The assumed self-extinguishing qualities of the cladding material were countered by the particular conditions under which it was used and by the domino effect of other adverse circumstances. This example leads to consideration of the period of exposure to the consequences of risk events. The risk trigger (source event) here was not the actual fire itself but the decision to use an inherently unsafe cladding material. The impact of that decision was not immediate and followed only after the occurrence of a subsequent event.

2.4.6 Uncertainty in Periods of Exposure to Risk Consequences

Project risk management tends to assume that the consequences of risk events flow seamlessly and immediately upon their occurrence. The example of the apartment building fire suggests that this is not inevitable. Uncertainty in event occurrence and impact magnitude can also be accompanied by uncertainty about when such impacts are felt. This is akin to the concept of patent and latent defects in quality assurance management. For patent defects, the condition is obvious and immediate, and the consequences can be determined with relative ease. For latent defects, the condition is neither obvious nor necessarily immediate, and because of this their consequences are difficult to forecast, let alone determine quantitatively.

Risk events where the consequences do not flow immediately, but are experienced as a single or continuous impact at some time in the future, are sometimes known as 'long tail' risks since the impact is temporally separated from the event by a period that may stretch over several years. They are literally like a 'sting in the tail' for the risk taker; they tend to have characteristics of low probability and high impact, and are among the most difficult project risks to manage, largely because of the high level of uncertainty associated with them.

Take the hypothetical case of a highway bridge project where, at the time of construction, a precast concrete beam unit (not known to be defective) had been erected to span between two piers. The original design parameters had included a safety factor of 2.5 for operational loading, but a defect occurring during the manufacture of the beam unit had reduced this to 1.25. After 10 years without incident, a double tractor pulling an abnormally heavy load had broken down on the bridge, and was passed by another abnormally heavy transport vehicle, resulting in a brief loading factor of more than twice the safe load on the defective part of the bridge structure. The structure failed, with severe consequent damage and loss, but fortunately with only minor injuries to people. However, failure, as a consequence of the latent defect, could have happened at any time during the previous 10 years or even at any time in the remaining 150 or so years of the operational life of the bridge, hence the uncertainty in risk exposure.

Since all projects take place against a continuum of time, the *dynamic* nature of risk and its constituents should be considered.

2.5 The Dynamic Nature of Risk

The probability of occurrence and the impact, or indeed both, of particular risks may change over the third element: time. Indeed, few risks continue unchanged indefinitely, although some may disappear completely after a specific time period and are 'closed off' for the purposes of risk management. This dynamic quality is particularly true for project risks, and examples from several areas of risk suffice to show this.

Many projects are vulnerable to *economic and financial volatility*. However, the economic and financial climate prevailing at the time the project was conceived, and which influenced its associated decision making, may change during any or all of the ensuing design, development, delivery, and even operational phases. A simple example is provided by the changes in loan interest rates that occur over time. While an upward or downward trend may be predictable, interest rates at a particular time in the future are hard to forecast, even in the short term, as also are the turning points that precede them. This dynamic tends to support a preference for fixed-interest debt finance for projects, rather than variable-rate loans that might be available at a lower rate, since the latter involve greater uncertainty for the borrower.

The fluctuating *seasonality* of weather (or, perhaps more accurately, its *un-seasonality*) may contribute to risk where project delivery or operational activities have to be undertaken outdoors.

Changes in *productivity* (flowing from intended or unintended changes in the factors that contribute to productivity) influence risks associated with the outputs of planned project activities.

In the area of *legal change*, unforeseen amendments to laws, codes of practice, and regulations may serve to alter the likelihood of occurrence and consequences of risk events associated with them.

The dynamic nature of risk may be regarded as interferences occurring along the time axis of the uncertainty–certainty polygon shown in Figure 2.2. We have already referred to the reality of the lack of perfect symmetry in the underlying concept for this figure. The temporal frame simply amplifies the space for dynamic volatility in any or all of the constituents of risk bearing decisions.

Uncertainty and the dynamics of risk are also heavily influenced by the *psychosocial* construct of risk.

2.6 Psychology and Perceptions of Risk

It is appropriate to reflect on expanding the concept of risk as a social construct into a *psychosocial* construct.

In Section 2.2, we explored risk from the perspective of the society in which we live and work, and then placed definitions of risk in the context of projects. Projects are discussed more fully in Chapter 3, but we know that they are carried out by people (and usually by people representing organisations).

The threat and opportunity polarities (Section 2.3) of project risk derive from the perceptions of the people working on projects and in particular the people engaged in project decision making, who will be influenced by the uncertainties associated with their decisions (Section 2.4). People also have to take into consideration the dynamic characteristics of the project risks they manage (Section 2.5).

In addition to the cognitive demands of such perceptions and the project decision making that underlies them, *feelings* about risk come into play. Paul Slovic, one of the foremost researchers and writers in this field, uses a psychometric paradigm to explore and test theories that risk 'is subjectively defined by individuals who may be influenced by a wide array of psychological, social, institutional and cultural factors' (Slovic 2010). Part of this subjective interpretation of risk is the way in which people allow their negative or positive feelings about a decision or activity to influence their assessment of the risks associated with it. Slovic regards this as the *affective* power of risk operating as a 'heuristic' (rule of thumb) model. Thus, we may consider particular risks in terms of the degree of dread they engender in us (threat risks), by the amount of excitement or pleasure they generate (opportunity risks), and sometimes even by the level of ambivalence we have towards them (the so-called 'neutral' state).

The affective power of risk perceptions transitions to the project domain in three ways, each related to project decision making. An individual may transfer emotionally influenced perceptions of project risks to the represented organisation through the project decisions he or she makes. Organisations may transfer their corporate perceptions of risk back to the immediate project decision makers through policies and procedures established from the risk experiences of forebears or determined by known or anticipated behaviours of competitors. Societal perceptions of risk translate to projects through their capacity to influence either or both organisational and individual decision making.

The latter route is also known as the social amplification of risk (Kasperson *et al.* 1988), whereby in the transfer of technical risk information to society, the psychological, sociological, and cultural perceptions of risk exert influence such that public attitudes towards risks become amplified (in the sense of becoming stronger and louder) or sometimes attenuated (weakened or diluted), thus affecting risk behaviours and responses.

Organisational aspects of affective perceptions of risk will be dealt with more fully in Chapter 3 (Projects and Project Stakeholders). We will expand upon individual perceptions in Chapter 9 (Risk Response and Treatment Options), while societal perceptions are reconsidered in Chapter 12 (Cultural Shaping of Risk). Suffice to say here that affective perceptions of risk can significantly influence levels of risk *awareness*.

2.7 Risk Awareness

'Awareness' may be understood as a state of knowing about, or being cognizant of, something. However, while this emphasises a cognitive process, awareness is often regarded today as being a combination of intuition, instinct, experience, attitude, emotion, and the individual contributions of at least some of the five human senses, as well as the product of formal and informal knowledge acquisition.

Think of the occasions when you attend a business meeting or enter a social situation such as a hotel bar. Immediately, you begin to be sensitised to the 'vibes': looking and listening for visual and verbal clues and cues (and sometimes those that are olfactory) so that you can quickly gauge the atmosphere and perhaps the potential for a positive reception to any contribution you had thought of making to the gathering.

In another context, developing awareness is not only a vital part of learner-driver training, but also a core aspect of advanced-level driver courses. The difference, of course, is in the level of awareness expected for each, but the focus in both is upon drivers becoming sufficiently aware of the conditions that could affect safe driving: weather, traffic, road surface, the behaviour of other drivers, pedestrians and their intentions, accident avoidance routes, and so on. The learner-driver is usually presented with a less busy driving environment so that awareness can be developed gradually and safely in parallel with mastery of the vehicle's controls. For the advanced course, drivers are deliberately exposed to difficult conditions and are normally expected to demonstrate their awareness with a running verbal commentary on what they see and hear and how they interpret and intend to act upon this from a driving perspective.

In a similar way, we can say that risk awareness is a state of being constantly sensitised to potentially risky situations – most often those that pose a threat but not excluding situations that might give rise to opportunities for gain. In an organisational setting (i.e. for a project stakeholder), it is developed largely through knowledge acquisition and experience, but the level of risk awareness will always be tempered by the myriad of less tangible sensitivities of the individuals in the organisation who make project decisions on its behalf.

An important point here is that, while sensitisation to risk is desirable, it may eventually be followed by desensitisation. This is a constant danger, not only for project risk management but also for any management system within an organisation. We will return to this issue in Chapter 17 (Planning, Building, and Maturing a Project Risk Management System).

Just about every technique and tool described in this book has, besides its specific purpose, the aim of raising awareness about risk. Our intention is that you will be constantly reminded of this as you read, and we do not apologise for any repetition!

Risk awareness in an organisation is substantially increased and improved by seeking to achieve common understanding about the organisation's risks. Adopting an appropriate method of *classifying* risks is an important step in this process.

2.8 Classifying Risk

Classifying risk is important in risk management as it enables a project stakeholder to consider risks within an organised and coherent framework. It also provides the opportunity to explore whether or not a particular class or type of risk is amenable to a particular type of treatment.

Ideally, classifying risk should also provide us with a universal and uniform risk language. This is important in fields such as project management, where it is often necessary to communicate risk information to a wide variety of project stakeholders. A universal risk classification system, complete with uniform terms, would allow us to establish a common understanding of different risks, and provide a desirable basis for effective knowledge transfer, not only from one project to another within and across an organisation, but also between project stakeholders, across industries, and beyond. Risk awareness would thereby be enhanced individually, organisationally, and generally.

In reality, this ideal state has yet to happen. Little uniformity exists in the classification of risk, even within recognised discipline areas. Different approaches are found to classifying what are often the same types of risks. Risk terms are frequently diverse.

Problems in classifying risk include issues relating to different cultural perceptions and confusion between sources and causes (trigger events) and between effects and impacts (consequences). Cultural aspects are discussed more thoroughly in Chapter 12 (Cultural Shaping of Risk).

In addressing the latter problem, we would argue that ultimately any risk classification system should be capable of grouping risks by the type of source event (i.e. what can happen that will give rise to consequences, and where the uncertainty associated with either will affect project objectives). Our premise is that a consequence must be preceded by an event, and thus first grouping types of events is a more logical starting point. Grouping risks according to their consequences is less logical and probably more difficult, since, as we noted in Section 2.4, a single risk event may have multiple consequences. A source-event approach to classification provides baseline categories beyond which other arrangements for risk classification can be devised, particularly for projects.

Following this line of argument, we first present a generic risk classification system based on types of risk event sources. Thereafter, we will suggest classifications systems based on organisational structures, project life-cycle phases, and a customised hybrid approach. Other approaches can be followed, but we prefer the basic simplicity of the generic system as a fallback, stand-alone system that is applicable beyond the domain of projects.

2.8.1 A Generic Source-Event Risk Classification System

In a generic source-event risk classification system, the top tier of separation is between *natural* systems and *human* systems risks. This is shown in Table 2.2.

Natural systems risks arise from sources beyond human agency. Human risks arise in terms of human involvement in some capacity. This distinction is particularly useful for exploring options for responding to risk. Humankind has little or no capacity to influence the occurrence of natural risk events, but some limited ability to deal with their consequences. For human systems risks, there is far greater ability to influence both their likelihood of occurrence and their impacts.

This binary separation is not without criticism. In a report (with over 50 contributing authors) about risk perceptions and management, the Royal Society in the United Kingdom proposed three primary categories of risk: natural, human, and technical (Royal Society 1991). As technical systems are a feature of human activity, technical risks can be subsumed under human risks.

Critics also point to the impact of humans upon natural systems (i.e. the effects of human-induced climate change). Without wishing to enter this polemic debate, it is possible to show that the natural/human distinction in risk classification is still valid.

Table 2.2 Generic source-event risk classification.

Natural risk systems categories	Human risk systems categories
Astronomical	Cultural
Climatological	Economic
Geological	Financial
Biological	Health
Ecological	Legal
	Management
	Political
	Sociological
	Technical

For example, the impact of a hurricane may be aggravated in a region where overgrazing by cattle or deforestation through deliberate land clearance has taken place (both human-engineered situations), but essentially it is the hurricane event (as a product of a weather system) which gives rise to the risk, so it is still a natural one in terms of its source. Similarly, while adverse ground conditions are a perennial risk for construction projects, this risk is also a natural, albeit latent, one. The potential for risk is simply activated by the decision to build on a particular site and the subsequent commencement of excavations. Yet again, the damaging effects of a soil wash-away may be exacerbated by the removal (during bulk earthworks excavations) of natural watercourses, but the source of the risk is still the storm which precipitated the wash-away. These examples suggest that it is one or more of the elements of risk – probability, consequence, and time – which may be affected by changing circumstances, not the type of risk event itself.

2.8.2   Natural Systems Risks

As Table 2.2 shows, the system subcategories of natural risks are: astronomical, climatological, geological, biological, and ecological. Typical risks arising within each category are shown in Table 2.3. The examples do not constitute an exhaustive list, and many others will be found in nature. Furthermore, while we have used just one or two words to nominate risk examples, this is inadequate for good risk management; a fuller, more precise risk statement is required for each risk. This will be discussed in Chapter 7 (Project Risk Identification Tools).

A brief amplification of natural risk categories follows.

Astronomical risks relate to conditions, movement, and interactions of bodies beyond planet earth, and are frequently sourced to the sun or moon. However, while

Table 2.3 Examples of natural category risk events.

Natural risk systems categories	Examples
Astronomical	Solar radiation
	Tidal variation
	Meteorite strike
Climatological	Hurricane
	Typhoon
	Tornado
	Lightning strike
Geological	Earthquake
	Volcanic eruption
	Geomorphological state
Biological	Human biology
	Animal biology
	Marine biology
Ecological	Plant environment

a meteorite shower would be an extraterrestrial natural risk, a potential collision between man-made space debris and earth would be regarded as a human (technical) risk since it would arise from human agency.

Climatological risks are weather-related and, besides those indicated in Table 2.3, would also include events such as rainstorms, snowstorms, windstorms, freezing cold, baking heat, drought, and so on.

Geological risks arise from conditions relating to the earth's crust. It should be noted that, in this regard, tsunamis are not geological risk events per se, but are a potential consequence of under-sea volcanic eruption.

Biological risks relate to events occurring in living organisms, including human, animal, and marine creatures. However, where effects on these systems are triggered by human action or intervention, then they are sourced as human systems risks.

Ecological risks emerge as a result of interactions between living organisms and their natural environments. The same caveat applies here as to biological risks.

2.8.3 Human Risks

Risks arising out of human agency are more difficult to categorise than natural risks, due largely to the overlapping and interrelational characteristics of many humanly devised systems. Nevertheless, it is possible to indicate examples of human systems risk subcategories. These comprise cultural, economic, financial, health, legal, management, political, and social systems, and they are shown in Table 2.4 together with a few risk examples.

Cultural risks emanate from the threat of potentially negative interactions between groups of people espousing different identity, living, or religious values. At a government level, of course, such risks are treated politically, but this does not deny their cultural nature. However, in Chapter 12 (Cultural Shaping of Risk), we argue against culture being regarded as a separate category of human risks, instead proposing it as a shaper or driver of other types of risk. The categories are therefore not fixed.

It is not always easy to make a clear distinction between *economic* and *financial* risk categories, and for much of project risk management to do so may be unnecessary. One way to separate them may be to distinguish between factors of production and factors affecting the cost of project finance or the security of project cash flows. Thus, examples of economic risk might include labour or material supply issues, while changes in interest rates or credit ratings, or the potential unavailability of project loan finance, would be regarded as financial risks. It should be noted that, while cash flows, rentals, fees, and charges are included as examples of financial risks, in many instances they are not risk source events but rather the consequence of other risk events.

Risks arising through surgery, medical treatment, or epidemia due to human causes may be categorised as *health* risks.

In the *legal* category are found risks arising from requirements to observe clauses relating to private contracts, deeds, and agreements, as well as other legally enforceable instruments such as statutes, regulations, and codes. It should be noted that, at the pre-enactment or pre-promulgation stage of public instruments, some risks would have to be regarded as politically sourced.

Managerial risks relate to events attributable to action (or lack of action) by management. Project examples include poor productivity or excessive wastage, inadequate project quality, and human resource problems such as inappropriate or inadequate staffing. Occupational health and safety risks may also be included under this category. Human communication risks would also fall into this category, but events relating to communication systems per se are considered as technical.

Table 2.4 Examples of human category risk events.

Human risk systems categories	Examples
Cultural	Religions
	Cultural customs
Economic	Materials supply
	Labour supply
	Equipment availability
	Inflation
	Tariffs
	Fiscal policies
Financial	Exchange rates
	Interest rates
	Credit ratings
	Capital supply
	Cash flows
	Rentals
	Fees and charges
Health	Epidemia
	Surgery
Legal	Contract clauses
	Regulations
	Codes of practice
Management	Communication
	Productivity
	Quality assurance
	Cost control
	Human resource management
	Safety management
Political	War
	Civil disorder
	Industrial relations action
Social	Criminal acts
	Civil torts
	Substance abuse
Technical	Design failure
	Equipment and systems failure
	Estimation error
	Collision and accident
	Communication system failure

Risks arising out of government action (or inaction) can be classified as *political* risks. Among these, war and abrogation of international treaties are straightforward examples. Political risks may also arise from opposition to government threats or actions, as in cases of civil disorder or industrial action. The actions of lobby and protest groups fall into this category. In most instances, the risk falls upon the target, either the politicians or protesters. However, innocent third parties can be caught up in the crossfire. A company leasing office space in the same building as a foreign consulate, for example, might find its business activities adversely affected by the actions of groups wanting to demonstrate against events occurring in that consulate's home nation. Similarly, the picket lines of striking workers rarely discriminate between the purposes and motives of people wishing or needing to cross them.

Behaviour which is unacceptable to society generally may be categorised as *social* risk (or perhaps more precisely as antisocial risk). This category might include criminal acts such as theft, robbery, assault and murder, vandalism, sabotage, arson, espionage, and treason; civil torts including trespass, slander, and libel; and substance abuse such as drunkenness and other drug-induced antisocial behaviour. Some citizens might want to include public graffiti as a social risk, but it is important to remember that the values set by a society determine which behaviours are regarded as antisocial. These values are culturally determined and may change substantially over time for a particular society or differ between societies.

The *technical* category of risk involves events attributable to a humanly designed technical system. Examples range from actual design failure or shortcoming, to equipment and system breakdown or inadequacy. While these are easy to understand as technical risks, the inclusion of accidents and collisions in this category may be more difficult to accept. Given that there is no deliberate human intent or presence of substance abuse (which would place them as the consequences of risk events in the social risk category), then collisions and accidents can more properly be seen as a failure (or absence) of technical systems intended to prevent them. However, they may also be seen as the consequences of design failure.

The anomalies we have pointed out in the generic source-event risk classification system confirm some of the difficulties of this approach and perhaps explain why it has not enjoyed universal application. Generally, however, it can be regarded as robust and reliable, and in the main it is a useful and practical approach to classifying risk.

2.8.4 Risk Classification Based upon Organisational Structure

Another way to classify risks is by the structure of the stakeholder organisation that is exposed to them. Table 2.5 shows a typical structure for a commercial organisation that regards its products as projects to be managed and sold to customers.

The list of organisational units does not pretend to be exhaustive and will vary from organisation to organisation. Governance, operational, and administrative units can be identified in other types of organisations including public authorities, universities, schools, hospitals, and other public services and utilities. These too can be used as a means of classifying risks.

The basis for this approach is rarely related to the source of risk events and is most often predicated upon the units that the organisation considers best placed to manage particular risks. This form of risk classification has advantages and disadvantages from a risk management perspective. It directly involves such units in organisational and project risk management. Used effectively, it forces each part of an organisation to consider the risks it faces. Staff in the units will become familiar with the terminology of the risks and probably more knowledgeable and experienced in ways to deal with them. 'Ownership' of the risks may be more easily achieved, and appropriate organisational cultures developed. Against these benefits, however, classifying risks according to organisational unit in this manner tends to encourage a focus on risk consequences rather than source events. It also leads to a 'silo' risk management environment where specific risks become the territory for specific units. Other parts of the organisation may not even know of them, and the overall level of organisational risk awareness suffers accordingly. Similarly, the effectiveness of risk knowledge capture and management in the organisation may be impaired.

Table 2.5 Risk classification by organisational structure.

Governance and structure
Board
Human resources
Quality management
Occupational health and safety
Purchasing
Design and planning
Production
Transport
Financial administration
Information and computer technology (ICT) and communications
Marketing and sales
Customer and public relations
Research and innovation
Legal affairs
Project management

Our view is that, while the organisational unit approach to risk classification can be adopted, it should not become the sole method used for describing risks in an organisation.

We will return to issues of organisational structures in Chapter 3 (Projects and Project Stakeholders) and again in Chapter 16 (Strategic Risk Management).

2.8.5   Risk Classification Based upon Project Phases

Life-cycle application phases are suggested in AS/NZS 3931 (1998) as an approach to categorising risk. The phases are identified in that document as the: concept and definition phase; design development phase; construction, production, transportation, operational, and maintenance phase; and disposal phase. In later chapters of this book, and for different purposes, we present examples that contain slight variations on this list.

Little obvious advantage accrues from using project stages or phases as a *primary* approach to risk classification, other than perhaps showing stages in a project where unique risks arise or stages where the same risks tend to occur. For some projects, this approach might serve to provide a clearer focus for the time aspect of risk.

2.8.6 Customised Hybrid Approaches to Risk Classification

Yet another approach seeks to separate *external* and *internal* risks. This too is more suited to specific contexts, since a risk might be classed as external in one situation, but as internal for another. This is part of a systems boundary issue for risk management that we discuss later in this book.

Table 2.6 shows categories in a customised external/internal risk classification system developed by a parastatal organisation in South Africa. It is a hybrid approach, as the categories are a mixture of terms relating to people, systems, and processes. Natural risks are simply regarded as events external to the organisation. For this organisation, the risk classification system is not intended purely for project risks, and each unit or department in the organisation is required to apply it to the work it undertakes.

Table 2.6 Customised hybrid approach to risk classification.

External risk subcategories	Internal risk subcategories
External stakeholders	People
Economic	Decision making
Political	Structure
Regulatory	Systems
Competitors	Information and knowledge
Technology	
Natural events	

Table 2.7 lists examples of some of the risks the organisation has identified in each of the subcategories for internal risks. Just as we noted for the examples listed for the generic risk classification system shown in Tables 2.3 and 2.4, the lack of sufficient

Table 2.7 Typical internal category risks for a customised classification system.

Internal risk subcategories	Examples
People	Behaviour
	Training
	Competency
	Motivation
	Communication
	Information and knowledge
	Sufficiency

(Continued)

Table 2.7 (Continued)

Internal risk subcategories	Examples
Decision making	Speed and appropriateness
	Processes for governance
	Suitable information
	Lack of communication
Structure	Accountability
	Functionality
	Change of structure
Systems	Suitable
	Reliable and available
	IT Systems
	Methods of work
	Procedures
Information and knowledge	Sufficient information and data
	Suitable and coherent information and data
	Ownership of intellectual property

descriptive information for the risk example entries in Table 2.7 is likely to be a problem for practical risk management, and particularly for project risks, within this organisation. This issue is addressed in Chapter 7 (Project Risk Identification Tools).

The major drawback with this customised and hybrid example of risk classification is that it does not appear to be systematic and could easily give rise to confusion about the most appropriate category to be used for classifying some risks.

2.8.7 Multisystem Risk Classification

The different approaches to risk classification described in this section suggest that a single method may not fully serve all the risk management requirements in an organisation. While we advocate a generic risk source-event approach, we reiterate our preference for it as a stand-alone or fallback system – if other methods are not successful, then this one should work.

However, modern information management systems are such that it is no longer necessary to rely on one risk classification system alone. Project risks that an organisation faces can be classified and recorded in several different ways at the same time. The information can be stored at the project level and at an organisational level. At the latter level, the information can be set alongside other risks (i.e. beyond projects) that the organisation faces, so that risk management can be more comprehensively applied.

In multisystem risk classification, identified risks can be tagged with several category descriptors. Risk information can then be input into the organisation's risk knowledge database and arranged, sorted, and retrieved in a variety of ways to suit different purposes. This approach is discussed more fully in Chapter 11 (Project Risk Knowledge Management).

Despite our preference for the generic approach to risk classification, we concede that it is vital to adopt a system that will not only be understood but also be meaningful across the entire organisation. A vital ingredient towards achieving this goal is *risk communication*.

2.9   Risk Communication

For project risk management to be effective, information about risk must be communicated to relevant participants in the process, so that shared understanding can be achieved. 'Risk communication' refers to the whole gamut of information, knowledge, opinions, advice, and suggestions about risks in general and also particular risks that may have to be shared with relevant people within an organisation, shared between organisations, or distributed to the public at large. It is grounded within the general theory and principles of communication and follows much of that larger field's precepts and models.

In this book, we have deliberately chosen to explore risk communication as a topic much later (Chapter 19, Communicating Risk). Delaying such discussion in this way enables us to consider risk communication retrospectively from the perspective of the risk management processes presented in the preceding chapters. The whole book is intended to communicate knowledge about project risk and project risk management. Chapter 19 will enable readers to assess how well our aim has been achieved.

2.10   Summary

In this chapter, we have attempted to establish a theoretical understanding of risk sufficient to serve as the platform upon which the propositions of project risk management can be presented. The chapter has ranged widely over different and important topics about risk. Each of them might warrant a complete book, but we hope our coverage has not only provided you with enough of the essential foundational information and knowledge about risk to proceed confidently with the remaining chapters about project risk management, but also stimulated a lifelong interest in this important area of project management.

From the array of definitions of risk available, we have chosen to adopt, albeit with some qualification, that of ISO 31000 (2018): 'the effect of uncertainty upon objectives'. The qualification we suggest is to frame the effect as positive or negative, rather than as a neutral state. If there is no discernible effect, there is no risk to the project.

Project risk is always associated with uncertainty, but uncertainty is not inevitably associated with risk. If the uncertainty has no tangible effect on project objectives, then it is not a risk. Exploring where uncertainties lie in project risks is helpful for dealing with them.

The elements of risk (likelihood of occurrence, consequences, and period of exposure) are likely to change over time, in both nature and degree. Thus, risk is dynamic.

A social construct of risk has been expanded into a psychosocial construct, reflecting not only that risk is experienced and perceived by people but also that those perceptions are also influenced by their feelings about risk and its dimensions.

All these aspects of risk influence our awareness of risks. The degree of sensitivity to project risks, developed individually and organisationally, influences and determines the level of commitment to managing them.

Classification adds to the effectiveness of processes in every area of management, and this is no less true for project risk management. While we have not been able to demonstrate the overwhelming benefits of a single universal approach to classifying project risks, we have described several different methods and their advantages and disadvantages. A generic risk source-event system is preferable, but modern information management technology allows multiple classification systems to be used conjointly with ease.

Effective communication leads to shared understanding, and risk communication thus plays an important part in project risk management.

Before launching into the actual processes of risk management, however, we must first clarify our understanding of projects, and this is the subject of Chapter 3.

References

Australian Standard/New Zealand Standard (AS/NZS) (1998). *Risk Analysis of Technological Systems – Application Guide* (AS/NZS 3931). Homebush, NSW: Standards Australia.

Australian Standard/New Zealand Standard (AS/NZS) (1999). *Risk Management* (AS/NZS 4360). Homebush, NSW: Standards Australia.

International Organisation for Standardisation (ISO) (2018). *Risk Management – Principles and Guidelines* (ISO 31000). Geneva: International Organisation for Standardisation.

Kasperson, R.E., Renn, O., Slovic, P. *et al.* (1988). The social amplification of risk – a conceptual framework. *Risk Analysis* 8 (2): 177–187.

Knight, F. (1921). *Risk, Uncertainty and Profit*. New York: Houghton Mifflin.

Oxford Dictionary of English (2005). *Oxford Dictionary of English*, 2e. Oxford: Oxford University Press.

Project Management Institute (PMI) (2013). *A Guide to the Project Management Book of Knowledge*, 5e. New Town Square, PA: PMI.

Royal Society (1991). *Report of the Study Group on Risk: Analysis, Perception, Management* (Chairman: Professor Sir Frederick Warner). London: The Royal Society.

Slovic, P. (2010). *The Feeling of Risk: New Perspectives on Risk Perception*. London: Earthscan Ltd.

3

Projects and Project Stakeholders

3.1 Introduction

Projects are the means and lifeblood of development. Traditionally, they have been closely associated with physical development through engineering, infrastructure, and building projects, and architecture, engineering, and construction (AEC) is correctly described as a project-driven industry. Physical 'hard' projects are also found in the mining, shipbuilding, automotive, aeronautical, and aerospace industries. More recently, however, 'soft' projects have emerged in fields such as finance, information technology (IT), telecommunications, media, health, justice, and education. We have embraced a world of projects. The connotation of projects has expanded and is much more widely used in contemporary management – to the point where almost every organisational decision is regarded as a 'project'.

In this chapter, we look at projects: their nature, objectives, constituents, and life-cycle phases. The processes of project implementation, and organisational structures for projects, are described. Project stakeholders, and their relationships to projects, are discussed in terms of stakeholder organisational structures and organisational decision making. We also explore what might make a project 'risky', but in this book we do not address the much larger topic of project success – that in itself warrants a complete book on its own. Much of the discussion adopts a generic approach to projects, but examples are drawn from various industries and applications. Many examples are presented here, and our suggestion is that you return to some of them as you work through the later chapters (Chapters 5–10) that deal with project risk management practice. They could be useful for practising risk management techniques.

3.2 The Nature of Projects

A project is a deliberate undertaking. It arises from an intention to create or do something specific. The underlying motivation is therefore one of expending effort to achieve a desired outcome. That outcome might be tangible and functional (e.g. a building or engineering structure). It could be physical and artistic (e.g. a painting, sculpture, or installation). It could represent the mounting of an exhibition, such as a blockbuster collection of famous artists like the Impressionists. Or it could be something more

Managing Project Risks, First Edition. Peter J Edwards, Paulo Vaz Serra and Michael Edwards.
© 2020 John Wiley & Sons Ltd. Published 2020 by John Wiley & Sons Ltd.

intangible (but still functional) such as a new smartphone 'app', a new online commercial product, or a new procedure for claiming welfare benefits. Events, such as pop concerts and sporting competitions, are also regarded as projects to be managed.

Dictionaries typically provide a quite terse definition for 'project' as simply 'a plan or scheme'. The Project Management Institute (PMI) more expansively offers: 'a temporary endeavour undertaken to create a unique product, service or result … definite beginning and end' (PMI 2013: 3).

Our view is that the latter definition can be misleading, and you may be better off sticking to the simpler dictionary version. Insistence upon uniqueness and definite beginnings and endings may give rise to unnecessary confusion. Confident identification of each of those characteristics might be difficult, tiresome, and not strictly necessary. Generally, each of us has a pretty good idea of what projects are in our individually familiar contexts, whether that is in industry, in public administration, in the creative and performing arts, or simply as a personal endeavour. Our plans and schemes are undertaken in the service of fulfilling objectives.

Interestingly, projects in the eighteenth century and earlier were synonymously associated with 'projectors', the people associated with promoting them. Often this term was used pejoratively, with the promoters regarded as confidence tricksters. Now its use is much diminished, other than in connection with the graphical display of images on a screen, while 'project' has not only survived but proliferates in modern language use. That said, contemporary companies project the images they wish to be seen by customers and the general public, using carefully developed projects aimed at social networks and other communication media that themselves have been initiated on a project basis.

As we shall see in this chapter, in undertaking projects, time and resources are required and consumed, but first the need for a project itself must be determined in terms of the objectives sought for it.

3.3   Project Objectives

Establishing project objectives is important not only in terms of clarifying our understanding of a project and the need for it, but also for setting out the criteria against which project success can be assessed. It is also important for project risk management, given the ISO 31000 definition of risk presented in Chapter 2: 'the effect of uncertainty upon objectives'.

The determination of project objectives also provides the foundational logic for our contention that, where multiple organisations are involved in delivering projects, a completely common set of objectives will rarely be found. Some objectives may be shared by some of the participating organisations, but each organisation will have its own objectives to consider. Sometimes these will clash with those of other parties involved in the same project, thus yielding a fertile area for risk. If project stakeholders have differing objectives, then differing uncertainties will arise in achieving them, and therefore each organisation will face different risks. If each project stakeholder faces different risks on a project, then each will need to manage its own risks. Thus, a common or shared project risk management system is not only impractical but also inappropriate for a multi-stakeholder project.

Often, project objectives are considered to be associated only with procurement (i.e. the delivery of the project), but they may also be operational or strategic. Figure 3.1 shows this diagrammatically as a hierarchy.

Figure 3.1 A hierarchy of project objectives.

3.3.1   Procurement Objectives

Project procurement objectives are couched in phrases that relate to fulfilment criteria which can be evaluated to determine the successful delivery of the project from inception to operational readiness. Typically, these criteria are expressed in terms of time, cost, and quality – the 'iron triangle' of all construction projects. Sometimes project scope (the extent of what is to be delivered) augments or replaces quality, especially in design-build procurement systems where design is integrated with construction and not treated as a separate professional service. Procurement objectives may also include objectives that are less tangible and sometimes more difficult to measure, such as safety and environmental management. Examples are shown in Table 3.1 for a high school building project that includes alterations and additions to existing buildings as well as new facilities. The example assumes a client (employer)–contractor arrangement whereby the client is the relevant government education authority.

Note how some objectives are almost open-ended in terms of their expectations, and also how they are framed in a way that implies that their fulfilment is entirely the responsibility of the contractor. This is not untypical of construction projects and tends to reinforce our earlier contention that a risk management system common to, and shared by, all project stakeholders is impracticable. In this example, the contractor might prefer to: amend Objective (1) to allow for authorised extensions of time; replace Objective (2) with one relating to profit expectations on the contract; amend Objective

(9) to allow short-term obstruction necessitated by the immediate demands of construction work; amend Objective (10) to ban inappropriate behaviour by students, staff, and visitors towards construction workers; and possibly add an objective (on the contractor's part) to foster positive long-term relationships with the client (in this case, the education authority) in the hope of obtaining further project work. For a stakeholder which is a subcontractor on the project, Objectives (1) and (2) would be modified to match the terms of its involvement. The subcontractor might also prefer to include objectives relating to prompt payment by the contractor, unhindered access for its work, and use of the contractor's plant and equipment (e.g. cranes and scaffolding) when needed. A relationship-fostering objective with the main contactor could also be considered.

Table 3.1 Procurement objectives for a public high school project.

Procurement objectives

1. To complete all construction work within x months from contract signing.
2. To complete all work within the agreed contract price ($\$y$), subject only to authorised variations.
3. To complete all work to agreed specifications and fit for purpose.
4. To match alteration work as closely as possible to the finishes and fittings in existing buildings.
5. To schedule work to existing facilities so as to minimise temporary closure and disruption during school terms, and to give timely warning to school authorities of proposed schedule changes that could affect the operation of the school.
6. To execute all work safely, in accordance with regulations, and avoiding any injury or harm to students, staff, parents, and other authorised visitors to the school. Temporary fences, barriers, and warning notices must be properly maintained and monitored.
7. To execute all work without causing damage to the internal and external environments of the school.
8. To ensure that all construction waste and harmful materials are removed promptly from the site and disposed of safely.
9. To ensure that construction plant, equipment, and vehicles do not obstruct required access routes around the school.
10. To ensure that the behaviour or actions of any employee of the contractor and subcontractors do not cause nuisance to school staff, students, and property; and to bar construction workers from entering any part of the existing premises unless expressly instructed to do so for the sole purpose of carrying out work.
11. To ensure that all temporary storage areas and facilities for workers are maintained in a clean and hygienic condition, and are removed promptly when no longer required for use.

Most of these objectives (other than those relating to relationship building) are likely to find their way (via a client briefing process) into requirements formally expressed as clauses in contract and subcontract agreements, but they tend to begin their existence as project objectives. Some objectives may not be stated expressly but are implied from customary expectations for particular projects and industries.

In terms of representing metrics for project procurement success, it should be easy to distinguish objectives that are 'hard' and straightforward to measure from those that are not. Objectives (1) and (2) in Table 3.1 are readily quantifiable. Assessing performance for 'soft' objectives is more difficult.

Different procurement arrangements could entail subtle changes to several of the objectives listed in Table 3.1. For a private school, the board of governors might replace the public education authority as the client or project sponsor; the latter will almost certainly still have a jurisdictional role to play, but in a less direct project stakeholder capacity. A design-build delivery system for the project might involve different procurement objectives compared with a more traditional separated client-design team– contractor arrangement; and the procurement objectives for a PPP (public–private partnership) project would differ even more, since such a project would almost certainly involve bundling several school projects into one 'mega-package' in order to create a more attractive commercial proposition for private sector bidders.

In complete contrast, consider a dedicated website design, deployment, and maintenance project to be undertaken by a professional IT company on behalf of its client, a small regional city café business. The client's primary objectives for the project are really operational – it wants to use a contemporary marketing medium to reach a wider customer audience. This aspect of the café example will be discussed more fully in Section 3.3.2, but the café owners' procurement objectives are likely to be associated simply with the time and cost elements for the delivery and maintenance aspects of the website project.

3.3.2 Operational Objectives

Operational objectives address the question of what the delivered project is required to do. What is its intended function?

For the high school project example, regardless of whether the school is run publicly or privately, the operational objectives sought by the client stakeholder might be to:

- Increase the pupil place capacity of the school by x number of students.
- Enable the introduction of new curriculum specialisms in the school (e.g. performing arts).

There may well be others.

For the café website project, some operational objectives are listed in Table 3.2. Again, there may be others.

This café business uses a Facebook page to augment its proposed webpage promotion, particularly for announcing changes in daily specials, dates for events, and details of community activities. As a deliberate policy, the café refrains from comparisons with competitors, despite this type of enterprise being universally highly competitive. The café owner's view is that its reputation is more valuable in attracting customers than engaging in destructive competitive practices, hence the emphasis on marketing its strengths. Note also that, apart from a brief allusion in Objective (3), the café does not stress food safety and hygiene since these are strictly regulated by law, with inspections, and breaches subject to prosecution and public notice. Food can be prepared and served carefully and safely. It can also be prepared and served carelessly, yet still be safe!

Table 3.2 Operational objectives for café website project.

Operational objectives

---

1. To operate and maintain an attractive, inviting, effective, and user-friendly Web presence.
2. To fully inform potential and existing customers about the café and what it offers, in terms of:

   - Genuine café ambiance, emphasising friendly and efficient service and unique and fascinating curio décor.
   - Interesting and tasty range of gourmet foods (prepared by professional chefs) and drinks.
   - Availability of daily specials.
   - Indoor and outdoor table settings.
   - Table reservations for party bookings.
   - All fresh food ingredients sourced locally and from regional growers.
   - Availability of external catering services with individually prepared menus.
   - A variety of sources and condiments made and bottled on the premises and available for sale at the café, online for home delivery, or for commercial order and delivery.

3. To emphasise careful food preparation and environmental management practices.
4. To showcase new products.
5. To announce upcoming weekend events such as live music and custom classic car displays.
6. To indicate support for the local community.

3.3.3   Strategic Objectives

Strategic objectives are concerned with what the client/sponsor wants to achieve *through* the project. They are generally of a more long-term or medium-term nature.

For the high school example, the strategic objectives sought by a public education authority are likely to differ quite markedly from those of the board of governors for a private school. The former might be seeking to increase and improve the provision of high school education in a region where demographic analysis suggests development and sharp population increase will lead to growing demand over the next 30 years. The latter, in undertaking the alterations and additions project, might be hoping to exploit evidence of increasing affluence in the local suburb and surrounding areas, or reach out to international markets, in order to attract more students (through the quality of its facilities as well as the education it offers) and increase its turnover and profit. It may also be trying to cement any competitive advantage it perceives it has in the provision of private secondary education in that region.

As for the café owners, their strategic objective may be to use the Web presence to widen the target audience for potential customers and thus grow their business, perhaps with a view to selling it profitably 5–10 years in the future.

In most instances, relatively few strategic objectives are established for projects, compared to operational and procurement objectives. Since they are usually aimed further into the future, they are also likely to be surrounded by greater uncertainty. It therefore makes sense to avoid setting too many strategic objectives, but rather to regard them as reflecting a provisional master plan for the relevant stakeholder.

Later, in Chapters 6 and 7, we shall see how project objectives can be used as a tool for identifying risks. Suffice it to say here that risks associated with strategic objectives can strongly influence 'go/no-go' project decision making, since, if the attainment of a strategic objective is likely to be threatened, it might be better to simply abandon the project.

Having explored procurement, operational, and strategic project objectives, a pertinent question to be asked is: Can they be shared by different participants in a project?

For the high school example, it is unlikely that the contractors and subcontractors involved in delivering the project will be interested in sharing anything other than some of the procurement objectives with the client, unless the work forms part of a PPP project arrangement. Even then, while the PPP contractor might be part of the private partner consortium in the procurement phase, in many instances this is a temporary, relatively short-term arrangement and the construction contractor is subsequently replaced in the consortium by a facilities management contractor.

In the case of the café website project, since the Web developer is charged with ongoing maintenance as well as creation of the website, this organisation is likely to share most, if not all, of the procurement and operational objectives with the café owner client. The Web developer may also share the strategic objectives, particularly if one of its own objectives is to engage and strengthen a longer term relationship with this client and eventually initiate a relationship with a new buyer for the business in the future. Another reason might be to use the success of the website project to showcase the achievements and capabilities of the organisation.

This serves to reinforce our earlier argument, in that a mutually *inclusive* project risk management system is only possible if all the relevant stakeholders share the same objectives for the project. If any stakeholder has different objectives, then it must maintain its own mutually *exclusive* risk management system. Later in this chapter, we will see that this has implications for project participants acting in a professional agency capacity on behalf of a project client.

Our final example in this section is the renowned Singapore Gardens by the Bay (SGBB) attraction in Singapore. Table 3.3 lists the procurement, operational, and strategic objectives for what are really nested 'projects within projects within a megaproject'. This SG$1 billion endeavour (from inception to first opening) has proved highly popular with tourists and residents alike, to the point where many international visitors now claim it as their main reason for travelling to Singapore.

The concept of 'nested' projects is used deliberately. Even for the procurement phase, distinct projects can be identified, including the international competition for the development of a concept and master plan; ground preparation (the 101 ha site is on reclaimed land); the two conservatory domes; associated administrative, ticketing, and catering facilities; outer gardens and giant 'tree' features; recycling installations for

Table 3.3 Client objectives for the Singapore Gardens by the Bay (SGBB) projects.

SGBB project client objectives
Procurement objectives
1. Time-related.
2. Cost-related (budget compliance; achieving value for money).
3. Quality-related.
4. Transparency of procurement process (corporate governance).

(Continued)

Table 3.3 (Continued)

SGBB project client objectives

Operational objectives

5. 'Bring the world to Singapore and showcase Singapore to the world'.
6. To develop a national garden and premier horticultural attraction for local and international visitors.
7. To create a showcase of horticultural and garden artistry that presents the plant kingdom in a unique way.
8. To provide an entertaining display, while educating visitors about plants seldom seen in this part of the world, with species from cool, temperate climates as well as tropical forests and habitats.
9. To create a lasting architectural wonder.
10. To mount short-term temporary displays as special draw-cards that will also enable local and elderly residents to see particular plants and their environments without having to undertake long and costly international travel.
11. To ensure that SGBB operates in an economically, financially, socially, environmentally, and ecologically responsible and sustainable manner.
12. To maximise recycling and energy conservation activities.

Strategic (global) objectives

13. To create a centre for horticultural excellence and research.
14. To contribute to the branding of Singapore as a distinctively global city.
15. To make SGBB a premier and easily accessible leisure attraction.

National objectives

16. To contribute to a 'green' environment for Singapore.
17. To create something of national pride for Singapore.
18. To create a 'People's Garden' which all citizens can identify as their own.

---

generating electrical power from garden waste (with a collection net beyond the SGBB boundaries); harvesting of rainwater and treatment of waste water; initial planting (with many large and costly exotic specimens); and furnishing with sculptures and artefacts sourced from many countries.

Several expansion projects have already been undertaken (e.g. a children's activity and display area with restricted adult entry), and more are underway, such as the Presidents' Garden on the eastern side of the river estuary surge barrier. SGBB mounts several major exhibition event projects each year, such as the Dutch Spring Tulip and Japanese Sakura displays, in the ticketed dome areas. These are managed through a three-year rolling planning cycle. The objectives for these short-term nested projects align closely with those framed for the original procurement project. The 'nested' project concept thus conflicts to some extent with the 'definite beginning and definite end' definition of the PMI and fits more comfortably into the dictionary definition. Interestingly, staff who have moved from the procurement phase to the operational phase of SGBB continue to refer to it as a project. Coincidentally, its forerunner in the west of England, near to St. Austell, is actually called the Eden Project, possibly as a marketing stratagem but also to denote a continuing characteristic to this commercial and environmental undertaking.

The time, cost, and quality procurement objectives listed in Table 3.3 are not articulated in detail here, but each has specific sub-objectives relating to whichever nested project is undertaken. Procurement objective (4) is interesting in the context of Singapore, where transparency in governance and administration is considered an important concern of public policy in order to combat corruption. The objective also relates to an implicit government culture of not allowing major projects to fail. To reduce this risk, most major public projects are initially managed by highly qualified and experienced public servants, seconded from their home departments, who stay with the project during the critical procurement phase and who sometimes then stay on into the operational phase. A few become even more permanently associated with such projects as they transit from wholly public sector ownership into semi-privatised governance enterprises but with the land remaining in public ownership.

The operational objectives for SGBB in Table 3.3 clearly support an overarching aim to create a major international tourist attraction that will contribute to the national economy, with due attention paid to environmental responsibility. The latter objective continues through to the strategic and national objectives statements.

The distinction between strategic and national objectives for SGBB is quite subtle, since the latter are deliberately framed to include sociopolitical objectives. It is important to the Singapore government that citizens also feel 'ownership' in this international attraction. Of the total SGBB land area, some 95% is allocated to gardens and features that are free for public entry, and many free public events are hosted in the open-air auditorium.

Through the examples presented in this section, it is easy to see how some project objectives can be regarded as 'hard' and quantifiable in terms of achievement, while others are 'softer' and less easily evaluated. This has implications for project risk management in terms of assessing the effects of uncertainty upon objectives.

Consideration of the different types of objectives for projects enables us to examine the phases of projects through which objectives are fulfilled.

3.4 Project Phases

Projects are temporal undertakings. They take place over time, and the way in which projects are perceived determines where the time markers should be placed. For the most part, project management tends to focus upon project activities from inception (when the initial concept is actualised) to a point where the intended function of the project can begin (it becomes operational). Intrinsically, these are the 'definite' beginnings and endings stipulated by PMI (2013: 3), with an implicit connotation that the time period involved is unlikely to extend beyond a few years. However, this restricted view cannot be applied to all projects, as our SGBB example has shown, and a more nuanced and open-ended perspective may be necessary.

For example, an open-cast mine project may have a relatively short design, planning, and construction period, followed by an operational life of perhaps 30 years. At the end of its operational life (whether physically or economically determined), the mine pit and surrounding land must be remediated. However, much of the decision making and planning for this (including the contractual obligations pertaining to it) are put in place during the pre-operational phase. Latency characteristics such as this mean that a longer term approach must be taken to risk management on many projects.

In contrast, other projects have far shorter life cycles. Events projects are good examples of this. With these projects, the planning process might take several months for an event held perhaps on a single day.

For a Formula One (F1) Grand Prix motor racing event, to be held annually on a temporary street circuit, planning might take eight to nine months (usually commencing before the finish of the preceding year's event). Roadworks, traffic diversion, signage, temporary security fencing and barriers, the erection of temporary pits, administration and hospitality facilities, and all the other activities necessary to create the actual circuit and event environment might take three months to put in place. Car racing, including practising, lower tier races, promotional side events, and the culminating F1 event on the final day, would occupy one week. After that, another five or six weeks might be the maximum period allowed by the local public authority to dismantle the temporary facilities and restore the whole area to its former use. In a sense, this represents a continuous cycle of successive similar, but individually and temporally unique, projects.

Perhaps an even starker example can be found in commercial telecommunications projects. It is not unusual for a new cellular telephone 'product' project to have a concept, business case, design, development, and roll-out phase of three weeks or less, followed by an operational life of perhaps only a few months before the service is replaced by another new and 'improved' product to match or beat the offerings of other service providers in a highly competitive market.

These examples suggest that a more integrated, phased approach to projects is advisable: one that does not automatically apply a deterministically measured temporal frame but recognises that a more holistic project view may have to be taken. The holistic perspective can be disaggregated to reflect the characteristics and requirements of each individual 'sub-project' in the greater whole.

The project phases suggested here are procurement, operation, and disposal. These are shown in Figure 3.2, which will be used as a template for further conceptual development. Note how the phases partly match the typology of objectives discussed in Section 3.3. The alignment is not exact, however, with respect to strategic objectives and the disposal phase.

Furthermore, we have deliberately adopted 'phases' as our preferred terminology, rather than 'stages'. This is not done to exclude the possibility of staged procurement (e.g. constructing different sections of a building project in a deliberately sequenced manner), staged operational life (e.g. bringing a gas multi-turbine power station on line as each turbine is commissioned), or staging the disposal phase (e.g. ensuring that delicate artefacts in an exhibition project are removed before the dismantling of more robust exhibits is undertaken).

Using project 'phases' in this way allows us to reserve 'stages' for distinctive periods of activity within phases. The project procurement phase covers all the activities required to initiate, justify, design, plan, construct, test, and commission a project before it becomes operational. The operational phase deals with everything that the project needs to fulfil its intended function, including staffing, maintaining, repairing, and replacing

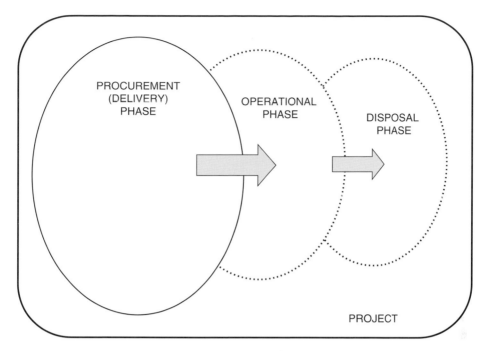

Figure 3.2 Project phases.

parts. The disposal phase occurs when the project function becomes obsolete or has been completed, or when ownership and operations are transferred to someone else.

Note that phases may be linked by commissioning (procurement-operational) and decommissioning (operational-disposal) activities, and thus phase separation is often blurred rather than finitely distinct.

Using this phase concept, we can consider the composition of projects.

3.5 The Composition of Projects

All projects comprise four elements linked by a fifth. These are:

1. Tasks/activities
2. Methods/technologies
3. Resources
4. Organisation
5. Decision making.

The project elements are shown diagrammatically in Figure 3.3.

Tasks and activities identify *what* must be done to bring a project to fruition. In project management, they can be represented through 'programming and scheduling' techniques such as bar charts, work breakdown structures (WBSs), and critical path networks (CPNs). In the procurement phase of a building project, for example, tasks and activities would include everything from the design concept, financial feasibility

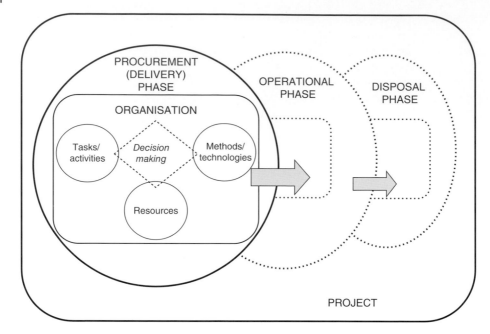

Figure 3.3 Project elements.

studies, estimates and budget preparation, detailed design and specification, and tender preparation and award, through to construction work, services installations, fitting-out, and eventual commissioning for operational readiness.

Methods and technologies propose *how* the project tasks and activities are intended to be carried out and are usually more fully explained in 'project methods statements'. A multistorey apartment building might employ steel columns and beams with bolted connections to form the structural frame (structural steel technology), with tower cranes as the method for materials hoisting (materials movement technology), and pumped ready-mix concrete (reinforced concrete technology) poured over steel bar reinforcement onto temporary metal deck plates used to form the intermediate floor slabs. The deliberate staging of sections of the work would also be a methodological consideration.

Resources comprise all that is needed to carry out the project tasks and activities and to support the proposed methods and technologies. The great variety of requirements for any project can be categorised under headings such as capital, labour, materials, and equipment.

Organisation is the overall responsibility for determining who will carry out the tasks and activities and when, where and when equipment for particular technologies will be required, and what supervisory and coordination arrangements are needed.

Decision making is the 'glue' or essential link between the other four elements, taking place within each element as well as between elements.

An important consideration for project management generally, but especially for project risk management, is that decisions made about elements in the project procurement phase can affect aspects of the operational phase and may even influence the disposal phase. Similarly, decisions made in the operational phase may affect the disposal phase. This is shown in Figure 3.4. Here, the regular project boundary seen in Figures 3.2 and 3.3 has been replaced by a delineation that is much less neat, less certain, but more realistic in terms of the 'messiness' found in many projects.

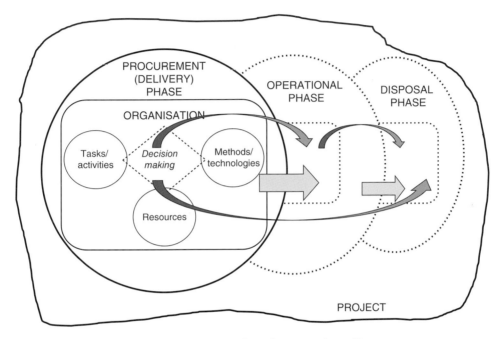

Figure 3.4 Project interphase decision making effects.

For most projects, because these interphase decisions relate to matters that will or could arise in the distant future, their effects are likely to be more uncertain than those pertaining only to the procurement phase.

While phases, elements, and stages provide underpinning for project structures, they also largely determine project implementation processes.

3.6 Processes of Project Implementation

All projects arise from a decision to respond to a perceived need or opportunity or, in the case of many purely creative endeavours, from an inner compulsion to express an inspiration or idea in a communicable form.

For example, a need for a new school is perceived through analysis of relevant demographics and population forecasts. An opportunity may be perceived to exploit a gap in the market to satisfy demand for a particular product or service. An author may recognise an historic or contemporary real-life event as the inspiration for a new novel. A poet may inwardly 'hear' a phrase or sentence fragment as the seed for a new poem; a songwriter might not only hear something similar, but also be inspired by a musical riff to compose a melody. Artists and sculptors follow an inner vision to produce an intended outcome. The decision to respond is the behavioural intent that follows the cognitive perception. Most projects follow a path of ideation, inception, concept development, project development, planning, resourcing, and project implementation, followed by deployment and operation. Thereafter, maintenance and upgrading are inevitable for many projects, and some go through a subsequent disposal process.

A useful and alternative mnemonic way to remember all this employs eight 'P' verbs to describe project processes:

- Propose (suggest the idea, recognise the need)
- Propound (explain, justify, and explore the proposal)
- Prepare (establish requirements and design solutions)
- Proceed (implement the project procurement activities)
- Protect (manage the threat risks associated with the project)
- Propel (ensure the project delivery process is completed successfully)
- Produce (arrive at a productive operational outcome for the project)
- Prolong (establish to what extent the project will be ongoing).

In many instances, each 'P' will be close to many decision points for the project.

3.7   IT Project Example

An example of project implementation is presented here as a real-life Web-based IT project. This project will be used again in Chapter 4 to explore the application of project risk management.

The key stakeholder in the IT project is a commercial group comprising several fashion retail companies (business units) that sell branded merchandise in branch stores and online. The online presence is the responsibility of a small project management unit that develops, maintains, and periodically refreshes the website of each company in the group.

The project is to develop and deploy a brand product personalisation service for one business unit to offer to its online customers. Strategic objectives for the project (for the business unit and the holding group) include: growing the brand's online business, increasing market penetration, maintaining customer brand loyalty, and maintaining a competitive edge. The operational objectives sought by the business unit include increased sales revenue by attracting more unit sales and by charging for the personalisation service. For procurement objectives, the Web development unit wants to complete the project in as short a time as possible, integrate the new service seamlessly into the existing online payment system, minimise downtime for the existing website, and create a template that can be used to replicate the service with other companies in the group. The project comprises a concept development stage, a project development stage, and a deployment and operational stage.

3.7.1 Ideation and Concept Development

Inception of the project commenced with the idea of providing a unique online service, over and above actual product unit sales, which would not only distinguish the brand label's presence among other online competitors, but also attract more online customers since that sales method incurs lower operating costs than direct store-based transactions. The idea arose from recognition of the Web presence as an opportunity to be exploited.

At this point the idea, which came from the company sales manager, did not specifically refer to the introduction of an online product personalisation service, but rather to providing some sort of unique personalised service to online customers. Development of the idea came later, where it was presented and brainstormed at a company sales meeting. This produced a proposal to create a product personalisation service, whereby online customers could order garment purchases to be embroidered in appropriate places (e.g. chest front, cuffs) with personal initials or names. It was recognised that the service would be exclusive to online sales, since the garment personalisation must be carried out at the factory/warehouse prior to delivery. For store-bought items, this could not be done at point of sale, and trying to do so might deter customers.

Informal discussion in this way acted as a contextualising and stimulating process, and ideation moved rapidly to concept development, such that the business unit was quickly able to formulate a draft proposal to be discussed with the IT project management unit. This discussion established a high-level requirements brief, a time and resources plan, and a target budget. The expanded proposal was checked for feasibility and resource adequacy by the project manager. At this point, an initial go/no-go decision was possible. If the proposed project cleared this hurdle, the business unit, with the help of the project manager, could prepare a full business case for the Web application project, including its anticipated ongoing maintenance requirements.

The project received a 'go' decision from the business unit manager and the IT project manager, and the developed business case was then submitted for approval to the board of directors for the holding company. Assent to go ahead was received, and a steering committee appointed to provide oversight to the project.

3.7.2 Project Development Stage

After the board approval for the project, the project manager and brand business unit manager conferred again to consider timelines and resource allocation in greater detail.

From this point, the project manager dealt directly with the in-house Web designers and developers for the project. The requirements brief was clarified and expanded as project staff requested further information. A third party became involved as the Web-based IT project required a 'payment gateway' as customers needed to make online payment for their purchases. Company and customer expectations of payment security meant that the gateway development was best outsourced to a specialist provider. Since such providers are usually very large international finance-based companies, the project timelines had to be fitted around their availability rather than vice versa.

Proof of concept was undertaken by developing a small part of the project and testing it for feasibility and integrity, and initiation of the third-party involvement followed on very quickly after the proofing was satisfied. In almost all IT projects, the designers and developers have already begun their work of building and testing the back-end (what is seen by the brand business unit) and front-end (what is seen by the online customers) systems for the project.

Test case protocols were drawn up for the whole system, and exhaustive testing was undertaken. System modifications were carried out and retested until the project system's beta performance was deemed satisfactory in terms of user-friendliness and reliability.

3.7.3 Project Deployment and Operation

By this time, the project manager was already working on another IT project in parallel with this one, but now the project launch and deployment were imminent. This is a critical time for all IT projects.

Once the new Web application was deployed and made 'live', a short period of evaluation and fine-tuning took place, during which any time needed to take down the website for further adjustment was kept to an absolute minimum.

No formal project closure or signing-off took place, but the project manager was very quickly made aware, through the business unit, of any deficiencies in the live application.

3.7.4 Operational Maintenance

This IT project continued to received maintenance attention in terms of resolving minor 'glitches' in the Web platform. Upgrading of the payment gateway was implemented by the service provider, and the whole application has been migrated to a new Web operating system. As yet, there are no intentions to discontinue and dismantle the product personalisation service, which has proved popular with customers and is fulfilling the project objectives. Additional potential for this project, in terms of an opportunity risk, is described in Chapter 15.

After looking at the processes of project implementation, it is appropriate to explore how they are organisationally structured for projects.

3.8 Organisational Structures for Projects

Project stakeholders are an important consideration in project risk management. In Chapter 2, we discussed risk and uncertainty, their association with each other, and their strong relationship with project decision making. It is therefore vital to know who may be involved in a project and the powers they have, or the influence they can exert, over the decision making of others. Establishing the organisational structure for a project is an essential first step in exploring stakeholder relationships and powers.

Project stakeholders are often allocated simple labels that attempt to describe the nature of their powers and their relationships to a project. For example, we may hear of internal and external stakeholders, of primary (or key) and secondary stakeholders, of direct and indirect stakeholders. Such descriptors often fail to fully accommodate all the subtleties of real-life projects, and it is usually better to start with a visual representation of the organisational structure for a project.

Organograms are a useful tool in many areas of project management. They can show stakeholder status, contractual relationships, project governance structures, and communication vectors. Such diagrams may also provide a rough measure of the complexity of a project. Rather than trying to show all these aspects in one diagram, different layers can be created to emphasise a single aspect or combinations of a few of them at a time. Trying to show everything at once may create a visual 'clutter' that is counterproductive to the intended purpose.

Figure 3.5 is a conceptual illustration of stakeholders for a hypothetical project, with very simple identification and labelling. This organogram is not particularly useful other than identifying several stakeholders and confirming that they are associated with the project. It does not tell us who each shape represents, nor the nature of the stakeholder's relationship to the project and to other stakeholders. The identification codes could refer to a more expansive, detailed list. Different object shapes might refer to particular types of stakeholders, as also could colour shading if it were introduced into the diagram. Purely conceptual, the figure does not yield much of value for project management, and its use may be limited to a presentation intended to 'sell' the project as part of its justification.

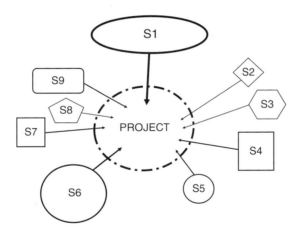

Figure 3.5 Project stakeholders.

In order to add more value to the conceptual diagram, it should provide greater information about the project stakeholder relationships.

3.9 Project Stakeholder Relationships

The management of project stakeholder relationships is more often than not based upon analysis of some level of the power/distance ratio, as this will largely determine *how* they should be managed. A high power/distance ratio for a project stakeholder means that it has strong capacity to *directly* influence project decision making and hence project risk. That stakeholder must be managed carefully, through direct and effective communication, frequent consultation, and high levels of information supply. A project client entering into a contractual agreement with a contractor provides a good example of two such stakeholders. By contrast, project stakeholders exhibiting low power/distance ratios tend to have far less capacity to influence project decision making, and usually fewer or largely indirect channels through which to express their views. They are often managed through minor or infrequent consultation processes and minimal provision of information through means such as short press releases. Tenants in a housing project or ticket buyers for an event project are examples of the latter situation. However, caution is needed in this type of stakeholder analysis. The principle is not absolute, and many a project has been led to abandonment or failure because of the collective efforts of low power/distance ratio stakeholders and the unexpected intrusion of an inquisitive press or strident adverse publicity on other social network media forums, for instance.

Figure 3.6 develops the previous stakeholder concept by adding more information about the stakeholder relationships. This diagram could represent a typical construction project where the design process is separated from the bidding and construction processes. S1 represents the project client, with S7, S8, and S9 representing designers (who might be a mixture of in-house and independent professional consultants), all grouped together as project stakeholders in a client–design team coalition.

On the other hand, S4 might represent the main contractor, and S2, S3, and S5 the subcontractors and suppliers to the main contractor, in a loosely bound contractor coalition. Project stakeholder S6 might be a regulatory authority independent of both coalitions, but with a high power/distance ratio in terms of knowing what can be constructed on the site and how it must be built.

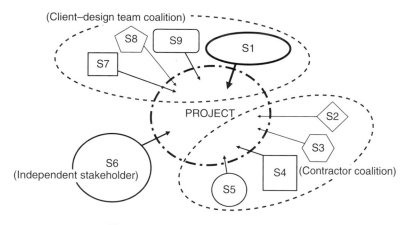

Figure 3.6 Project stakeholder coalitions.

Figure 3.6 begins to provide some information about stakeholder positioning in the organisational structure for a project, but it still falls short of being really useful. The next series of figures, taken from real-life projects, shows how structurally rearranging the organograms can redress this lack.

Figure 3.7 is taken from a refurbishment and extension project for a hospital in southern Africa. Several aspects of this organogram are worthy of attention.

Four of the project stakeholders are public authorities: three at the national level and one at the local municipal level. A fifth stakeholder could also, at least partly, have public sector association since the electrical supply organisation could be a parastatal (publicly financed, but operating as a quasi-private entity).

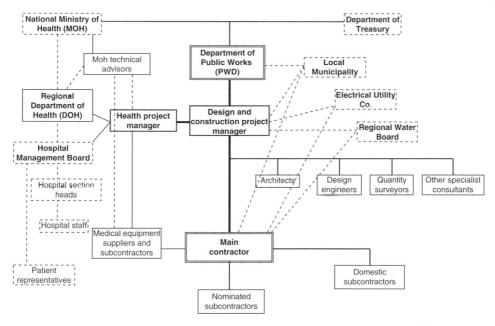

Figure 3.7 Hospital project organogram.

The Department of Treasury is quite unique in this and similar organograms. It is a high-power and high-distance stakeholder, since it is remote from almost all project decision making, but has immediate power to restrict funding. Even if funds are already committed, unspent amounts can be withdrawn simply at the political will of the government in power.

Relationships between government departments can be at best cautious and at worst quite acrimonious, and this can give rise to threat risks for other project stakeholders.

The organogram is generally structured as a governance diagram for the hospital project, but some vectors denote communication lines. For example, the local municipality, the electrical utility company, and the water board each have temporary contractual relationships with the Department of Public Works (or its project manager). These temporary arrangements eventually become permanent agreements with the Hospital Management Board when the completed hospital works are handed over. The three organisations each maintain communication links with the main contractor to deal with day-to-day issues during construction.

Surprisingly, the organogram does not make it clear which organisation is actually the project client. In this instance, the client stakeholder is likely to be the Regional Department of Health. The presence of a health project manager, Ministry of Health technical advisors, and specialist medical equipment suppliers in the organogram points to the potential complexity of this project. None of them is in a direct contract relationship with the construction contract governance arrangements.

The nominated and domestic subcontractors to the main contractor are shown as single groups in the organogram, as are the medical equipment suppliers and subcontractors. In reality, these are likely to comprise many individual organisations, each with stakeholder attributes.

Two final observations merit attention. Firstly, while the organogram does not represent a design-build integrated procurement system, it does not necessarily therefore depict a traditional separated design-bid-construct procurement system. After an agreed level of design is reached, all the design professionals could be 'novated' to the main contractor under a hybrid design-novate-build system, whereby the ongoing professional services and fee expectations of the consultants are transferred from the client to the main contractor after the construction contract has been awarded.

Secondly, regardless of the procurement system, all the consultants providing professional design, advisory, or project management services to the project could be permanent employees of the relevant stakeholder organisations (Department of Public Works, Ministry of Health, and Regional Department of Health); or they could be independent organisations who have tendered for the opportunity to provide professional services; or they could comprise a mixture of both. Such possibilities directly influence the level of project complexity and indirectly influence the degree of project uncertainty and associated risk.

Figure 3.8 is the organogram for a radically different type of construction project in southern Africa, this time for informal settlement upgrading. Housing settlements are common throughout the African continent. They start as bottom-up informal responses to social housing needs that largely lie beyond the financial capacity of governments to satisfy through provision of formal housing. Such settlements are invariably associated with overcrowding, poor water supply, and poor sanitation services. In the past, this has led to attempts to destroy the settlements due to fears about social discontent and the spread of diseases such as typhoid and cholera. Few such attempts have been successful, and, unless costly security arrangements are put in place, the razed settlements are quickly repopulated, not only by the former occupants, as rehousing schemes are often inadequate or non-existent, but also by newcomers to the site as rural-to-urban migration increases.

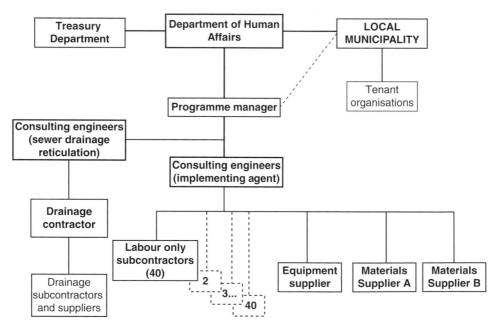

Figure 3.8 Settlement upgrading project organogram.

Latterly, governments and municipal authorities have begun to formulate radically different policies, belatedly recognising the inherent social stability in many well-established informal settlements. Upgrading schemes have been implemented to extend the provision of water supply and to introduce basic but more hygienic sanitation systems for dwellings. At first, such schemes involved the construction of basic ventilated pit toilets. Subsequently, portable toilets (with emptiable 'buckets') were used in areas with difficult ground conditions. More recently still, for stable settlements and where it is physically feasible, pit bucket toilets are being replaced with more conventional facilities connected to waterborne sewage disposal systems. This project example deals with the reticulation of sewer drains and the construction of more permanent outdoor toilets for almost 1800 dwellings in a relatively long-established housing settlement where many of the former 'shack' houses have already been improved by their occupiers. The upgrading project does not attempt to deal with issues of land title and occupation but addresses only the more urgent sanitation inadequacies of the settlement. Other issues can await more patient attention and resolution. Funding was made available through a specific government scheme.

Each toilet facility comprises a washdown closet and basin. Cement block walls, which are rendered externally, are built on a simple concrete slab floor. Roofs are mono-pitch and covered with corrugated cement or galvanised steel sheet fixed to timber rafters. Water piping is 12 mm diameter plastic tubing and fittings. The toilets are each connected to the newly reticulated drainage system.

The organogram shows that public stakeholder involvement is still evident, but to a lesser extent than in the previous hospital project example. The Treasury Department supplies scheme funds to the Department of Human Affairs but plays little part in project decision making. The transactions of both departments are, of course, subject to scrutiny by an Auditor General.

The local municipality communicates with the Department of Human Affairs on matters of policy and funding, but directly with the programme manager for day-to-day concerns. Tenants/occupiers of the settlement dwellings are regarded as important stakeholders by the municipality.

In this example, the programme manager and the consulting engineer (as implementation agents) are appointed after bidding for the opportunity to provide these services. The programme manager has overall responsibility for several similar projects, while the consulting engineer provides design services and acts as project and contracts manager for the toilet upgrading project. A different engineering consultant has design responsibility for the drainage reticulation and management oversight of the successful tenderer for this work.

Although visually a simple and straightforward project in the organogram, project complexity arises through the large number (40 at any one time) of labour-only subcontractors involved, each of which must be managed by the consulting engineer. Each subcontractor bids for a limited number of toilet installations (thus essentially giving rise to mini-projects), and this diversifies the risk of overall project failure. This strategy also encourages small business enterprise development with limited capital requirements and easier market entry. Bidders must show how much local labour they intend to use and what training opportunities they will offer to unskilled workers. The simple construction technologies involved do not require sophisticated tools and machinery, and occupational safety implications are thereby minimised. There are therefore economic and social upliftment benefits deliberately built into this project.

The equipment suppliers, Materials Supplier A (construction materials) and B (plumbing materials), bid for the supply contracts on an annual basis, thus introducing greater competition into this part of the project. These suppliers also have to demonstrate affirmative action and social upliftment in their project stakeholder roles.

Complexity (we expand on this topic in Chapter 13) is also added by the need for the drainage reticulation to proceed slightly ahead of toilet construction. Tenders for this work, on a competitive labour and materials basis, were invited from local small and medium enterprises. Economic and social upliftment clauses were also included in the contract agreements. As the drainage work and toilet construction essentially comprise 41 separate contracts, a high level of intercommunication has to be maintained, at least between the contractors and the two managing consulting engineers.

A different type of service installation project for a residential development is shown in Figure 3.9. In this case, a large greenfield site adjacent to a rural town has been set aside for township development that will include houses, shops, schools, churches, clinics, and other community and commercial facilities. The national electricity supply authority has already installed major transmission lines and transformer stations in this region, and now proposes to build a 130/11 kV step-down substation to service the new development.

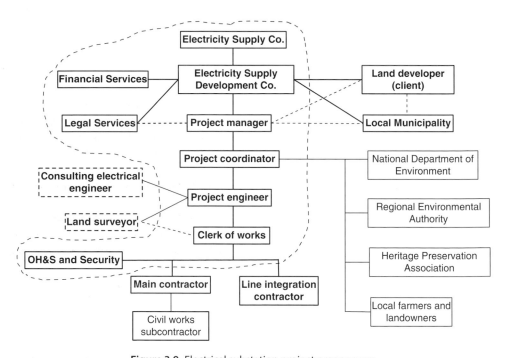

Figure 3.9 Electrical substation project organogram.

The dotted irregular line in the organogram indicates the various in-house roles of the electricity supply company. All other stakeholders are external. The land developer is the initial 'client' for the project and will be charged take-off power costs until sales or lease contracts are signed between the developer and eventual owners or occupiers.

As with the toilet upgrading project, two main contractors are involved: one for construction and electrical work within the substation, and the other for the connection work to the existing high-voltage transmission lines. The substation contractor has one subcontractor shown on the organogram for the civil engineering work required. Both main contractors will have additional subcontractors and suppliers not shown on this diagram.

Because this is a greenfield site in a largely rural and historic area, there is a high level of interest shown by environmentalists and conservationists, as indicated by the stakeholders on the right-hand side of the organogram. This adds political risk to the project (Chapter 14). Adjacent farmers are also concerned for their interests to be represented.

Despite the largely in-house organisational structure for the substation project, the demands for effective intra-stakeholder and inter-stakeholder communication links are high, with the dual-contractor arrangement adding to the complexity of the project. On the other hand, the electricity supply company is well-experienced in such work.

Yet another residential project example is represented in Figure 3.10, which shows the organogram for a 180-unit residential development constructed as one 5-storey block and one 12-storey block, with two basement floors of parking and $500\,\text{m}^2$ of leasable retail space on the ground floor. This project, to be built over a two-year period, is in an outer

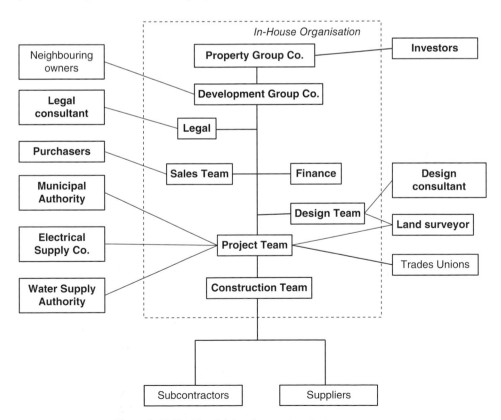

Figure 3.10 Residential development project organogram.

suburb of a major international city. Units will be sold to individual purchasers, and two large banks have agreed to provide construction finance as investors in the project.

Noteworthy in this example is the tight in-house management of the project, in a manner that is commonly regarded as 'enterprise project management'. Not unusual for this type of speculative development project, the developer is a subsidiary of a larger property company. This in itself is a mitigating response to the threat risk of insolvency of either company.

The in-house design team is augmented by the broader professional services offered by an external design consultant, and a professional land surveyor provides land information and setting-out services to the internal design team and later to the project team when construction begins.

The in-house legal team provides general legal administration to the whole group, while an external law firm provides services for unit sales contracts and advice about any legal issues relating to neighbouring property owners and other matters.

Concerns of neighbouring property owners are initially received by the development company, and if they are potentially matters of dispute, they are immediately routed through the internal legal department to the external legal consultant.

Matters relating to the municipal authority, the electrical and gas services supply company, and the water supply company are directed first of all to the project management team and thereafter to the internal design team or the construction team, whichever is the most appropriate unit to deal with them.

The project management team deals with the procurement of subcontractors and suppliers, with the construction team assigned responsibility for their day-to-day management.

Trade union matters are initially directed to the project team, but where necessary are transferred to the legal team, and even further to a specialist industrial relations lawyer.

Although not immediately apparent from the organogram, the key role in this example is taken by the project management team. It is only placed below the design team in the diagram because the latter often prepares conceptual sketches for proposals that do not eventually become firm projects. The internal project team is appointed as soon as the directors of the development group receive approval from the parent company to go ahead with a project. Thereafter, all project matters are notified to the project team, including unit purchasers' occupation dates and requests for individual variations to finishings and fittings. Issues with neighbours are also dealt with by the team if they involve disruption to work activities or temporary access.

While enterprise project management undertaken in this way can be highly effective and efficient, it requires clear internal and external communication that can quickly degenerate into a 'scatter gun' approach where email messages and attachments, for example, are sent to almost everyone with little regard for who *should* receive them.

Further examples of project organograms can be found in Case Study A (a PPP correctional facility; see Figure A.1) and Case Study B (a rail improvement project; see Figure B.1). Each of these projects involves a state government, and the organisational structures show how contemporary governments 'distance' themselves from day-to-day project matters. Up until the late twentieth century, permanent public works departments (PWDs) would have undertaken this responsibility. In less economically developed countries, this is still the case.

The examples of project organisational structures shown here demonstrate potential arrays of project stakeholders that can be encountered. They enable us to perceive one aspect of project complexity and give some insight into the governance structures for projects and into the locations of project decision making. In order to pursue the latter in greater detail, it is necessary to consider the organisational structures of individual project stakeholders.

3.10   Stakeholder Organisational Structures

Using the dominant one of the five basic components of all permanent organisations (the operating core, the strategic apex, the middle line, the technostructure, and the support staff), Robbins and Barnwell (1998) exemplify the basic structural alternatives for all organisations as:

1. Simple structures
2. Machine bureaucracies
3. Professional bureaucracies
4. Divisionalised forms
5. Adhocracies.

3.10.1   Simple Structures

In a simple organisational structure, the organisation tends to be vertically and horizontally centralised – with little or no decentralisation – often under the direct supervision of a principal who is the strategic apex of the organisation. Small trades-based subcontractors and sole principal or small partnership consultancies tend to fit this type of structure. Its advantages include: rapid decision making, clear lines of accountability, direct and unambiguous communication, and simple goal setting. On the other hand, simple organisations are usually vulnerable to unforeseen loss or change in personnel at the strategic apex, they may have growth limitations, and the sheer concentration of power may inhibit effective decision making.

3.10.2   Machine Bureaucracies

Organisations configured as machine bureaucracies tend to rely on standardised processes as a coordinating mechanism, with the technostructure as the key part. In this part are grouped all the technical knowledge and skills of the organisation, with separate departments providing support services. Such organisations tend to have little vertical, and limited horizontal, decentralisation. They can carry out standardised activities efficiently, often with economies of scale, but have potential for conflict to arise between the operating core and support staff. Such organisations may also be reluctant to innovate, and they may dislike change.

Figure 3.11 provides a somewhat hybrid example of a machine bureaucracy in the form of a large construction company eligible to undertake project work in the highest value band (unlimited project value) of the category range imposed by a public sector client.

In this example, one of the company directors has direct oversight of all projects undertaken by the company, while other directors have responsibilities in the various support mechanisms. It is the use of directors in this way that makes this a hybrid

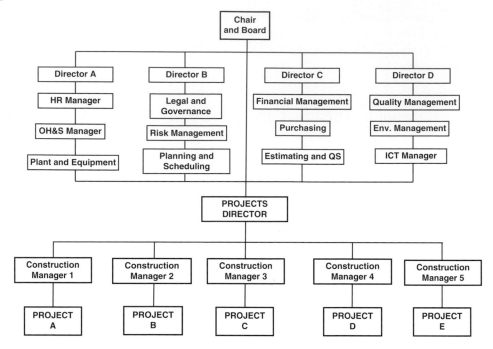

Figure 3.11 Construction company organisational structure.

example; in a strict machine bureaucracy, these roles would be filled by general managers. The horizontal decentralisation at the project level is not limited to five projects each in the highest value category, since company policy allows designated contract managers to deal with one maximum value project and up to two more projects in lower value categories.

The Estimating and Quantity Surveying (QS) section (under Director C) also deals with bid preparation and submissions. The Plant and Equipment section (under Director A) deals with company-owned equipment and also with operational matters for equipment leased in conjunction with the Financial Management section. Not shown in the figure are the foremen, work crews, site administrators, and site quantity surveyors under each construction manager.

From the project director downward, the organisational structure more closely resembles a simple structure, and therein lies its vulnerability. If realistic succession planning or crisis planning is not in place, the whole organisation could be jeopardised by loss of control at this level.

In this example, there is no clear identification of responsibility for ongoing business development activity, nor is there any inclusion of corporate social responsibility (CSR) involvement. As will be seen in this chapter, this organisation could be part of a divisionalised form as a regional branch of a national or international conglomerate. In its present form, the organisation's structure looks to have grown organically (i.e. with sections added on as needed) rather than having been systematically planned.

A simpler example of this type of structure is shown in Figure F.1 (Case Study F, an aquatic theme park project).

3.10.3 Professional Bureaucracies

Professional bureaucracies are similar to machine bureaucracies, but the main difference lies in the identification and implementation of professional knowledge and skills as the key operating (project decision making) part of the organisation. They also tend to be more flexible with regard to vertical and horizontal decentralisation. Standardisation of processes is decentralised across the organisation, and task specialisations occur. There may be minor potential for sub-unit conflict, but more often an obsession with 'professional rules' can be found.

Figure 3.12 shows an example of an organisation structured as a professional bureaucracy. It too is in a somewhat hybridised form. This example is taken from a large professional engineering consultancy that offers different specialised consultancy services. Currently the consultancy has two senior partners, with the core operating under the direction of the other three partners (with a fourth planned). The core is multistranded, with each strand comprising several teams under the leadership of a partner. The teams are each led by associates or senior engineers, and comprise junior qualified engineers, engineering technicians, and trainees. The hybrid nature of this organisation lies in the way the strands and teams operate. In many engineering consultancies, the partners each direct project teams wherein the teams carry out similar engineering design work on different projects. In this organisation, the strands comprise different engineering specialisms.

Thus, Strand A (under Partner A) has Team A1 dealing with civil engineering, earthworks, and roads; Team A2 deals with bridges and tunnels; while Teams A3 and A4 focus mainly upon structural engineering for building projects. Strand B has four teams

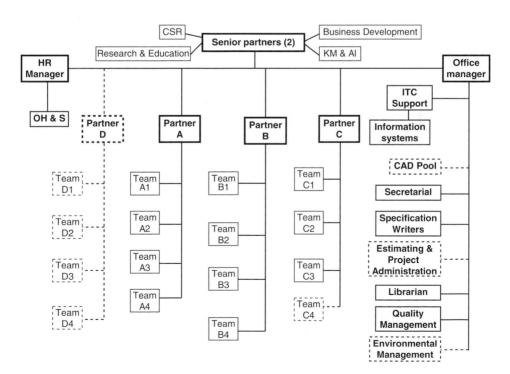

Figure 3.12 Engineering consultancy organisational structure.

that deal with mechanical engineering, marine engineering, mining, and rail engineering. Strand C covers services engineering including electrical; heating, ventilating, and air conditioning (HVAC); and hydraulic engineering. It is about to commence operating an environmental engineering team to meet perceived demand for this service. When this happens, the existing environmental management section will move into the new team section under Partner D.

A part-time associate partner (brought back from retirement) has reporting responsibility direct to the senior partners about business development opportunities, and this has led to the proposal for another specialist strand (D), yet to become operational, that will expand the consultancy range into township planning, architectural design, and project management. This is quite a radical change since it transforms the consultancy from multi-engineering specialists into multidiscipline consultants. Under this change, the Estimating and Project Administration section will move from the control of the office manager to be under the direction of Partner D and align more closely with the new Project Management section. The proposal is one of several arising from an independently commissioned review of the whole of the organisation's activities and structure, part of which will almost certainly lead to corporatisation of the current partnership arrangement. Other areas of expansion explored, but that have not yet reached the planning stage, include fire-safety engineering (already partly dealt with in the services engineering strand), pipeline engineering, safety management, and risk management. The proposal is that the latter two will become part of the proposed project management unit, rather than continuing to be rather weakly represented in each of the strand teams as they currently are. In this way, they could be offered as independent billable services to clients.

The dotted panel for the CAD (computer-aided design) Pool, composed of computer-aided drafting technicians, under the oversight of the office manager also signifies change occurring under contemporary trends. Previously, this pool serviced the drafting requirements for all teams. Latterly, however, qualified engineering graduates have been recruited with accredited CAD skills as part of their degree programmes, and the whole organisation has moved to an enhanced high-performance workstation configuration within each design team. Existing drafting technicians have been encouraged and assisted to gain engineering technician qualifications and have been transferred into the teams where they are best suited. The CAD Pool is almost finished with. This change process has been almost seamless, and the intention is to repeat it soon with the specification writing pool.

The teams operate not only as engineering specialist contributors to projects led by other teams, but also as design engineers and project managers for their own projects. Thus, the electrical engineering team may provide input to a project managed by the structural engineering team, or may provide independent professional services to an external project where none of the other teams is involved. Here, too, change is underway, as Strand D is planned to include a Project Management team or unit. This not only will provide project management services for all projects undertaken by the consultancy but also will seek and bid for opportunities to offer project management services as a business specialism outside the organisation, initially in a limited number of the partnership's specialist engineering areas but eventually across all areas.

In addition to exercising oversight of all the activities of the company, the two senior partners deal with more fluid aspects. Business development and organisational change comprise one area, and this also involves exploring strategic expansion opportunities through national and international alliances and mergers. Involvement in CSR activities is another. More speculative areas are also pursued at this level. Recognition of the need to engage more closely with knowledge management (KM) and artificial intelligence (AI) has led to deeper investigation of both, with the collaboration of the organisation's librarian and IT Support unit and by using external consultants. Perhaps the most interesting area is that covered by Research and Education. An associate partner has been seconded from one of the specialist teams to explore, on a two days per week basis, innovative engineering solutions, focusing particularly on biomimicry whereby the activities of natural organisms can provide clues for solving engineering problems such as structural stability, wind flow, weather resistance, water harvesting, high-temperature toleration and resistance, and so on. The investigative activities are also aimed at sponsoring and co-funding university research projects where this is feasible, while the education aspect aims at involving partners and staff directly in the curricular activities of local universities and technical colleges.

Comparison between the two examples shows that the construction company is probably less flexible in its organisational structure than the engineering consultancy and is likely to be less amenable to change. The engineering consultancy is more proactive in this regard than the construction company. The latter is also very dependent on the capability of the project director yet shows no clear evidence of backup capacity, whereas the engineering consultancy has diversified specialised areas across the organisation such that each strand is likely to have at least some resilience and redundancy in the event of shocks to the system. Both examples serve to show where the locus of project decision making exists and how decision support, approval, and oversight are carried out.

3.10.4 Divisionalised Forms

In the divisionalised form of organisational structure, a larger meta-structure is imposed over a conglomeration of separate machine or professional bureaucracies. The separate entities may be quite disparate in that each sub-organisation could be involved in quite individual activities, and the meta-structure acts as a coordinating mechanism for the whole group. Alternatively, semi-autonomous regional organisations conducting the same type of business might be brought together to exploit strengths and opportunities at a national or international level. Thus, separate construction companies would operate semi-independently in different regions of a country but be brought together as a group under a national banner with some shared group support services such as IT, finance, and legal administration. Staff might be seconded from one company to another to facilitate particular project undertakings, but usually under a temporary rather than permanent arrangement. Professional consultancy organisations group together in a similar semi-autonomous way at regional levels and internationally.

Autonomy can be an advantage in organisations with divisionalised form, but it may also be disadvantageous. For example, decision making is more rapid within each single organisation, but is often found to be less consistent across the group.

The success of divisionalised organisations relies heavily upon the competency of the middle management that undertakes most of the decision making in each sub-organisation. However, this form of organisational structure does leave the apex free for 'big picture' strategic planning and management, and also provides an onward career path opportunity for senior managers in the sub-organisations.

Tracking decision making can be elusive in the meta-organisation, but this is rarely needed in project risk management since the projects are mostly undertaken autonomously by the sub-organisations in the group.

3.10.5 Adhocracies

Adhocratic organisational structures are essentially those that do not quite fit any of the other structural forms. They are predominantly temporary in nature and often comprise a number of separate and independent organisations brought together by mutual intent. Project organisational structures are thus typical of adhocracies, and examination of any of the example structures offered earlier in this chapter will show how well they fit this description. Any decentralisation in adhocracies occurs selectively and intentionally, and for many the apex is actually bipolar (e.g. client and contractor).

The strengths in most adhocracies lie in their flexibility and responsiveness, although this is not always a feature of project organisations brought about by a series of formal contract agreements. Adhocracies rely heavily on the specialist advice and input of each participating stakeholder for decision making. There may be conflicts and power struggles within adhocratic organisations, and sometimes power ambiguity, for example where one stakeholder finds its expected decision making capacity is actually diluted by the power held by other stakeholders.

Adhocratic organisational structures tend to operate in a dynamic and complex environment (thus matching the characteristics of risk) and can be difficult to coordinate. Perhaps the best examples of this form of organisational structure are found in the film industry, where many comparatively small companies make unique creative contributions to each production.

For many project managers, a 'matrix' project organisational structure is preferable, whereby the matrix shows project stakeholders listed down the left-hand column. Along the top row, two alternatives are possible. Either the required functional roles are entered, or the detailed project stages. Thus, the matrix indicates who is doing what, or it indicates the project stages during which each stakeholder will be active. As such, its usefulness is limited to two possible 'layers', neither of which portrays any type of relationship *between* project stakeholders, and the matrix acts more as a memory jogger in both formats.

Having explored the nature and types of stakeholder organisational structures, we can now turn to *how* they may be managed.

3.11 Modes of Organisational Management

Parkin (1996) usefully offers two contrasting modes of organisational management: the *administrative* mode and the *enterprise* mode.

In the administrative mode, organisational management seeks to establish an orderly approach in terms of policies, structures, hierarchy, functions, control and adaptability, and external relationships.

The contrast with the enterprise mode lies in the latter's less tangible but greater pre-occupation with meaning, interests, power, politics, and symbolism. The administrative mode will try to establish a common set of objectives and then work closely together to achieve them; while the enterprise mode recognises individual and group interests, and the organisation becomes a loose coalition or alliance of these. The administrative mode seeks harmony and removal of conflict (i.e. with everyone following the rules), whereas the enterprise mode accepts conflict as natural and often desirable. The administrative mode prefers clear lines of authority, leadership, and control; but the enterprise mode recognises the power of individuals in resolving conflicts of interest.

This is not necessarily a mutually exclusive 'either/or' situation, but rather one of the 'dominating mode' of approach to management found in any organisation. However, the question arises as to whether or not an organisation can switch its dominant mode easily and without major disruption to staff and processes. This dilemma is frequently in projects where design is a major input. The creative traits of designers, such as architects, tend to make them more amenable to enterprise modes of management where they are able to argue passionately for their cause. While this may be reasonably accommodated within a project stakeholder organisation where design is the major activity, it does not always fit easily in the organisational structure for the project itself, particularly where project managers attempt to impose a more controlled administrative mode. In practice, design consultancies have tended to adapt to what could become an explosive, conflict-ridden project organisational environment by using different staff for pure design activities and for the project implementation role.

Having considered the organisational structures for projects and project stakeholders, and the ways by which they can be managed, our attention now turns to project stakeholder *decision making* since this plays a large part in managing project risks effectively.

3.12 Project Stakeholder Decision Making

Decision making occurs in every aspect and at all stages of all projects. It is the essential precursor to how and why projects are conceived and implemented.

In management, Parkin (1996) describes decision making as the 're-traceable expression of the process of organising'. While this may be so for the management process, for decision making at the intrapersonal level, such auditing may be less practicable. As individuals, we do not always know precisely why we make the decisions we do, often not even at the time of making them, let alone when we try to trace backwards to them at some point in the future. What this means is that the cognitive inputs to intrapersonal decision making are not always memorable.

Because of the nature of projects, the process of decision making may be intrapersonal, interpersonal, intra-organisational, and inter-organisational. It is also a cognitive and behavioural process. If we leap forward momentarily to Chapter 11 (Project Risk Knowledge Management) and Figure 11.1, we find knowledge depicted as steps on an upward gradient, comprising: data, information, knowledge, understanding, and wisdom. Data are simply assemblies of observable facts or opinions. Information becomes evident when these assemblies are arranged into concepts. Knowledge is gained when the concepts are fitted into appropriate contexts. Understanding occurs when the context derives from a particular philosophy, and wisdom is the capacity to choose an appropriate course of action or know how to act if there are no alternatives to choose from. The first four steps are cognitive; the last is both cognitive and behavioural.

To explore the decision making process, we offer an example taken from a construction context. It represents the situation of an independent professional consultant (architect, engineer, quantity surveyor, etc.) commissioned to prepare information and advice for a client for a proposed construction project. The decision process shown here conceptually is applicable across a wide range of consultancy fields. In this case, the process shown in Figure 3.13 follows upon earlier client decisions about selecting and appointing professional consultants for the project and about preliminary project briefing information.

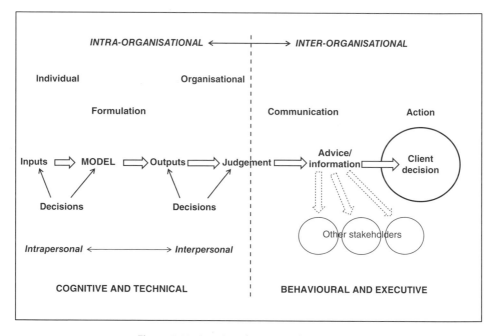

Figure 3.13 A project decision making process.

The 'model' referred to in Figure 3.13 might be, for example, the cost modelling tool of a quantity surveyor, a computer-based Monte Carlo simulation model, a structural design calculator, an architectural spatial allocation model, an electricity demand load calculator, a pipe sizing model, a reference set of legal precedents, or a software code algorithm. These models are representational; they are intended to represent real-life systems. We use them on every project to inform our decision making.

Project decision making is often guided by codes and specifications and is largely conducted by discipline-based professionals in the construction industry. Control of the knowledge and information needed for such decision making actually constitutes the power source of each professional discipline, and in some countries their roles are legally protected. The codes, specifications, and regulations, and professionalism itself, are all aimed at minimising threat risks to society, but they can also tend to stultify the creative exploitation of opportunity risks.

For the example shown in Figure 3.13, the decision process commences with a series of cognitive choices opened up by posing technically based project issues as questions:

- Is a model needed?
- What type of model should be used?
- How reliable and accurate is the preferred model?
- What data/information inputs are required to service the model?
- How will such data be sourced?
- Will any data transformation be required?
- How reliable and accurate are the data/information?
- What uncertainty is associated with the data/information?
- What model output format is required?
- What uncertainty is associated with the model output values?

The decision process so far might be carried out intrapersonally by one staff member or interpersonally by a few staff in a small team. It comprises intellectual (cognitive) effort, followed by judgement (also cognitive) and then deliberate action (behavioural) based upon that judgement. Although 'model' is used in the singular here, in practice the modelling exercise might require the combined use of several different models.

The outcomes of the model are then assessed within the consultancy organisation (usually interpersonally rather than intrapersonally) by applying criteria such as appropriateness, accuracy, and reliability, thus demonstrating understanding. The implications for the project are considered by applying professional wisdom. These implications may be discussed with other project consultants on an inter-organisational basis.

In practice, the model outcomes are now framed in terms suitable for communicating inter-organisationally as recommendations or advice to the project client, who receives and considers the advice intra-organisationally. The client subsequently accepts or rejects the advice and issues appropriate instructions for action inter-organisationally to other project stakeholders.

The example suggests that project decision making may comprise a series of mini-decisions, each requiring cognitive inputs and behavioural actions. This approach can lead to 'hurdle' or 'gateway' decisions that will determine whether or not a project proceeds, is radically changed, or is discontinued.

The cognitive and behavioural aspects of decision making generally are explored in greater detail by Russo and Schoemaker (1989) and by Parkin (1996), and the key elements are listed in Table 3.4.

Table 3.4 Key elements to decision making.

Key elements to decision making	
Russo and Schoemaker (1989)	Parkin (1996)
Framing	Problem recognition
Information gathering	Information and analysis
Systematic conclusions	Judgement (concluding)
Learning from feedback	Deciding/acting

While the earlier authors appear to leave out the important action element, too much should not be made of this since they infer that action will automatically follow the drawing of systematic conclusions, although that is not inevitable for projects. Parkin (1996), however, fails to note the importance of feedback (i.e. post-decision evaluation in a learning cycle perspective of decision making), and our example too fails to show this. From a project perspective, also missing from these lists is the appropriate 'buy-in' by other stakeholders, as any lack of acceptance on their part imperils the effectiveness of a decision.

From the construction industry example of Figure 3.13, the decision framing or problem recognition element would be to establish the decision context by identifying what is to be modelled and why it should be modelled. Information gathering and analysis are represented by deciding upon, sourcing, and collecting the required data; inputting them into the model; and analysing the model outputs. Systematic conclusions are drawn from considering the implications of the model outputs for the project and communicating these in a form that is understandable to the client. The deciding/acting element is represented in the project client's subsequent direction to stakeholders. Learning from feedback occurs post-decision and should be achieved by deliberate observation of the efficacy of the decision taken.

In any situation, barriers to good decision making can arise, and Russo and Schoemaker (1989) offer 10 instances of such barriers:

1. Frame blindness
2. Lack of frame control
3. Misreading historical evidence
4. Overconfidence in personal judgement
5. Over-reliance on heuristics
6. Over-reliance on intuition
7. Hasty conclusions
8. Lack of ongoing monitoring and control
9. Inadequate recording of decisions and outcomes
10. *Groupthink* failure.

Frame blindness can occur where a project stakeholder does not have a clear understanding of project aims and objectives (or does not understand the decision issues in their relevant context). Effective project briefing processes can help to prevent this barrier from arising.

Lack of frame control occurs when project aims and objectives are allowed to shift over time (or if the context of the decision issue is allowed to drift). Good project management – including firm handling of the client – is required to avoid this possibility.

Misreading historical evidence can include choosing too selectively from past data and information for model inputs or failing to recognise the true extent of variability and uncertainty associated with such data. This barrier can be overcome, at least to some extent, by using more heterogeneous (as compared to homogeneous) data, and by treating the uncertainty stochastically rather than deterministically. Range estimates might have to replace single-point values. There is also the possibility of an 'information order effect' having an undue influence on project decision making. The 'primacy effect' occurs when a project decision is affected more heavily by earlier information, while the 'recency effect' means that the decision is influenced more by later information. Although intuition would suggest that the latter effect is preferable, the quality of the more recent information should first be considered. Also, from a project risk perspective, project stakeholders exhibiting one effect may end up in conflict with stakeholders exhibiting the other effect.

Overconfidence in personal judgement, instead of consulting colleagues for a second or third opinion, can be detrimental in making decisions. There is obviously a threat risk in relying on too much intrapersonal, rather than interpersonal, effort.

Heuristics are 'rules of thumb' often gathered through many years of experience on projects. Occasionally, they constitute abbreviated essentials extracted from larger and fuller sets of procedural guidelines. The latter form is preferable, and their effectiveness must be demonstrable, as relying upon experience alone can sometimes be a fickle and dangerous approach.

Frequent use of 'gut feel' (intuition) in project decision making should be treated with cautionary suspicion. Wherever possible, it should be backed up by more solid evidence.

Hasty conclusions can occur where project time constraints are extremely severe or where the available decision making time is too short. Time pressure must always be balanced against the need for good decisions.

Lack of ongoing monitoring and control is a failure to track progressive outcomes of decisions that have been made, or is neglect to take remedial action for outcomes that are going off-course.

In a similar vein, *inadequate observation and recording of decision making outcomes* leaves an organisation vulnerable to repeating poor decisions in the future.

Finally, *groupthink failure* in project decision making can be attributed to several factors. It may occur if a project team is not sufficiently differentiated, thus leading to the lack of a sufficiently wide pool of opinion in making decisions. This can happen in organisations where poor or covert organisational cultures are allowed to prevail. It may also happen if the organisation is too hierarchically structured and the decision making team is too subservient to an overdominant leader. Decision making processes involving high levels of information processing or high levels of information integration lead to decision complexity, especially if the decision process is constrained by time pressures and the group is dominated by an overzealous leader.

We can see from all of this that good decision making plays a vital part in achieving project success, and that barriers can actually be the causal events for threat risks, but does bad decision making inevitably lead to project failure? Is it the prime ingredient in 'risky' projects?

3.13 'Risky' Projects

What makes projects 'risky'? It should not be automatically assumed that riskiness in projects is most strongly associated with threat risks. Perhaps the strongest indicator of riskiness arises in projects where there is a high probability of failure matched by a high probability of success – a bipolar risk situation. The quadrennial Olympic Games event is almost always considered a risky project, mainly because of the huge financial and logistical demands placed upon the host nation over a comparatively long time period preceding the actual short duration event. Yet the Games are always expected to be spectacularly successful – and they always are!

Or, retrospectively, consider a project such as the Sydney Opera House: seriously over time in its construction, wildly beyond its original budget, highly innovative in its structural and acoustic technologies, compromised by last-minute changes in terms of its fitness for purpose, and dogged by political vulnerability throughout the procurement phase. Yet this iconic project has been hugely successful, operationally and strategically, for almost 50 years. At the time of its inception, did anyone ever think of it as a risky project?

What is it that sets risky projects apart from 'normal' projects? Here, admittedly, threat risk factors dominate. Projects are risky when decision making is accompanied by high levels of *residual* uncertainty. These are most frequently associated with procurement time, cost, or fitness-for-purpose parameters. Thus, a shopping centre project which *must* open on a particular date, in order to be successfully kick-started by the annual pre-Christmas spending spree, has a critical completion time criterion. To actually meet the required date may involve high levels of progress uncertainty all the way through the construction period – and even up to a week or two before the scheduled opening.

Similarly, a small event project, such as a production by an amateur dramatic society, may face committed production costs beyond its financial capacity, and have to rely upon its 'angels' – enthusiastic supporters who will underwrite the financial outlay. But there is no complete certainty that ticket sales revenue will eventually repay them.

In terms of fitness for purpose, a new television studio project may have demanding standards for acoustics and TV camera floor tracking evenness that are likely to severely tax the capabilities of the contractor responsible for building it.

We might think of all these examples as risky projects. There are other contributory factors.

New technology requirements, or technologies that are new to the user, can also signal a risky project. The technology learning curve may be steep with uncertainty.

Project decision making carried out under highly politicised circumstances will almost certainly make for a risky project, as also will situations where project stakeholders other than the primary or key stakeholders enjoy abnormally high power/distance ratios.

Projects undertaken in countries with unstable governments or ruled by absolute dictators obviously carry high levels of threat risk, and participants should look for a commensurate level of reward, with internationally enforceable means of repatriating profits.

The possibility of encountering corruption makes any project risky, and our advice is to abandon such projects as soon as the first whiff is detected, unless you are the instigator!

Leading on from this, trust plays a large part in all projects, regardless of the system of procurement employed. The *bona fides* of project stakeholders is expected and hoped for, but inevitably this will not always be the case. If *mala fides* (bad faith, bad intent) has been observed from a particular project participant on other projects, do not assume that this leopard has magically erased its spots. Treat the next project involving that stakeholder as risky, and, if you go ahead with it, take suitable precautions.

Other projects that should be regarded as risky are those where there is a high level of uncertainty that primary objectives can be fully achieved, especially where it soon becomes clear that they were unrealistic in the first place and perhaps only established originally to support what might otherwise have been a shaky business case.

Stakeholders inexperienced in project management, or stakeholders engaged in projects completely unfamiliar to them, will always make a project risky, but ensuring good project management is implemented will lubricate their learning curves substantially.

It is important to remember that, as described in Chapter 2, we have asserted that risk is a psychosocial construct, and therefore the riskiness of projects lies largely in the perceptions of the beholder.

There are no formal metrics for riskiness, and even restrictive time, cost, and quality criteria do not automatically make a project risky, since, if they really were impossible to meet, the project could not go ahead on the basis of them.

Another pertinent question might be: In what phases are projects likely to be most risky – the procurement phase, the operating phase, or some combinations of them? No authoritative answer is possible, but the clue might be to look closely at the phase where you perceive the greatest uncertainty will be found.

We have suggested some criteria for assessing project riskiness that may work for you, but your evaluation will almost certainly be subjective rather than objective. Experience and good risk awareness will be your best guides.

3.14   Summary

In some ways, this has been a difficult chapter to write, but we hope you have found it interesting to read. We have had to remind ourselves constantly that we are writing a book about managing project risks and not a book about project management. There are many of the latter available, and we have not attempted to cover all the same ground. Providing an understanding of projects through a risk management lens has been our aim.

Topics dealt with in this chapter have included the nature of projects and their composition. We have argued for a project perspective beyond the procurement phase, to include the operational and even disposal phases. The progressive life cycle of projects has been considered.

All projects are structured in some way, often as adhocracies, and the organisational structures of individual project stakeholders illuminate governance complexities. Organograms are a useful visual method for exploring the intricacies of organisational structures. They can be layered to reveal different perspectives including stakeholder types, governance priority, power, proximity, and communication vectors.

Stakeholder organisations exhibit different modes of management. Those with a more freewheeling enterprise style of management may find it difficult to interact with organisations, including the project organisational structure itself, that operate in a more formal administrative mode. Any conflicts arising from encounters between differing modes of organisational management provide the seed beds for threat risks.

The review of organisational structures was undertaken in the service of informing an exploration of project decision making, as this activity is strongly associated with project risk through the uncertainties it releases. Decision making is found to have cognitive and behavioural elements and may be undertaken intra- and interpersonally, and located intra- and inter-organisationally in projects.

Finally, we have attempted to provide some guidance about the subjective perceptions of the riskiness of projects.

From this point, the following chapters focus upon risk management for projects and the detailed activities to be followed in a project risk management system.

References

Parkin, J. (1996). *Management Decisions for Engineers*. London: Thomas Telford.

Project Management Institute (PMI). (2013) A Guide to the Project Management Book of Knowledge, Vol. 5. PMI, Newtown Square, PA. *ISBN* 9781935589679.

Robbins, S.P. and Barnwell, N. (1998). *Organisation Theory: Concepts and Cases*. Upper Saddle River, NJ: Prentice Hall (Includes Mintzberg's 1979 theory.).

Russo, J.E. and Schoemaker, P. .J. .H. (1989). *Decision Traps: Ten Barriers to Brilliant Decision-Making and How to Overcome them*. New York: Doubleday.

4

Project Risk Management Systems

4.1 Introduction

The concept, meaning, and implications of risk were introduced in Chapter 2, and projects formed the substantial content of Chapter 3. Now, we bring the two topics together by exploring how project risks can and should be managed.

In doing this, it is important to remember that, as we noted in Chapter 1 and reinforced in Chapter 3, each project stakeholder should manage its own risks. Trying to implement a risk management process specific to a project and yet common to all of the stakeholders associated with that project is unlikely to be successful, given the ISO 31000 (International Organisation for Standardisation [ISO] 2018) definition of risk as 'uncertainty and its effect on objectives'. The reality is that, for the same project, while each stakeholder may share some objectives in common with other project participants, other objectives will be mutually exclusive.

Stakeholders acting in an agency capacity on behalf of a project sponsor (client) should be acutely aware of this dilemma. An agent (i.e. a consultant) can be thought of as wearing two risk management 'hats': one for the client's risks arising from the aspect of the project for which the agent has responsibility, and another hat for his or her own risks in terms of the agency processes of participation in the project.

We can illustrate this with an example. A project client commissions an architect to design a new sports and leisure facility. Along with other project objectives the client has established will be those perennial favourites relating to time, cost, and quality. The architect, as agent for the client, will strive to ensure that at least the third of those, quality (or fitness for purpose), is fulfilled to the client's satisfaction, and may also have at least some management responsibility for achieving the others. In carrying out that design and management responsibility, however, the architect will almost certainly have practice objectives: to establish an ongoing relationship with that client, to generate appropriate and timely fee income from the project, and to enhance his or her professional reputation through carrying out noteworthy work. Wearing 'two hats' in this way may well give rise to additional risks if any of the project objectives falling under one hat come into conflict with those under the other. The same dilemma faces other client agents acting in a professional consultancy role such as project managers (PMs), engineers, interior designers, and so on. However, even where a PM operates within a host organisation (typically in a project management office [PMO] or enterprise project management capacity), the 'two hats' phenomenon may still be present, albeit in a

Managing Project Risks, First Edition. Peter J Edwards, Paulo Vaz Serra and Michael Edwards.
© 2020 John Wiley & Sons Ltd. Published 2020 by John Wiley & Sons Ltd.

subtler way. While delivering projects for the host organisation, the PM may also be seeking to enhance the reputation, expansion, and future security of the PMO itself.

Indeed, even the principal stakeholder (i.e. the project client or sponsor) has to recognise the dual-risk dilemma. Whilst directing management effort towards its project-related risks, the organisation cannot ignore those relating to other, everyday activities. In effect, some organisations may operate two risk management systems (RMSs): one organisationally focused and the other project focused. This is not necessarily a bad approach – after all, it does mean that risks are being deliberately managed somewhere in the organisation! However, it is essential that the organisation is aware of what it is doing and that, at least at some higher level, a measure of risk management integration is taking place – even if only to capture risk knowledge. General intra-organisational risk management is not the intended focus of this book, however, and the content is directed mainly towards stakeholder-to-project risk management issues and processes.

After discussing risk management standards and guides generally, we introduce a systematic approach to risk management based on the principle of a learning cycle as a dynamic process. This is where knowledge and learning accumulate over experiences gained from dealing with risks on successive or multiple projects.

The stages in the cycle are introduced and briefly explained as a platform for the individual treatment of risk management processes that will subsequently be described in more detail in later chapters of this book.

The system is generic, not only for all types of projects but also for all types of stakeholders. Thus, no matter what your project background is, you should be able to grasp the overall approach and tailor it to suit your particular circumstances. Examples drawn from two different project contexts are used to illustrate this.

The tasks of building, operating, maintaining, and improving an effective RMS are given greater attention towards the end of this book.

4.2 Risk Management

We all manage risk. As individuals, we tend to do this cognitively and intuitively most of the time and rarely exercise a deliberately systematic and purposeful approach. My 'project' objectives may be to take a healthy morning walk, buy a daily newspaper and some milk while I am out, and achieve all of these safely. On most days, I may also take the family dog to share the exercise with me. Part of my project route includes having to cross a busy road. As I stand at the kerbside, waiting to cross the road, my eyes take in the traffic conditions and my brain simultaneously and intuitively converts the visual data into information (amount, proximity, and speed of traffic) that will enable me to decide when it is safe to cross. If the dog is with me, my brain tells me how that decision should be modified. Additionally, when the crossing is safely achieved, my brain will store parameters of the experience in my memory and use them for reference in similar 'project' conditions in future.

Although cognitive intuition might suffice for a person, it is hardly likely to be adequate for an organisation seeking to achieve specific objectives in a project, especially where decision making is quite diversified in the organisation. Many types of projects are becoming increasingly complex, the rate of technological change has become more rapid, and expectations of accountability have become more demanding. It is now necessary to cope with volatility and turbulence in almost every aspect of organisational activity. Status and reputation have to be protected. National and international incidents have led to greater recognition of local and global vulnerability to many issues of contemporary society. All of this has contributed to the development of a broader conceptual view of risk (particularly threat risk) among communities of all kinds and to the recognition that more formal procedures are needed to manage risk.

Put simply, risk management is a process of dealing with risks. It may be implemented reactively or proactively and informally or formally, as shown in Figure 4.1.

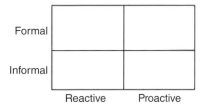

Figure 4.1 Approaches to managing project risks.

A risk is managed reactively when something happens on a project and the relevant stakeholder takes action to deal with the post-event consequences. Risks are managed proactively when the stakeholder gives timely and deliberate consideration to the possibility of particular events occurring during its involvement with the project, where the consequences of those events could affect the stakeholder's objectives for the project, and thus makes plans beforehand to deal with potential eventualities.

Informal approaches to managing risks imply that the actions are ad hoc, decided upon on the spur of the moment and unrelated to any pre-planning. Formal approaches, in contrast, derive from policies and procedures that have been considered well ahead and are appropriately planned for, resourced, and budgeted.

In contemporary society, with its project-driven nature and turbulence, we cannot afford to simply rely on our ability to react to risks after they happen, but must take a more proactive approach.

From Figure 4.1, it follows that risk management approaches that lie towards the upper-right quadrant will probably be more systematic than those falling towards the lower left. By extension, we would argue that risk management effectiveness is likely to follow the same trajectory. Note the deliberate choice of words here in 'probably' and 'likely'. From this point on, few terms in this book will be intended to denote conditions and situations that are absolute and certain. We have already seen how the cloak of uncertainty lies heavily over risk. To some extent, it must also lie over risk management. Our consolation argument is that, by trying to deal with project risks formally, proactively, and systematically, you will be putting yourself, and the project stakeholder that you represent, in the best position to expose the uncertainties inevitably associated with any project at a time well in advance of the point where substantial commitments, in terms of finance and other resources, have to be made. You will be better prepared to manage the eventualities that could arise. In doing so, you may also secure a valuable competitive advantage in the environment in which you operate, particularly if that is as a commercial business. Another important point emerging from Figure 4.1 is that there is no single ideal system of risk management, but those that are more proactive and formal are likely to be more effective than those that are not.

4.3 Risk Management Systems

A systematic approach to risk management increases the capacity of an organisation to handle risks at all levels. Where it is implemented uniformly, it promotes internal transparency and common understanding of the business activities of the organisation, and facilitates the establishment and growth of a commitment and culture towards managing risks. Having a formal RMS in place simplifies the organisational capture, transformation, and transfer of risk knowledge. It provides the means for assessing best practice, for benchmarking risk management performance, and for establishing a platform that permits the benefits of modern information and communication technology to be exploited.

An effective RMS delivers further benefits. Within a stakeholder organisation, congruence between the organisation's objectives and the project objectives can be confirmed, or any conflict between them quickly identified. Trust and confidence between the levels of management, and between management and operatives, are improved. Responsibilities are clarified, and ownership of responsibility is more willingly accepted. Potential problems are exposed to a wider internal audience. The organisation's capacity to deal with new risks is enhanced. More focused risk training is possible. Creative approaches to handling risks can be encouraged and developed. The nature and scale of potential crises are better understood, allowing better visualisation of more distant time horizons. Anticipation replaces surprise, and informed thinking takes over from reliance on luck. Above all, a systematic approach to risk management should encourage decision making within an organisation that is more consistent, more controlled, and yet at the same time more flexible.

A good RMS will allow an organisation to look forward to the future for each project, while maintaining a convenient capacity for using the wisdom gathered from its previous project risk experiences. This might entail the adoption of a dual-system approach. On the one hand, there would be dynamic, project-specific RMSs as the vehicles for managing the risks of each project. On the other hand, there should also be an organisational meta-RMS, containing a risk register structured as a repository of organisational knowledge of all risks on all projects.

Ideally, there should also be a third system in place. This would have an organisational macro-focus, to embrace risk information associated with the internal tasks, technologies, and resources necessary to sustain the organisation itself as an ongoing concern (i.e. to deal with the likelihood of events occurring which could [positively or negatively] affect the continuing life of the organisation). For the purpose of this book, however, we will concentrate here upon *project* risks and *project* RMSs.

At a project level, the project risk register (PRR) will comprise the means for identifying competing interests and employ techniques for weighing up inadequate information about the project. It will take note of risks already allocated or risk treatments already in place. It will provide for the recording of decisions for action and the assignment of responsibilities. The systematic nature of the PRR will be inherent in the definitive processes it incorporates, the approaches to data analysis it employs, the treatment regimes it embraces, and the usefulness of the information and knowledge that it captures.

A good starting place to explore systematic project risk management is in the development of internationally recognised standards for risk management.

4.4   Risk Management Standards and Guides

The international standard for risk management, ISO 31000 (ISO 2018), promotes a framework for risk management that is substantially similar to those offered by many texts. The development of this standard lies in its origins in Australia and New Zealand, where the Australian Standard/New Zealand Standard (AS/NZS) standard 3931, *Risk Analysis of Technological Systems – Application Guide* (1998), provided a starting point. This standard was aimed principally at the chemical engineering industry, and its focus was mainly upon analytical techniques for assessing risks in the operating phase of production facilities. AS/NZS 4360, *Risk Management* (1999), a more generic guide to risk management, followed quite quickly and was subsequently amplified in a useful companion handbook that elaborated upon the basic principles and techniques (AS/NZS HB 436; 2004). Growing international interest in the field of risk management prompted Australia to publish two Chinese language versions of AS/NZS 4360.

Almost thirty Australia/New Zealand standards and handbooks are now in publication, most of them aimed at risk management for a particular industry or application, including those for: information technology (IT) security, Legionnaire's disease health risks, security systems, occupational health and safety, organisational governance, outsourcing, weed control, legal systems, healthcare, business continuity, risk financing, risk assurance, sports and recreational organisations, not-for-profit organisations, and environmental management. Other countries have followed suit, particularly in terms of risk management standards tailored to suit specific industry needs.

While embracing the unique terminologies of different fields, and thus presenting risk management in a way that may be more familiar and intelligible to the users, the adoption of more specialised standards has brought with it its own threat risk: the possibility of failing to ensure that basic concepts and principles of risk and risk management are clearly expounded to, and grasped by, each user. It also means that the potentially greater solution space for risk treatment options, made available through a more generic risk management approach, is instead likely to remain limited to solutions conventionally embraced within a particular industry context. Our experience has been that the wider the reading about risk and risk management in other areas of application, the greater will be the number of options available to deal with the risks encountered in your own project context.

Standards Australia and Standards New Zealand (SA/SNZ; note the slight changes to host names) have also published three companion handbooks for ISO 31000. SA/SNZ HB 436 (2013), clearly acknowledging its ancestry to the original risk management standard, is an updated version of the earlier handbook. SA/SNZ HB 89 (2013), however, probably owes allegiance to AS/NZS 3931 in that it offers guidelines on risk assessment techniques. SA/SNZ HB 327 (2010) offers advice for communicating and consulting about risk, and a guide to ISO 31000 describes a vocabulary for risk management (ISO Guide 73; ISO 2009). There is also an application guide to managing risks in projects (AS/NZS IEC 62198:2015; SA/SNZ 2015).

The information given in this section was current at the time of writing. It is important to note that new and updated versions of risk management standards and guides will continue to be published in future.

The Project Management Institute's (PMI) *A Guide to the Project Management Book of Knowledge* (2013; hereafter, *PMBOK Guide*) offers a six-point plan for project risk management:

- Plan risk management
- Identify risks
- Perform qualitative risk analysis
- Perform quantitative risk analysis
- Plan risk response
- Control risks.

While useful, with many detailed subsections included, this plan omits essential starting and finishing points: establishing the project context, and capturing risk knowledge.

The ISO 31000 (2018) guide is also a good starting point as a partial framework for systematic risk management. It comprises five stages:

1) Establish the context
2) Identify risks
3) Analyse risks
4) Evaluate risks
5) Treat risks.

These stages are intended to be continuously accompanied by two other activities: communicate and consult; and monitor and review. ISO 31000 (2018) emphasises the importance of contextualising risk, but only in the 2018 revision does it (belatedly) acknowledge the need for risk knowledge capture, referring to this process as 'Recording and reporting'.

In fact, few texts about risk management include the first contextualising stage, but it is important for good risk management. Some authors combine Stages 2–4 under an umbrella heading of 'risk assessment'. Many fail to offer specific guidance about the two parallel activities advocated by ISO 31000.

Terminology differences among risk management standards and guides are seen as an issue for those seeking to establish a universally accepted risk vocabulary (e.g. ISO Guide 73; 2009). In our view, this is a somewhat forlorn hope, as it is rarely achievable and is always susceptible to different interpretations, especially if translation into languages other than English is attempted. Even among English speaking countries, subtle differences in terminology may arise (e.g. word processors offering the choice between UK English and American English). We suggest that the processes of risk management are more important than the terminology.

While existing standards and texts can take us a good way along the risk management journey, we consider that they are deficient in two important respects. Firstly, if they consider projects at all, they tend to look at them in isolation; whereas most stakeholders, while they may be concerned with only one project at any time, tend to have a sequential or even continuous engagement with projects where one follows closely upon another. The second deficiency derives from the first. In the risk management frameworks offered by some standards and most texts, there is an absence of deliberately planned action to capture risk experience from a project and subsequently harness it as a valuable risk learning and knowledge resource to be used on future projects. Even recording that a particular risk event did not happen on a project can be useful to know. If historic project risk information is not gathered, analysed, and exploited, a resource is lost that could have been critically important for project risk management to become fully and effectively integrated into the stakeholder organisation. Furthermore, just relying on the risk wisdom of 'oldies', the experienced personnel in the organisation, becomes a threat risk in itself if they are headhunted by competitors or are hit by a bus while crossing the road on the way to work! In Section 4.5, we consider how these deficiencies can be resolved in a more systematic manner.

4.5   A Cycle of Systematic Project Risk Management

Systematic risk management is a dynamic process of activities intended to deal with project risks. When the process includes the deliberate capture of risk knowledge from a project, it becomes a learning activity in addition to its management function. If the process, together with the acquired knowledge, is then applied to subsequent projects, it becomes a *learning cycle*. Figure 4.2 shows how this is achieved, and indicates important activities and system requirements for effective risk management.

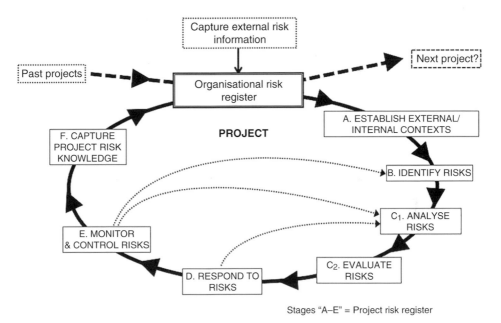

Figure 4.2 The dynamic cycle of project risk management.

The cycle of systematic risk management assumes that a stakeholder comes to a new project with some archival level of organisational information that encapsulates its knowledge (and hopefully its wisdom too) about project risks. Thus, while the project and its uncertainties will be new, they will not necessarily be unique in terms of their risk characteristics. Similarities to previous projects are likely to exist, and the existence of a searchable database allows that to be confirmed or disproved. The database, through retrieval and transformation of the knowledge it contains, reduces or even removes undue reliance upon the personal risk experience and tacit knowledge of individuals within the stakeholder organisation. The processes of organisational knowledge management are dealt with more fully in Chapter 11.

We call this compendium of stakeholder risk knowledge the 'organisational risk register' (ORR). It is deliberately arranged as an organisation-wide repository of risk information and is distinguished from documents intended to record risk management activities for individual projects. As shown in Figure 4.2 (see tasks B, C, and D), the project-specific document is the PRR. It is also important to note that, besides recording the risk management decision making and activities for an individual project, the PRR may also serve an evidentiary purpose should the organisation's risk management be subjected to formal enquiry at some point. The PMI's *PMBOK Guide* (2013, p. 314) refers to a similar document as the 'Plan Risk Management' component of a 'Project Management Plan'. Whilst we concede the need to understand the importance of placing risk management within a larger project management system, the Plan Risk Management proposed by the PMI covers only five activities: risk identification, qualitative risk analysis, quantitative risk analysis, risk response planning, and risk control. Our proposed RMS incorporates much more.

A brief explanation of the proposed cycle is given here. More detailed explanation of each stage, and the techniques involved, is provided in subsequent chapters.

Clearly, the timing of each stage in the project risk management cycle will be influenced by the nature of the project itself and the project phases that ensue. While Activities A–D may occur within a relatively short time frame – weeks or months – those from D to F may span several years; although it is worth noting that, for a typical telecommunications service provider embarking on a project to develop a new product, the whole cycle may be constrained to just a few weeks as competitive 'time to market' pressures influence the project timelines.

It is also important to appreciate that the graphical simplification in Figure 4.2 is deliberately intended to aid conceptual understanding of the risk management cycle that we propose. It depicts each stage as distinct and separated in time. In reality, several activities are iterative or can be overlapped to some extent. Risk identification (Stage B), for example, is a distinct activity, but it can be followed almost immediately by analysis of the risks (Stage C_1) as soon as they are identified. Activities B–D (risk identification to risk response), besides creating the information that populates the PRR, may be repeated several times during the early design period. Stage E (monitor and control risks) is undertaken almost continuously as the project delivery process moves from pre-planning to implementation and thence to completion. Stage F (capture project risk knowledge) could be undertaken at any point where parts of the project have been finished and are no longer susceptible to risk. Besides desiring conceptual simplicity, our reason for showing distinct activities is that generally they do happen sequentially, that specific tools and techniques are usually needed for each stage, and that for some projects they may involve different people or teams.

For each new project, the process starts with accessing the ORR (see Chapter 11) to retrieve any information about similar types of project, similar locations (if it is a physical project), and similar clients if that is a relevant situation. Depending upon the structure of the ORR, it should also be possible to adopt a template upon which the new PRR can be based.

4.5.1 A: Establish the Context

As noted in Section 4.4, we consider the context establishment stage to be essential to good risk management, although it is often ignored in texts on risk management and frequently omitted by organisations that (mistakenly) believe that they are already fully aware of the contexts within which they operate.

Contextualising a project firmly establishes its role and significance to the organisation (the internal context) and its place in the wider environment or society (the external context). Thorough contextualisation will also expose the internal and external drivers which will shape or influence the project risks that are identified. These considerations are expanded in Chapter 5.

4.5.2 B: Identify Risks

For risks to be managed systematically, they must first be identified. Otherwise, they are unknown (or at least unrecognised) but possible future events that, should they eventuate, may have to be dealt with in a purely reactive manner.

We advocate a team-based workshop (or workshops) approach to project risk identification. This is discussed in Chapter 6, followed by several risk identification techniques in Chapter 7.

4.5.3 C_1: Analyse Risks

This stage marks the first of the actual assessment activities in the risk management cycle.

Generally, a project stakeholder, besides wanting to know the risks it faces, also wishes to gain an appreciation of how severe the threat risks are or how beneficial opportunity risks might be. Often, this analysis is done simply to gauge the importance of each risk relative to that of other risks, so that response treatment options can be prioritised efficiently in terms of their resource requirements. Less often, more precise analysis is needed in order to ensure that financial conditions and commitments of the stakeholder are not inadvertently compromised. The analysis, among other things, involves examination of the likelihood of the risk event occurring and the nature and magnitude of the subsequent consequences of that event for the project (and thus the stakeholder). Techniques for qualitative and quantitative risk analysis are explored in Chapter 8. The emphasis in that chapter will be mainly upon qualitative assessment.

4.5.4 C_2: Evaluate Risks

Risk evaluation is also elaborated in Chapter 8. The evaluation process focuses on the level of risk in terms of magnitude of the potential severity or benefit. This process also considers each identified and analysed risk against the parameters of any strategic risk management criteria that the stakeholder may have established. Such evaluation is important in that it can constitute a major decision 'gateway' for that stakeholder. A risk that cannot pass through the strategic risk management gate may be a critical factor in an organisation's decision to proceed no further with its involvement in a project. The threat risk may be too severe or the benefit of an opportunity risk too small. Doubt may surround much of this evaluation, since the process involves risk uncertainties. Clearly unacceptable risks, those that are beyond doubt, will trigger the 'no go' decision. Others will be allowed to proceed to Stage D and be re-evaluated after they have returned through an iterative post-treatment risk analysis stage. Many stakeholders will rank their identified risks in the evaluation stage, in terms of perceived degree of severity or extent of opportunity benefit.

4.5.5   D: Respond to Risks

Also known as the risk treatment stage, at this point each evaluated risk (usually in highest rank order) is explored in terms of possible treatment or response options. The aim here is to minimise severity, or maximise benefit, as far as is practicable for the stakeholder.

We consider this to be one of the more creative aspects of risk management, since for any risk there are likely to be several, if not many, options to mitigate threats or enhance opportunities. This is also why we advocate reading about risk beyond the confines of your own project context. Looking for effective risk treatments, whether in human society or in nature (biomimicry), is rarely wasted effort in risk management. Responses to risk are considered in Chapter 9.

Exploration of treatment options may require additional analysis, hence the dotted arrow in Figure 4.2 from Stage D back to C_1. Risk responses are discussed in greater depth in Chapter 8, but it is important to note two things here. Stage D is essentially a decision process: decisions will be made about whether or not individual risks should be treated and, if so, how this is to be done. Those decisions are heavily influenced by the decision maker's attitude towards risk. Response treatments may be associated with uncertainty about their degree of effectiveness.

One outcome of this stage, bearing in mind the possible iterations of Stages B–D, will be a partially complete PRR for the stakeholder.

4.5.6   E: Monitor and Control Risks

Completion of the project stakeholder's risk response decision making often coincides with (but should certainly not lag) tangible project activity. The project moves from a planning phase to implementation activities deliberately intended to bring it eventually to a stage where it can begin its operational life. This may mean the commencement of construction work on a building site, the development of code algorithms for a software application development project, the trial roll-out of a customer service procedure, or the addition of new bed spaces in a hospital operating ward.

Project risk management may take on a different character at this point, for what could be a considerable period of time. The frequency of risk management activity undertaken by the stakeholder may be aligned with other progress, cost, quality, safety, and environmental review procedures implemented for the project. Different staff may be assigned risk management responsibilities, and reporting channels to senior management established. The aim is to monitor the risk situations embraced in the PRR and to review and control their ongoing status. Since the project is now proceeding on its intended course in reality (as opposed to the more abstract processes of project planning), the potential imminence of risk events is closer, and the team delegated to manage those risks must be constantly aware of dynamic changes in the project environment.

While it should be possible to 'close off' some risks on the PRR as their associated 'real-time' activities are completed, changing project circumstances may give rise to the emergence of new risks or different parameters for existing risks. These should be subjected to analysis and evaluation, hence the dotted arrows in Figure 4.2 looping back from Stage E to Stages B and C_1. Chapter 10 considers the risk monitoring and control process in greater depth.

4.5.7   F: Capture Project Risk Knowledge

An organisation committed to implementing systematic risk management on its projects will have directed valuable resources towards doing so. Common sense suggests that harvesting the risk knowledge arising from the risk management process would be a worthwhile process. Yet few project stakeholders do this with any thoroughness, and even fewer look for information about risk beyond their particular project environments.

While the capture of project risk knowledge is shown towards the end of the risk management cycle in Figure 4.2, it is not necessarily a 'one-time' activity. Ideally, knowledge capture should be carried out as soon as possible after the completion of particular components of a project, while participants' memories are still fresh. Often, this can be done through 'lessons learned' project debriefing workshops, with relevant information entered into the ORR. Risk information gleaned from other sources (external to the project and the stakeholder) can be dealt with in a similar way, to create a comprehensive and accessible body of knowledge about risk and risk management for the organisation. Knowledge management and its implications for organisational risk management are considered more extensively in Chapter 11.

Having briefly described the processes of systematic project risk management, we now turn to the alignment of these processes with the project life cycle.

4.6   Project Stages and Risk Management Workshops

As we discussed in Chapter 3 and saw in Figure 3.2, projects may comprise more than one phase. Within each phase, identifiable and sequential stages may occur. In the procurement phase, for example, the stages might include: inception, conceptualisation, design development, bidding, implementation, and completion or delivery.

How risk management is overlaid across, and integrated with, the project stages is unique to each project and depends upon the nature of the project and the demands it imposes upon the project stakeholder. Nevertheless, it is possible to offer some general guidelines, and these are described and illustrated by way of examples. The first is a hypothetical case drawn from the construction industry, while the second is based upon a real IT project. The timing of project risk management interventions is considered again in Chapter 6 as a precursor to risk identification.

4.6.1   Construction Project Example

In the first example, the project stakeholder (acting initially in a pre-appointment capacity) is a construction contractor which has responded to a call for expressions of interest (EOIs) to design and build a new manufacturing facility for a large pharmaceutical company (the client stakeholder). On the advice of its consultants (also stakeholders), the client has decided to adopt a design-build (DB) method of project procurement. This method, also known as D & C (design and construct), integrates the design and construction processes under a single point responsibility (the contractor), as compared to traditional building procurement methods where the building design is undertaken by separate consultants on behalf of the client, the design documentation is then made available for competing contractors to prepare and submit tenders, and the winning bidder subsequently proceeds with construction under the supervision of the client's consultants. DB procurement attempts to achieve greater diversion of threat risk from the client towards the contractor, whilst at the same time fostering more efficient project delivery through the earlier involvement of the contractor. It is not our objective here to compare building procurement methods and we use the DB example just to illustrate a parallel trajectory for project risk management, in this case for the contractor as a project stakeholder organisation.

As a further assumption for this example, the client has indicated in the call for EOIs that, if the winning bidder delivers the project successfully, it will be invited to negotiate for contracts to construct similar facilities in other locations, notably South America, Africa, and South East Asia. This may act as a 'sweetener' relationship management objective for the contractor.

For the contractor, the project broadly divides into two stages: design-bid, followed by build. Figures 4.3 and 4.4 illustrate this with brief reference to the activities undertaken in each stage, the growth in project information and concomitant reduction in uncertainty, the potential for identifying risks, and the points where focused (as distinct from general) risk management activities could be targeted. To avoid too much complication, we have excluded the operational and disposal phases of the project, although in reality

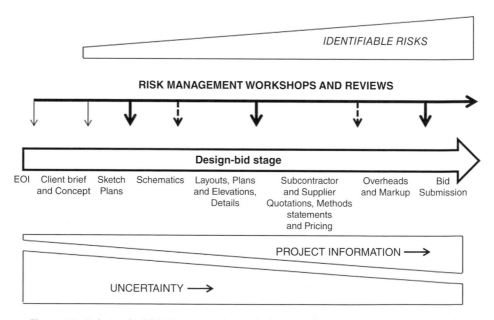

Figure 4.3 A design-build (DB) project's design-bid stage: information, uncertainty, and risk management.

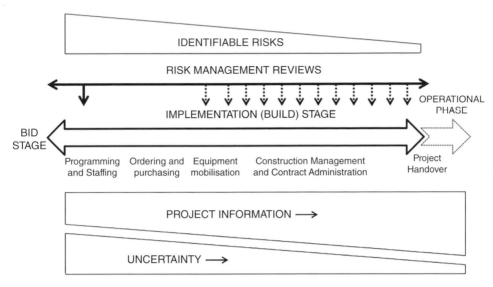

Figure 4.4 A design-build (DB) project's build stage: information, uncertainty, and risk management.

even the contractor stakeholder will pay some attention to these, particularly in light of the relationship objective noted here.

The task descriptors below the main arrows are not intended to be exhaustive but suggest activities typical to each stage. Timelines in the figures are obviously not to scale, and it should be noted that the gradients indicated on the information, uncertainty, and identifiable risk polygons are rarely consistent and never predictable! For a project such as this one, the design-bid stage might take three or four months. For an 'off the shelf' DB design – a single-storey warehouse with standard bay sizes, for instance – two or three weeks might suffice for design, with additional time for obtaining permits and so on. Also, while the identifiable risk level indicators in Figures 4.3 and 4.4 do not distinguish between threat and opportunity risks, in practice the stakeholder's risk management activity will almost inevitably focus upon threat risks.

4.6.1.1 The Design-Bid Stage
Figure 4.3 shows the contractor's activities after receiving the call for EOIs for the construction of the pharmaceutical manufacturing facility.

The contractor will try to obtain at least some amplification of the client's brief and requirements before proceeding with any graphic design, but at least a design concept will be established. Architectural, structural engineering, services engineering, and other specialised design inputs will commence, either from in-house resources or by outsourcing them to professional consultants. Site investigation will be carried out. Cost planning (not shown in Figure 4.3) will attempt to set budget parameters. The amount of design information will increase rapidly over this stage, together with other information needed to obtain quotations for subcontract work and materials supplies. These will be harnessed to the development of methods statements (broad explanations about how the facility will be constructed), thus allowing more refined pricing of the work against the established budget. Eventually, this pricing reaches a point of sufficiency where overhead costs and the bid markup can be added, and the bid is ready for submission.

Important points to note from Figure 4.3 are that neither the amount of available information, nor the level of identifiable risks, commences at zero. There will always be some information available to the contractor at the beginning of a project, and some risks will be self-evident.

The polygon for uncertainty is deliberately drawn larger than that for project information. There is a paradox here, as the increasing level of project design automatically increases the amount of information that is generated. However, the design process itself advances though a series of design decisions, many of which are complex and interrelated. The decisions themselves generate inherent uncertainty that can often prevail well into succeeding project stages.

Thus, the two thinner downward arrows shown on the left-hand side of Figure 4.3 represent potentially separate mini-workshops for project briefing and context setting.

In the first of these, the essential requirements and objectives for the project are considered by assembling project briefing information.

Project risk management activity for the design-bid stage commences with establishing the internal and external project contexts (Figure 4.2, Stage A) for the stakeholder (i.e. the contractor in this example). This process is described more fully in Chapter 5. In the example, it may be carried out beforehand as a separate workshop activity by senior executives in the contractor organisation or by the project risk management team at its first workshop.

The full involvement of the project risk management team really commences with a risk identification workshop (Figure 4.2, Stage B), as represented by the first heavier solid downward arrow in Figure 4.3.

The team will retrieve relevant material from the ORR and will probably set up the template for the PRR. This workshop will undertake risk identification, together with analysis, evaluation, and response activities (Figure 4.2, Stages B, C, and D; see Chapters 6, 7, and 8), focusing as much as possible on issues that will inform subsequent design decision making. This workshop may take place over a period lasting from half a day up to two or even three days. It is important in that, besides commencing the stakeholder's risk management process for the project, the workshop is also instrumental as an exemplar for subsequent management. The performance 'bar' should be set as high as possible, within reasonable constraints of practicality, and the PRR will rapidly become well-populated.

Between the formal risk management workshops (indicated by solid downward arrows in Figure 4.3), ad hoc review meetings (shown as broken vertical arrows) will take place as and when they are needed to consider architectural, engineering, and cost-related aspects of design. Each review will entail updating the PRR.

At a point somewhere just prior to inviting subcontractor and supplier quotations, we advocate a second formal risk management workshop to review and augment the previous processes and to identify any risks associated with the outsourcing of work. Another update of the PRR follows this.

Further risk reviews will be arranged if required, but the third formal risk management workshop should consider the sufficiency of all that has been undertaken so far. This workshop should precede, but may coincide with, the decision making with respect to the overhead costs allocation and profit markup to be added to the bid. Again, the PRR is updated.

The thoroughness of risk management undertaken in the design-bid stage of a DB project is always constrained by time and cost. Severe limitations on the time available for this stage influence the adequacy of the project design, diminishing the generation of appropriate information and increasing the level of uncertainty and, by definition, the number and magnitude of threat risks. The search for opportunity risks may be non-existent or inadequate. Restricting the cost of design, and the cost of risk management effort, has the same effect. While reducing these costs may be in the interests of minimising sunk costs in a project bid that might not be successful, any mismanagement of threat risks during the design-bid stage could flow on, with adverse consequences, into subsequent project stages and phases if the bid is successful.

4.6.1.2 The Build Stage

The longer build stage of DB procurement for the project example is shown in Figure 4.4.

The successful bidder now commences construction work on the site, and this work will continue for as many weeks, months, or years as it takes to complete the facility to the point where it can be handed over to the client for occupational use. Few of the plethora of activities are shown in the diagram. In reality, the initial programming and scheduling process may reveal upwards of 4000 separately identifiable activities for a project such as the proposed pharmaceutical manufacturing facility. This provides us with some clues about the potential size of the PRR.

The next formal risk management workshop should take place as early as possible in this stage. We advocate a formal workshop at this point, despite the fact that it might be quite close to the one held just before bid submission. The main reason for this is that project reality now replaces project possibility. Also, the composition of the risk management team may change and incorporate more site-based personnel, some of whom may need to be brought 'up to speed' with the project risk management processes. New risks, associated with the programmed work activities, may be identifiable and previously identified risks revisited in the light of scheduling decisions.

After this first workshop in the build stage, the PRR will be developed into a comprehensive management tool. Further formal risk management workshops may not be required unless major changes to the scope, timing, or contractual or financial status of the project are anticipated. Instead, risk reviews may be instigated at regular intervals timed to coincide with, or immediately follow, project progress reviews. Existing risks are reviewed, new risks identified, and each subjected to analysis, evaluation, and response treatment. This risk monitoring and control process is described in more detail in Chapter 9.

Risk knowledge capture, while shown at the point of project completion (for the contractor stakeholder), may actually take place at any time during the build stage when particular activities finish and the risks relating to them are no longer applicable. Ideally, the knowledge capture process (covered in Chapter 10) should be more extensive and thorough than just a project debriefing or 'lessons learned' meeting, although it may be partly based upon that.

Note that, in Figure 4.4, the levels of project information, uncertainty, and identifiable risk potential continue beyond the contractor's handover of the completed project to the client. This is largely due to the contractor's short-term responsibility for remediating defects and the assignment of component and services systems warranties to the client. Any liability for latent defects will constitute a longer term threat risk situation for the contractor.

This example shows how the cycle of systematic risk management can be applied to the temporal processes of delivery over several months, at least through the procurement phase, for a construction project.

4.6.2 IT Project Example

The second project example is IT-related and is based upon the project referred to in Section 3.6 (Processes of Project Implementation) of Chapter 3. The stakeholder in this case is a retail conglomerate comprising several distinct retail 'brand' companies (or business units), all served by the IT Department of the holding group. The stakeholder organisational structure most closely resembles a divisionalised form.

The IT project management services are undertaken within the department as part of its activities, but are not formally constituted as an IT PMO. The major part of the IT Department's activities relate to the website needs of each brand business unit. Other IT-based services, such as stock control and purchasing, are largely dealt with by IT managers in each brand company, with needs-based assistance provided by the group IT Department.

As described in Chapter 3, the project is to develop and deploy a brand product personalisation service to offer to online customers. Project activities are illustrated in Figures 4.5 and 4.6. For IT projects of this nature, the period from inception to deployment might range from 4 to 12 weeks.

Figure 4.5 shows the flow of activities during the conceptual development of the proposed project, including ideation.

No formal risk management was undertaken during this stage, and the risk management activity shown in Figure 4.5 is thus notional. The PM, however, used experience and judgement to build schedule flexibility and a cost contingency buffer into the concept proposal.

The IT project development stage is illustrated in Figure 4.6.

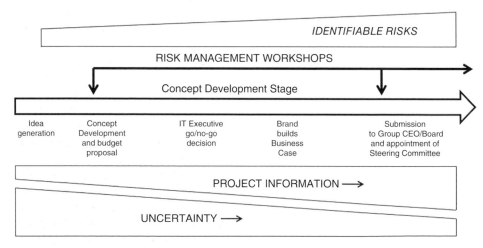

Figure 4.5 An IT project's concept development stage: information, uncertainty, and risk management.

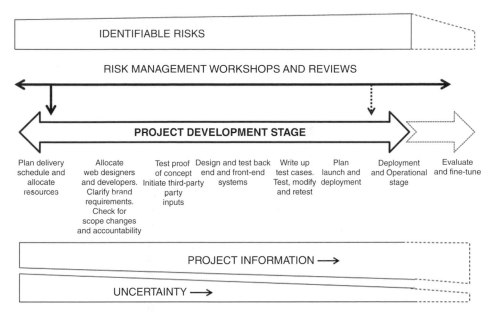

Figure 4.6 An IT project's development stage: information, uncertainty, and risk management.

No formal project closure or signing off took place on this project. Since no formal risk management was undertaken, no formal risk knowledge capture was carried out, and any risk-related information was simply added intuitively to the PM's store of experience. The organisation's project risk 'wisdom' thus primarily resides in one person.

Bearing in mind that the whole project duration may be no more than a few weeks, fitting a formal risk management approach to the project processes cannot result in an extensive treatment for projects such as this example represents. For the concept development stage, two workshops are suggested, each at key points. The first risk management workshop could be included with any meeting arranged to instigate the concept and budget proposal, with perhaps only the business unit manager and the PM participating. A second workshop might be held just prior to board submission of the developed proposal, when project information availability nears its peak. The PM, brand business unit manager, and preferably a senior manager or parent board representative should participate in this workshop.

A third risk management workshop should take place during the project development stage, when task decision making is probably at its highest, and with technical experts participating as well as the PM. A risk management review meeting might be worthwhile just prior to the system testing process, to ensure that 'all bases are covered', and formal risk knowledge capture should be undertaken soon after the project website enhancement has been deployed.

In this example, the constraints of fast-delivery IT projects are revealed in terms of applying systematic risk management. Adopting such an approach can be done, but would add considerably to the workload of the PM since the workshop and review activities must be allied to appropriate risk documentation processes. However, it is questionable if the company in the longer term can afford to continue without formal risk management in place. The vulnerability of the organisation to the continuing well-being and capability of its IT PM suggests that this in itself is a threat risk. Also, note that we have framed the IT project example from a threat risk perspective. Until the organisation has a more established culture of systematic risk management, processes of opportunity risk management are unlikely to be exploited successfully.

4.7 A Project Risk Register Template

Before completing this chapter, it is useful to consider a typical format for a PRR. Tables 4.1 and 4.2 provide a template that we will refer readers back to in later chapters. The template is easily constructed as a spreadsheet and is only separated into these two tables to meet page size limitations and maintain legibility. Table 4.2 follows immediately from the right-hand side of Table 4.1.

Table 4.1 Project risk register template (part 1).

Project (threat) risk management system (RMS) template						
Organisation			**Key**	**Key**		**Key**
Project			Rare = 1	Insignificant = 1		Negligible
			Unlikely = 2	Minor = 2		Low
			Moderate = 3	Moderate = 3		Medium
			Likely = 4	Major = 4		High
			Almost certain = 5	Catastrophic = 5		Extreme
			Note: Scale intervals must be defined for the RMS owner.			
	Risk identification			**Risk analysis**		
No.	Objective (or task/ technology/ resource/ organisation)	Identified risk (risk statement)	Risk category	Likelihood (1–5)	Impact (1–5)	Quantitative. analysis (calculate severity from objective likelihood and impact data)	Severity use code table
001							
002							
003							
004							
005							
006							
007							
008							
009							
010							
Etc.							

Table 4.2 Project risk register template (part 2).

	Risk response				Monitoring and control				
No.	Existing treatment (or risk allocation)	Missing controls	Treatment plan (response)	Reassessment (revised severity)	Responsibility (by whom?)	Action (by when?)	Monitoring (who? When? How?)	Treatment cost estimate ($)
1								
2								
3								
4								
5								
6								
7								
8								
9								
10								
Etc.								

Note: Consider risk duration factor?

Table 4.1 (Part 1) corresponds to Stages B (Risk Identification), C_1 (Risk Analysis), and C_2 (Risk Evaluation) of the risk management cycle shown in Figure 4.2. Entries for this part of the PRR are discussed more fully in Chapter 7 (Project Risk Identification Tools) and Chapter 8 (Project Risk Analysis and Evaluation).

Table 4.2 (Part 2 of the PRR template) covers Stages D (Risk Response) and E (Risk Monitoring and Control) in the cycle shown in Figure 4.2. Data entries for this part are dealt with in Chapter 9 (Risk Response and Treatment Options) and Chapter 10 (Risk Monitoring and Control). Row allocation in the two tables has been restricted here to 10 risks, but when used as a spreadsheet the row entry capacity is almost limitless for practical purposes. The 'Severity' column data can be sorted to group risks of similar severity. Spread-sheeting the PRR in this way allows several additional analyses to be undertaken. Some of these are considered in Chapters 8 and 18.

Additional discussion of the PRR is pursued in Chapter 11 in terms of the knowledge transition into an ORR and further knowledge transformation.

4.8 Summary

This chapter forms a bridge between the theory and principles of risk and risk management and their practical application to projects.

Risk management has been defined as a process for dealing with risks, in our case in a project context and proactively from a project stakeholder perspective.

A cycle of project risk management is proposed as a systematic process that can also provide a learning cycle resource for the stakeholder and its engagement with subsequent projects. The RMS described here is generic and can be tailored to suit the needs of stakeholders involved in different types of projects and in different industry contexts.

Establishing and understanding the internal and external contexts for a project are shown to comprise an important first task in risk management, as these contexts shape the nature and magnitude of project risks for a stakeholder. Risk identification, analysis, and evaluation then amplify project risk information and influence risk response decision making. Prudent risk managers will seek to monitor and control risks during the ensuing stages of a project, and at suitable points will also seek to capture risk knowledge from the project as a valuable and growing asset for the stakeholder organisation.

Project examples from different contexts have been used to illustrate the application of the RMS. Much of project risk management involves interrogating the project and its contexts. We must not only ask questions, however, but also ask the right questions if our risk management is to be effective.

Finally, the PRR template provides a brief introduction to the documentary requirements of project risk management. The following chapters will focus and expand upon each stage within the cycle of systematic project risk management, beginning with context establishment.

References

Australian Standard/New Zealand Standard (AS/NZS) (1998). *Risk Analysis of Technological Systems – Application Guide*. AS/NZS 3931. Homebush, NSW: Standards Australia.

Australian Standard/New Zealand Standard (AS/NZS) (1999). *Risk Management.* AS/NZS 4360. Homebush, NSW: Standards Australia.

Australian Standard/New Zealand Standard (AS/NZS) (2004). *Risk Management Guidelines Companion to AS/NZS 4360.* AS/NZS HB 436. Homebush, NSW: Standards Australia.

International Organisation for Standardisation (ISO) (2018). *Risk Management – Principles and Guidelines.* ISO 31000. Geneva: International Organisation for Standardisation.

International Organisation for Standardisation (ISO) (2009). *Risk Management – Vocabulary.* ISO Guide 73. Geneva: International Organisation for Standardisation.

Project Management Institute (PMI). (2013) A Guide to the Project Management Book of Knowledge, Vol. 5. PMI, Newtown Square, PA. ISBN 9781935589679.

Standards Australia/Standards New Zealand (SA/SNZ) (2010). *Communicating and Consulting about Risk.* Homebush, NSW: Standards Australia.

Standards Australia/Standards New Zealand (SA/SNZ) (2015). *Managing Risk in Projects – Application Guidelines.* AS/NZS HB 436. Homebush, NSW: Standards Australia.

Standards Australia/Standards New Zealand (SA/SNZ) (2013). *Risk Management Guidelines – Companion to ISO 31000.* SA/SNZ HB 436. Homebush, NSW: Standards Australia.

Standards Australia/Standards New Zealand (SA/SNZ) (2013). *Risk Management – Guidelines on Risk Assessment Techniques.* SA/SNZ HB 89. Homebush, NSW: Standards Australia.

5

Project Risk Contexts and Drivers

5.1 Introduction

Chapter 4 introduced a staged cycle for project risk management that commences with establishing the appropriate context for managing risks on a project. Context setting is important in risk management but is frequently overlooked or even deliberately ignored, usually on the grounds of familiarity – 'After all, we've done so many projects that by now we should know what they are all about, shouldn't we?' That may be so, but the old adage about familiarity breeding contempt is apposite here. Indeed, what could be worse than commencing a prudent process of risk management by immediately exposing ourselves to a threat risk of mismanagement? While many projects may be similar, and that helps enormously in risk management, each one remains unique. The factors that influenced the previous project may not be present (or not to the same extent) for the next. The Malvina Reynolds (1962) song 'Little Boxes' lampooned the repetitive and self-perpetuating characteristics of a tract housing development seen on a hillside whilst she was driving through Daly City, California; it attempted to characterise the occupants in similar fashion, but even she graciously conceded that the houses were painted in different colours and that the occupants worked in different jobs. Perceptions of the uniformly boring 'ticky tacky' do not diminish the presence of risk!

Contextualisation is the process of setting or exploring something in its particular surrounding and connecting environment. This enables us to consider our target of exploration more fully. It is like looking at the whole forest – and indeed where the forest is located and perhaps how it may be compared with other forests – before we determine how we should deal with particular trees.

More than one context is involved. Firstly, there is the obvious context of the project itself – what we can regard as the internal context. However, a project rarely takes place in cocooned isolation but, more often than not, in an arena (sometimes competitive) of many other projects involving multiple stakeholders, all influenced in one way or another by features, conditions, trends, and so on that are present in the physical world and in the societies that inhabit it. This extensive domain constitutes the external context for a project.

One aspect of contextualisation in risk management is the need to set a system boundary for project risk, since risk events arising in some parts of the external context may have consequences that impact the project and its stakeholders. The consequences of these external risk events interact with other characteristics of the project, and the

Managing Project Risks, First Edition. Peter J Edwards, Paulo Vaz Serra and Michael Edwards.
© 2020 John Wiley & Sons Ltd. Published 2020 by John Wiley & Sons Ltd.

stakeholders, to affect the internal context for a project. However, while external risk events may be part of the external context, they are not the whole of it.

The observable constituents of the internal and external project contexts act, severally and sometimes jointly, as risk drivers for projects; hence the importance of establishing context in systematic project risk management. Knowing the project context helps us to determine the manner, level, and timeliness of subsequent project risk management activities. We can broaden or narrow the context appropriately to suit the management requirements for particular risks.

5.2   The Contextualising Process

The process of establishing appropriate contexts for project risk management was identified as part of the systematic management cycle described in Chapter 4 and shown in Figure 4.2 (Stage A). This should be neither a hurried nor a curtailed process, since the outcomes can significantly affect our ability to identify and manage project risks effectively.

A workshop-based approach, using brainstorming techniques applied in a structured or semi-structured manner, is usually the best way of conducting the contextualising process. Given the position of this stage in the project risk management cycle, the workshop should be arranged as early as possible in the stakeholder's involvement with a project. Key decision makers in the stakeholder organisation, and particularly those with direct project responsibility, should attend, but small workshops are always better than large ones in terms of participant numbers. Note, however, that this is not a task for one person alone!

While a context establishment workshop could be incorporated into a project briefing session, as advocated by many proponents of project management, in practice it is usually better to arrange a separate (and slightly later) occasion for the process we suggest here. Context setting involves more than elucidating what the project and its purposes are.

The contextualising workshop could also be combined with the risk identification stage (Figure 4.2, Stage B), but in our experience this may be a false economy for three reasons. Firstly, it may shorten the time available to properly identify project risks. Secondly, the participants needed for good contextualisation may not be the same as those best equipped to identify actual risks. Thirdly, trying to combine the two tasks in the same workshop does not really allow sufficient time for the context setting outcomes to be properly summarised and assimilated in terms of their influence as risk drivers for the project.

A separate project contextualising workshop allows a risk awareness atmosphere to be generated and subsequently maintained in the ensuing risk management stages. Remember that this awareness may include an opportunity perspective, although it more often takes on a threat risk emphasis by focusing upon negatively framed contexts.

The workshop duration for context setting should not extend beyond an hour or two, and should be held free of interruptions and other distractions. It may be necessary to reconvene for brief meetings later on, if it becomes evident that additional information has to be obtained that is not immediately available in the workshop. While formal minutes of workshop proceedings are probably not necessary (unless required under the governance policies of the stakeholder organisation), secretarial help may ensure the full capture of ideas and opinions. 'Sticky notes' are useful as a quick means of recording initial thoughts, comments, and suggestions in a way that can be more coherently organised later on. Contemporary communication devices such as laptops, tablets, and smartphones will also do the job, but the visual and manual advantages of arranging the scribbles on those small yellow adhesive pieces of paper in some sort of rough order on a whiteboard are very compelling! The shared satisfaction is often self-evident. Audio-recording and even video-recording of the proceedings are also possible, but in our view may be technical over-elaborations of what is needed. However, we cannot ignore project exigencies. In some circumstances, evidentiary requirements may dictate the level of data capture for the contextualising process. The formality of contemporary risk management increases in step with growing demands for demonstrable evidence of its implementation.

Many organisations adopt an approach to context establishment for projects that in some ways resembles a short checklist. Case Study F (an aquatic theme park) depicts an example of this in Table F.2. The civil engineering company makes its use mandatory on all new projects. In our view, this is a good policy but one which could be improved by distinguishing clearly between internal and external project contexts and by using a more clearly structured format. To be fair, however, the company regards the policy more as a business case alignment tool than as a project contextualisation technique.

The outcomes of a contextualising workshop should be greater risk awareness on the part of those who will be involved in managing project risks, together with a more informed understanding of the factors that could shape those risks. Ideally, a brief written summary of the internal and external contexts for the project will be prepared, distributed to the relevant personnel, and captured in the project risk register. This will guide subsequent risk decision making, particularly in the risk identification and risk response stages.

5.3 Internal Contexts as Risk Drivers

All projects have an internal context that is usually associated with their particular circumstances, nature, and objectives.

Internal contextualisation should follow a fairly structured list of questions that interrogate the project:

- What is this project?
- What are the procurement, operational, and strategic objectives for the project (Chapter 3, Figure 3.1)?
- As a project stakeholder, do we have additional or different objectives?
- What are the environments of the project (procurement, operational, and disposal; see Chapter 3, Figure 3.2) in which we will be directly involved?
- What are the governance (organisational structure) arrangements proposed for the project?
- What do we know about other project stakeholders (Chapter 3, Figures 3.7–3.10)?
- What level of stakeholder management will we need to engage in?
- What financial arrangements are in place for the project?
- If it is a 'hard' project, how far is the location from our normal management base?

- How similar is this project to others that we have completed successfully?
- What unsuccessful aspects of past projects are likely to arise in this one?
- What is our 'project horizon': in other words, what will be the likely status of current projects, and the prospects for new projects, when we become fully engaged with this one?
- What technical, financial, resource, and management factors in our current organisation (Chapter 3, Figure 3.3) could affect our capacity to undertake this project successfully?

More questions could be added to this list, but the workshop should not be allowed to become excessively protracted. What we are trying to achieve is a 'richer' picture of the proposed project and how it might impact our organisation from a risk management point of view in terms of influencing or 'driving' the risks that we face. A corollary question to the answer of each question such as those in this list should therefore be:

- How could this aspect influence the level of uncertainty in the project risks we seek to identify?

Asking the corollary questions should expose internal drivers that will shape the project risks that our organisation will face and influence the way in which we can manage them.

For example, we may ascertain that the project involves adding more operating theatre facilities to an existing hospital and that the current facilities must remain in uninterrupted use as far as possible during the construction work. This internal context knowledge will influence the management of scheduling risks for the project, requiring greater activity programming flexibility as a response, as well as the possibility of having to pay higher rates for work to be carried out at inconvenient times. Safety and security risks are also likely to be affected.

Care should be taken to ensure that addressing the corollary questions does not allow the context setting workshop to drift into becoming a full-scale risk identification session. The 'richer picture' sought here is still more about looking at the whole forest, and not about examining the condition of each species of tree! At this stage, all we need to deduce is how that context might affect the likelihood of occurrence, the consequences, or the period of exposure for project risks generally rather than individually.

Once we are satisfied with what we have learned from this view, an even wider perspective should then be explored through the external contexts for the project.

5.4 External Contexts as Risk Drivers

At the beginning of this chapter, we noted the need to set system boundaries for project risks. This is important because it clarifies the extent of control we can exercise in project risk management. By and large, if events are likely to happen beyond the direct boundaries of a project, then risk management may be limited to dealing with the consequences of such events rather than addressing the events themselves. This is not inevitably so, of course. For weather risks, we may be able to adjust the timing of the project so as to avoid seasonal weather tendencies such as winter storms. For financial risks, we may be able to enter into loan finance arrangements that will correspond to prevailing interest rates that are forecast to remain low for the period of our involvement in the project. Generally, however, we are beholden to potential external situations and events that lie beyond our ability to fully control them.

Setting project system boundaries is shown conceptually in Figure 5.1. Risk events occurring in distinguishable external systems give rise to impacts and consequences that cross the arbitrary boundary of our project system and influence the nature or magnitude of the risks we face on that project.

For many projects, there will be a far greater number of identifiable external systems than shown in Figure 5.1. As with the process of establishing the internal project context, looking at each of these individually in the contextualising workshop process should be avoided. It is simply too much to attempt in the time available. The appropriate time for that is in the ensuing stages of project risk management, when risks have been identified and assessed and risk treatment options are being considered.

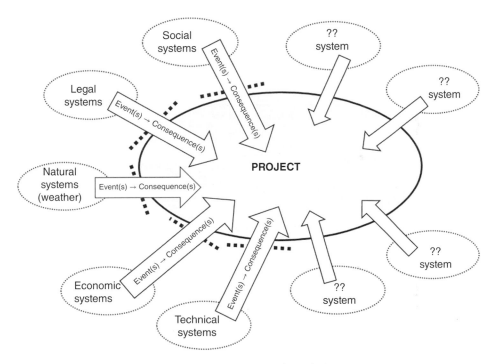

Figure 5.1 Project system boundaries.

For example, we may have identified a risk whereby market volatility in the price of materials essential for the project could leave our organisation exposed to project cost-overrun uncertainty. Our response to this risk may be to make forward purchasing arrangements for the materials in question.

The concept is useful, however, for enhancing our awareness of the external contexts for projects and the ways in which they can act as risk drivers. Thus, in establishing the external contexts, we may consider levels of price volatility for important resources needed for the project.

This awareness is actually better achieved by reducing the number of external contexts to four, and then considering each of these from the more general 'big picture' perspective that we advocate. The four external contexts are the physical, technical, economic, and social settings for a project. These are shown in Figure 5.2, but note how this also incorporates the earlier project concepts of Figure 3.3, with its 'messy' project boundary and internal contexts, and now places them within the purview of a similarly 'messy' external context.

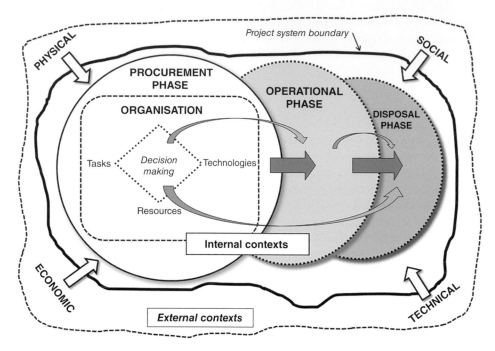

Figure. 5.2 Project risk driver contexts.

Geographically, the external project contexts may be as wide as needed, embracing local, regional, national, and even international aspects.

The exploration of each of the four perspectives usefully adopts the interrogative approach used for establishing the internal project context, although the questions are more generally framed.

5.4.1 Physical Contexts

Establishing the external physical context for a project is not the sole preserve of 'hard' projects (e.g. buildings and infrastructure such as transport facilities, water supply, and services distribution), nor even the commissioning and installation of art projects such as sculptures or the provision of temporary facilities for events projects. While hard/'outdoor' projects are inevitably subject to physical circumstances and constraints, 'softer' projects may be similarly affected. For example, a project to implement or improve customer relationship services in a national (or even international) chain store organisation will be subject to physical factors such as the number and whereabouts of store locations and the layouts of individual stores.

Given this caveat, however, it is hard projects, or more accurately the 'hard' aspects of projects, that most often give rise to the need to consider their physical external contexts. The contextualising question is therefore:

- What are the physical settings for this project that are capable of influencing the project risks that we face?

Without limiting the diversity of physical contexts that might be encountered, these could include:

1) Environmental issues
2) Prevailing climatic and seasonal weather conditions for the duration of the project
3) Proximity to rivers and waterways, and propensity to flooding
4) Proximity to essential temporary and permanent services (water, mains electricity, sewage, and waste disposal)
5) Remoteness (access, transport, and accommodation for workers and project staff; component and materials delivery)
6) Local traffic conditions (street access to site, parking restrictions, street closure capability, and traffic management capacity). This would also include similar constraints associated with maritime projects.
7) Geological constraints (or tidal conditions for maritime projects)
8) Continuity, access, and non-interruption requirements for existing services and facilities (roads, railways, airports, river crossings, hospitals, schools, etc.)
9) Proximity to sensitive areas (weapons firing ranges, defence installations, nuclear waste disposal facilities, high-security prisons, etc.).

Once each physical context is identified, the follow-up question to be answered is again:
- How could this context or these constraints and conditions influence the uncertainty in the project risks we seek to identify?

A real-life illustrative example is the expansion and refurbishment of an existing rail terminus in a major Australian city, using a Public-Private-Partnership (PPP) arrangement for the procurement and operation of the upgraded facility. The drafting of the PPP agreement took into account the need for continuing train operation during the construction phase, with local and interstate passenger train services to be shut down from midnight to early morning each day. Knowing this constraint allowed the private partner's contractor to make due allowances in the construction schedule. However, the continuing use context had failed to consider the 24-hour operational needs of freight trains run by a different company. Nor did it take into account the passenger service operators' practice of leaving empty regional passenger trains standing at the existing platforms overnight. The two factors severely impacted access to rail tracks and platforms for necessary construction work, aggravating the scheduling risks already faced by the contractor. The subsequent renegotiation of the PPP agreement and cost reimbursement to the private partner in the project delivery stage could have been avoided by more careful consideration of the physical, external project context.

5.4.2 Technical Contexts

In considering the external technical context for a project, we are not exploring the actual technical needs of the project itself. These are part of the internal project context and relate directly to the decision making associated with the 'technologies' element of a project (Chapter 3, Figure 3.3). The threat risks (and indeed the opportunity risks) arising from the use of particular technologies may be influenced by the wider 'technology climate' in which the project takes place.

This climate relates to issues such as the availability of suitable technologies; their reliability or proneness to failure; restrictions on rights to their acquisition, transfer, or use; the capacity to implement them; and the level of innovation, or familiarity, with particular technologies in the project setting. Some of these may be subject to international embargoes or restrictions aimed at specific countries.

The questions to be asked about the external project context are therefore:

- What technologies will be needed to deliver this project?
- What technologies will be used in the operational phase of this project?
- What constraints might arise in obtaining and implementing those technologies?

And the corollary question:
- How could any of these constraints influence project decision making?

Communications technology provides an appropriate example for this context. An information technology (IT) project may envisage the development of a smartphone application for particular payment purposes (e.g. parking fees in a commercial parking garage or payment of visa fees at an international entry point). While the smartphone technology itself may be readily available, the capacity to implement it reliably in different situations and locations may be constrained by many factors besides simple technical feasibility. Not the least of these would be the adequacy of cellular service coverage.

5.4.3 Economic Contexts

The wider economic context for projects also includes financial aspects. Matters to be considered here might include, *inter alia*:

- Market supply/demand conditions for resources (materials, labour, and equipment)
- Fiscal and monetary policy settings by governments and treasuries
- Financial markets, loan availability, and interest rates
- Economic performance
- Levels of national debt
- Unemployment statistics.

Any one of these is capable of influencing the occurrence of project risk events, the consequences of those events, and the ways in which we can respond to, monitor, and control them as risks for the project. Each is subject to considerable uncertainty in terms of forecasting future levels and conditions, particularly in the detection and modelling of likely turning points in graphs of trends and outcomes.

Appropriate questions for exploring the external economic context are:

- What are the likely market supply/demand conditions applicable to the resources and technologies needed to carry out this project?
- What are the relevant fiscal and monetary settings surrounding this project, and how likely are they to change?

The corollary question is again:
- To what extent will any future changes in these settings and conditions affect our project decision making and capacity to manage project risks?

For example, a civil engineering infrastructure project such as a major new harbour will almost certainly require the installation of large-diameter supporting piles for the construction of wharves, and dredging for new deepwater channels. The capital cost of equipment needed for such works is high, as also are the costs of transporting it around the world. The work is, more often than not, carried out by specialist contractors operating in an international market. Tenderers for the new harbour will therefore have to consider not only the availability of the specialist contractors, but also the likely proximity of the dredgers and floating pile-drivers to the required location when their input is needed. While such information will lack certainty at the time of tendering, a general picture of the market situation is essential, especially in terms of being aware of the possibility of other maritime projects coming on-stream internationally that will threaten the main contractor's ability to secure the necessary resources at the required time.

5.4.4 Social Contexts

Establishing the external social context for projects is usually a more wide-ranging process than the exploration of the physical, technical, and economic contexts discussed so far in this chapter. In fact, this perspective gathers up and embraces most of the external 'systems' referred to at the beginning of Section 5.4. Again, without limiting the actual diversity of contexts that might have to be considered, setting the external social context for a project entails looking at matters such as:

- Political environments
- Public administration systems
- Health systems
- Education
- Civil and criminal law
- Justice systems
- Public welfare systems
- Industrial relations and employment conditions
- Equality issues and affirmative action policies
- Public attitudes, mores, and social capital (considered as the nature and value of the abilities and relationships, existing and developing among people living and working in a particular society, that enable that society to function effectively)
- Culture and customary practices.

The level of exploration may be local, regional, national, or international, or even combinations of these.

Appropriate context setting questions would be:

- What social 'systems' are applicable to this project?
- What are the relevant issues arising in each of these?
- What changes are taking place (or can be foreseen) in them?

The necessary corollary is:

- How could these issues or conditions (or any changes in them) influence our project decision making and capacity to manage project risks?

Abundant examples of how external social contexts may affect project risk management could be drawn from any of the areas mentioned in this section, but two should suffice here.

For a public sector authority seeking to adopt a PPP procurement system for the first time for public asset and service delivery, a key factor in successful procurement will be the capacity and expertise of its public administration system to manage PPP projects, not only during the delivery phase but also throughout the operational life span of each project. The type of expertise required is likely to be different for each phase, and possibly for each type of PPP project. Issues relating to the recruitment, training, retention, and replacement of suitable managers will influence how the public partner deals with its risks on a project, not only in the short term but also for the considerably longer duration of the operational phase for each partnership agreement.

In the second example, consider the situation of a company seeking to establish a mining concession project in a foreign country. Many of the external social context areas listed in this section will come into play. What is the nature and stability of the foreign government? How easy will it be to interact with public administrators in that country? How will expatriate staff be affected by the law, justice, health, and education systems there? How will industrial relations conditions in that country affect the recruitment and employment of local workers? What local cultures and customs will have to be observed? What are the likely attitudes and reactions of different groups in the local and national population to the proposed concession?

Having described the important process of establishing appropriate contexts for risk management, we now consider how the contextual information and knowledge can be used.

5.5 Using Contextual Information

In the interrogative workshop approach to establishing the project contexts, the questions aimed at exploring them are unlikely to yield detailed information for specific risks.

However, the answers should greatly increase the risk awareness of a project stakeholder by raising the awareness of the relevant managers in the stakeholder organisation. Participants in the workshop process will become more alert to the myriad factors that can influence the nature and treatment of project risks.

It does not matter too much if some contextual aspects are overlooked at this stage of project risk management, although practice and experience will help to avoid too many gaps. Raising the general level of awareness acts as a stimulator for the subsequent identification, analysis, evaluation, response, and monitoring and control stages of project risk management.

A human physiological and psychological analogy supports this proposition. If I become more aware of my own physiology and its limitations (the skeletal, muscular, nerve, cardiovascular, and respiratory systems, etc.), and more aware of how I respond mentally and emotionally to different situations, then I can make more informed decisions about my life. While I will rarely be more informed than expert specialists in any of these areas, my overall self-knowledge will be improved. Of course, I could continue to make decisions on an uninformed basis, but surely that would be much riskier?

As noted in this chapter, the summarised information arising from the contextualising process should be incorporated into the documentation comprising the project risk register. This makes it easily accessible during subsequent stages of risk management.

Techniques for *risk identification* are presented in Chapter 6, but knowledge about the internal and external project contexts can provide guidance and augment other methods, such as checklists, at this stage of risk management. Our questioning approach again proves invaluable, for example: if we use this particular technology on this project, what could go wrong? Or: given the relative instability of the current government in that country, what particular sovereign risks arise for us if we undertake this project there?

Contextual knowledge influences *risk analysis* by moderating our estimates of the uncertainty associated with the probability of occurrence of risk events and the nature and magnitude of their consequences if they do occur (Chapter 8). For example, contextual information about the availability of electrical power supply and its propensity for outages allows us to factor in the additional costs of maintaining alternative and temporary or even permanent standby generation facilities for delivering and operating the project.

In the *risk evaluation* stage, the contextualised risk analysis outcomes may now entail assigning a project risk to a different severity category (Chapter 8), thus influencing any strategic risk management approaches for the organisation (Chapter 16).

Context information may affect the options available in deciding upon a *treatment response* to particular risks (Chapter 9). This, possibly augmented by additional information, might influence the way in which we plan to monitor and control those risks as the project proceeds (Chapter 10).

Finally, contextual information is likely to be valuable beyond the requirements of risk management for particular projects. Suitably organised and filtered, such knowledge can make a significant contribution to the store of risk knowledge available within an organisation. A suitable repository would be the organisational risk register (see Chapter 11).

5.6 Summary

More profound consideration has shown how important the initial stage, of establishing the context, is to project risk management. It informs and influences subsequent risk identification, analysis, evaluation, response, and monitoring and control activities. Given some modification, it may also contribute to a project stakeholder organisation's store of risk knowledge and wisdom.

Ideally, the internal and external contexts for a project should be considered in a workshop setting, using techniques that interrogate the project and its wider environment.

The major benefits of the contextualising process are: a 'richer picture' of the project and its setting, greater risk awareness for the relevant staff in the project stakeholder organisation, and more informed guidance for their risk decision making.

Once the internal and external contexts for a project are more clearly known, risk identification can proceed with greater confidence.

Reference

Reynolds, M. (1962). *Little Boxes*. New York: Lyrics and music copyright Schroeder Music Co.

6

Approach to Project Risk Identification

6.1   Introduction

Good risk identification is crucial for effective risk management, since unidentified risks cannot be managed systematically, yet they remain risks. An organisation wishing to implement, operate, and maintain a good risk management system (RMS) should be prepared to spend sufficient time and resources in considering how best to identify the project risks it faces.

Essentially, *threat risk* identification sets out to answer the question: What could happen that would adversely affect the satisfactory achievement of this objective, the completion of this task, the application of this technology, the acquisition of this resource, or the performance of this organisation? Since we proposed earlier (Chapter 2, Section 2.4, Risk and Uncertainty) that risk is closely associated with decision making, the question could be posed in another way: What could happen to make this decision a bad one for us?

Identification of *opportunity risks* employs a different questioning approach, and this is addressed more fully in Chapter 15, but all project risk identification is really a process of targeted exploration and discovery.

Note that the emphasis is on the occurrence of a risk event, rather than its consequence – upon cause rather than effect. This is not to say that impact plays no part in risk identification. Many people intuitively use it as a starting point in the process, as do some formal risk identification techniques. Cost or time overruns, for example, are often suggested initially as risks to projects, but a little thought will show that these are actually consequences of prior risk events, and the causal factors or situations must then be explored to explain the threat risk outcomes.

In this chapter, we will focus largely on the *approach* to risk identification by trying to establish reliable processes through which the threat risks faced by a project stakeholder can be identified with reasonable confidence. We also look at clarifying what we refer to as *trigger events*, particularly those occurring within the project system boundary proposed in Chapter 5 (Figure 5.1). In some instances, this will mean that the consequences of risk events occurring outside the project must be treated as events when they cross the project system boundary. In order to avoid or reduce confusion, this manipulation will be made clear for specific risk situations and for particular risk identification techniques.

Managing Project Risks, First Edition. Peter J Edwards, Paulo Vaz Serra and Michael Edwards.
© 2020 John Wiley & Sons Ltd. Published 2020 by John Wiley & Sons Ltd.

An attempt is made to categorise the large array of techniques available to project managers for risk identification, and to show where they are most appropriately used. Some individual techniques and tools will be examined in more detail in Chapter 7.

In this chapter and in Chapter 7, it is assumed that the earlier context establishment stage (Chapter 4, Figure 4.2, Stage A) in the cycle of project risk management has been completed satisfactorily. A further assumption continues from Chapters 1, 3, and 4: that a project stakeholder organisation is implementing its own RMS and identifying its own project risks.

Topics covered in this chapter include: the approach to project risk identification, timing of the identification process, and types of available tools and techniques.

6.2   Approach to Risk Identification

Perhaps more than any other stage of systematic project risk management, risk identification is best suited to a workshop approach and using brainstorming. However, the brainstorming should be structured in some way. While risk identification brainstorming could be used on its own in a completely unstructured 'freewheeling' manner, this would hardly accord with an organisation's desire to be systematic in its risk management. In fact, brainstorming is easier, and usually more successful, when it is guided in some way (i.e. applied within particular project 'frames' or perspectives). Freewheeling approaches are unlikely to yield consistently adequate results. Unless carefully managed, they can tend to drift away from the intended purpose, take too much (or too little) time, and leave participants feeling irritated and dissatisfied by the process and the results.

Workshop participants should comprise a small group of people who are familiar with the stakeholder's organisation and its projects, and are closely involved in the important decision making for various aspects of the project. As with the previous stage (context establishment) of project risk management, fewer participants are better than many for this type of workshop, but it is definitely *not* a task for one person alone! We repeat our contention that 'one-man band' approaches to project risk management expose any organisation to the serious threat risk of mismanagement. Ideally, the workshop participants will form a 'team' for the purposes of managing risks on the project.

Since the project risk identification stage could flow on fairly seamlessly into risk analysis (and possibly beyond that into risk evaluation and response treatment), a minimum duration for the initial workshop is likely to be four or five hours, depending upon the nature and complexity of the project and the extent to which it is familiar to the participants. Workshop logistics such as location, refreshment, rest intervals, and documentation are therefore an important preparatory consideration. The value of a good organisational risk manager is appreciated here, particularly for the documentation aspect, since it should be possible to set an agenda for the workshop, create a template spreadsheet to record risks (the project risk register [PRR]), and even provide participants with an appropriate information 'kit' beforehand.

The approach should be a 'narrowing' one, hence our reference in Section 6.1 to targeted discovery. Narrowing can be carried out progressively and quickly in order to:

1) Clearly identify the stakeholder.
2) Clearly identify the project.

3) Confirm the project contexts (objectives, etc.).
4) Clarify relevant external contexts (the risk drivers).
5) Select the project environment(s) to be considered: procurement, operation, and disposal.
6) Select the appropriate risk identification technique(s).
7) Focus on the associated project decision making, and then:
8) Ask the question(s): What uncertain aspect(s) of the decision(s) could threaten a satisfactory outcome (achievement of objectives) for this stakeholder?
9) (Note that for opportunity risk identification: What additional benefits beyond the achievement of objectives could this decision realise for this stakeholder?)

Activities 1, 2, 3, and 4 should have been completed earlier in the context establishment stage of the project risk management cycle. Now the workshop team will begin to 'drill down' into specific aspects of the project itself.

With their thinking guided to the relevant aspects, workshop participants are encouraged to brainstorm as many project risks as possible, preferably in short 'bursts' of intellectual effort each culminating in a roughly collected summary of recorded risks. 'Sticky notes' are again useful for this task, and risk identification at this point may comprise no more than a few words to describe each risk event. Alternatively, to minimise the 'paper trail', handheld, tablet, or laptop computer devices can be used, or audio-recordings made, but 'sticky notes' allow information ordering to be undertaken before it is captured more formally.

Later in this stage, precise risk statements must be prepared, but that is best done just prior to the risk analysis process, and we explain these statements more fully in Chapter 7. For now, written notes should suffice, and the workshop leader could use time during workshop rest intervals to transfer information from them to the PRR spreadsheet.

Inevitably, some of the rough notes will refer to consequences of risks rather than trigger events. This is not a matter of concern at this time, and the workshop should not spend valuable time in trying to correct them at this stage. That will be done later, when more precise risk statements are prepared.

A risk identification workshop (and more than one may be needed) should not spend excessive time on any particular aspect of the project, but should move fairly quickly through all the matters that need to be considered. It is usually possible to return to a particular issue later on, should participants think that additional risks might well emerge from further identification effort.

6.3 Workshop Timing

The timing recommended for project risk identification workshops is profoundly influenced by the nature of the project and the manner in which it is to be delivered. Since these are largely strategic matters, the arrangement and timing of project risk management workshops should be the responsibility of senior management in the organisation. While such arrangements may be established as a procedural policy, there should be sufficient flexibility to tailor them, in terms of timing and agendas, to suit the needs of individual projects. Experience plays a big part in this, hence adding to the importance of the organisational risk register (ORR).

Figures 4.3–4.6 in Chapter 4 showed the suggested timing for risk management workshops on two very different project examples. Now, Figures 6.1–6.3 illustrate in more detail how workshops might be arranged for two construction project stakeholders. Also, instead of a design-build contract arrangement, the hypothetical project represented is one procured under 'traditional' contracting arrangements typically found in the construction industry in the United Kingdom (and its colonial counterparts) from the early nineteenth century to today. Under traditional procurement, the design and construction processes are separated by an intervening tendering stage. The design phase for the construction project is carried out by an architect and other professional consultants commissioned by the client as project sponsor, and this arrangement was described as a 'coalition' in Figure 3.6 in Chapter 3. This team thus has an agency role towards the client (unless they are part of an 'in-house' unit in the client organisation). So Figure 6.1 assumes an individual project stakeholder perspective within the design team (e.g. an architect or engineer), and the nuances of this were discussed in Chapter 3. When tenders are invited for the construction work, the project stakeholder perspective in Figure 6.2 changes to that of a potential bidder. Thereafter, Figure 6.3 assumes that the tenderer's bid has been successful and that this stakeholder is proceeding with the construction work as the head contractor.

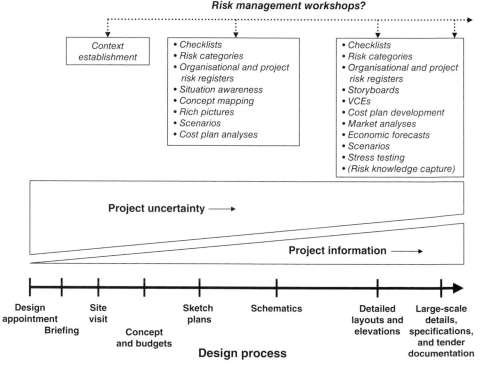

Figure 6.1 Design consultant risk management workshops in a construction project inception/design stage.

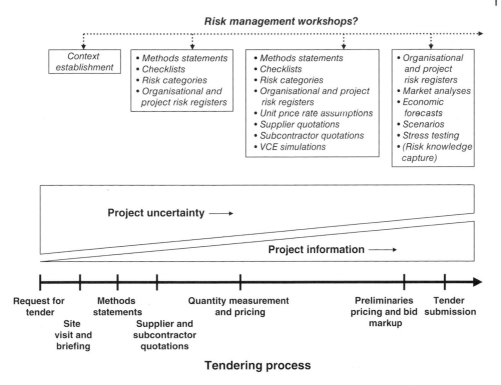

Figure 6.2 Bidder's risk management workshops in the construction project tendering stage.

Different procurement arrangements (both 'separated' and 'integrated' in terms of design) abound in modern physical development projects, and slight variations are found in almost every country (and even within countries). Describing and illustrating each one would be beyond the practical limits of a book such as this, as would attempts to portray risk identification workshop timing for many other types of project. The important thing to note from Figures 6.1–6.3 is that the second and third relate to a different project stakeholder (and thus a different project RMS) than the first diagram.

A good general principle to apply in determining the timing and frequency of risk management workshops is to follow the decision making for the project. When many key decisions are being made (i.e. about design options and delivery process activities), then the timing of risk identification procedures should match the decision intensity if possible.

We have emphasised that the RMSs for the two different stakeholders in the construction project example will be separate and different. The approach, however, will be broadly similar for both stakeholders, as it will for many types of project and for many types of project stakeholders.

Figures 6.1–6.3 include suggested techniques and methods that are appropriate for risk identification at each stage. Some may be more suited to particular project circumstances, but most can be adapted to suit requirements. The actual techniques are discussed in greater detail in Chapter 7.

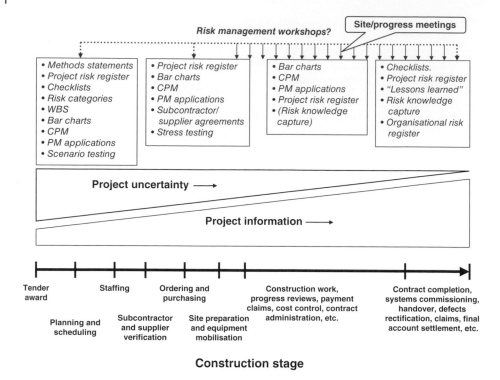

Figure 6.3 Contractor's risk management workshops and the project construction process.

The project inception/design stage inevitably starts with the least amount of project information available, and thus the level of project uncertainty is at its highest. However, information accumulates quite rapidly as design proceeds, and generally this means that risk identification workshops should be arranged to coincide with the accumulation process. It also means that, for each successive workshop, the risk identification techniques can be more detailed in nature. One technique, the use of scenarios, is an exception to this progression, and this will be explained later in this chapter.

Most of the design consultants involved in this stage of the project will employ unique discipline-based models for their work (the modelling process of decison making was described in Chapter 3). The decisions flowing from such modelling activity provide information that can be brought to the project stakeholder's risk management workshop for the purposes of risk identification. Indeed, discussion in the earlier workshops will be informed by these decisions and by contextual information about the project, and this knowledge will play its part, not only in influencing the stakeholder's subsequent risk analysis and risk response activities, but also in revisiting and reviewing identified risks in later workshops.

The outcome for the project design stage as shown in Figure 6.1 should be a PRR for the client/design team stakeholder, fully populated with identified, analysed, and evaluated threat risks for which appropriate response, monitoring, and control treatments have been determined. The PRR will provide the main platform and resource document for the project stakeholder's ongoing project risk management, which may be similar, but not necessarily identical, to that shown in Figure 6.3.

A different project stakeholder perspective – that of a bidder for the construction contract – is reflected in Figure 6.2.

For this stakeholder, the tendering process usually begins with a response to a request for tender (RFT) or expression of interest (EOI) notice issued by the project client, or by the lead consultant acting for the client, and is often followed by some level of briefing about the project that is made available to all those who have responded to the RFT.

The responding tenderer should then initiate what, at this point, will be a provisional implementation of risk management for the whole project (i.e. as if the bid were successful), commencing with context establishment.

While the design information available to the tenderer will indicate the nature and extent of the work to be carried out on the project, it must be interrogated and used to determine how and when construction activities will be conducted, who will do the work, and what resources will be needed. These construction process decisions (or methods statements) are used to inform the risk management workshops for the tenderer.

For this stage, the tenderer should hold a final risk management workshop as close as possible to the tender submission date in order to capture and consider the most recent risk information, particularly with respect to any up-to-date knowledge that could affect unit pricing for the work or the assessment of overheads costs and profit expectations. The separate PRR developed by this stakeholder at this point is likely to be a preliminary draft that will be more fully developed later on if the bid is successful.

The first of the workshops suggested for this stakeholder during this stage might have a duration of one to two hours. The second might take four to six hours, depending on the complexity of the project and the level of risk analysis attempted. The tender pre-submission workshop should not require much more than an hour.

Once the construction contract has been awarded, the successful tenderer then re-activates the earlier provisional risk management process. This is shown in Figure 6.3, which may represent a real project duration from a few weeks to several years.

If any conditions are thought to have changed, the successful contractor may first decide to recontextualise the project. Then at least two risk identification/risk review workshops will be necessary as the construction process moves from more detailed planning to commencement of operations on site. The formerly tentative PRR for this stakeholder now becomes an important reality for good risk management, but should quickly become far more comprehensive as project decision making accelerates and devolves downward throughout the stakeholder organisation.

If the tender is not successful in winning the contract, the provisional risk management work is not entirely wasted since the original draft PRR can be archived into the ORR and contribute to the body of risk knowledge for that organisation.

It is important to recognise here that techniques used mainly in other areas of project management, such as bar charts, work breakdown structures (WBSs), and critical path network (CPN) analyses, can be harnessed for risk identification and management purposes. The greater the decomposition of planned project activities, the greater the available access to the decision making associated with them, thus again following a principle of aligning project risk identification to project decision making. The benefit of adopting project management tools used for other purposes, and adapting them for risk identification, lies not only in the greater application value extracted from them, but also in the heightened risk awareness generated in the users and increased recognition of the role of systematic risk management in the project delivery process.

During the often lengthy period of construction work, risk management will largely comprise monitoring and controlling the risks previously documented in the contractor's PRR. This is usually best done as part of the contractor's agenda at regular site meetings (indicated by the thin downward arrows in Figure 6.3), and these occasions can be used to modify individual risk parameters, or to identify, assess, and treat new risks as the work progresses (see the dotted return arrows in Chapter 4, Figure 4.2).

While the capture of project risk knowledge is shown as a terminal, once-off stage in project risk management in Figure 6.3, it can also be carried out at any appropriate time by the contractor during the building contract to record outcomes for risks which have been 'closed off' by the completion of relevant risk bearing activities. This is discussed more fully in Chapter 11.

The risk identification processes shown here, for a construction project undertaken through a traditional 'separated' form of procurement and conducted in workshop environments, can be modified to suit the needs of other types of projects and other project stakeholders. Each stakeholder must decide the best timing and frequency for its project risk management workshops. Sometimes it may be convenient to conduct them in collaboration with other project stakeholders, but this practice should be regarded more as a procedure for allocating risks to particular parties and not as a more deliberative approach to managing them. The caveat about each project stakeholder facing different risks, and thus needing to manage them itself, remains relevant.

In principle, project risk identification involves consideration of what can affect the nature, timing, and resourcing of the activities necessary to bring a project to completion for a particular stakeholder; and of how that stakeholder fits into the wider project 'picture' and may be affected by other project participants. The risk identification techniques suggested for each stage in Figures 6.1–6.3 for the construction project example are also open to adaptation for other project circumstances.

A typology of the tools and techniques available for risk identification can be devised. Classifying them in this way promotes better understanding of their underlying power and timing, thus assisting in the process of selecting an appropriate technique.

6.4   Types of Risk Identification Techniques

Many risk identification techniques exist. Most are adaptations of management tools used for other purposes. This not only adds value to such tools, but also means that people in project stakeholder organisations are likely to be familiar with them already.

Table 6.1 presents a typology of techniques, with examples of each type.

The active ingredient common to every one of these techniques is brainstorming. Currently, we know of no fully automated techniques of risk identification, at least for projects, although some project management software applications can be customised to prompt with lists of risks at particular application points in the management process. Where they do exist, the prompt lists are similar to checklists and tend to be fixed at one point in time. Chapter 17 (Computer Applications) considers this situation in greater depth.

Whatever technique is used, at relevant intervals in the risk identification process the workshop participants have to respond to the (threat risk) question we raised in Section 6.1: What could happen in this particular context to threaten a satisfactory outcome for this organisation on this project? The process is therefore largely iterative

Table 6.1 A typology of risk identification techniques.

Type	Technique or tool
Activity-related methods	Methods statements
	Work breakdown structures (WBSs)
	Bar chart schedules
	Critical path networks (CPNs)
Analytical methods	Decision tree analysis (DTA)
	Event tree analysis (ETA)
	Fault tree analysis (FTA)
	Failure modes and effects criticality analysis (FMECA)
	Hazard and operability studies (HAZOPS)
	Safety hazard analysis (SHA)
Associated representative methods	Context setting
	Budgets
	Cost plans
	Quotations and purchase orders
	Contract clauses
	Other information resources (market analyses, economic forecasts, etc.)
	Checklists
	Generic risk classifications
	Organisational risk registers (ORRs)
	Procedural manuals
	Project stages and schedules
	Riskiness review models
Functional value methods	Value engineering (VE; also called value management [VM])
Matrix methods	Activity/risk category matrix
	Project phase/risk category matrix
	Project phase/activity/risk category matrix
Simulation and visualisation methods	3D visualisation (virtual environments)
	Concept mapping/rich pictures
	Storyboarding
Speculation methods	Scenario testing
	Stress testing
Structural and management methods	Organisational structure
	Governance and operational management procedures
	Project management applications

and, because of this, workshop leaders/facilitators should be wary of the potential onset of participant fatigue and ensure that sufficient rest intervals are included in workshops.

The typography of risk identification techniques is arranged alphabetically in Table 6.1. In the remainder of this section, they are discussed in the same order. In practice, however, the choice and use of any technique is strongly influenced by the stage (or stages) in the project where its application is most appropriate and effective. Some are best suited to project design and planning processes. Others are meant to reflect project delivery processes. A few deal with the operational aspects of projects. It is important to remember that, for effective project risk management, they must be applied proactively.

In this section, the various types of techniques are described in a general way. In Chapter 7, selected techniques are given more detailed attention with examples of their application.

6.4.1 Activity-Related Techniques

The activity-related risk identification techniques are overwhelmingly associated with project *delivery* activities. As noted, they are not risk-unique; that is, they are not devised solely for risk management, but are essentially used for other purposes that can be adapted, usually by the inclusion of the questions we have described.

Using these techniques in risk management relies on the availability of sufficient documented information about the task, technology, resource, and organisational requirements for delivering the project. While probably being closest to project decision making and thus closest to the risks themselves, these techniques are substantially information-driven and thus more suited to the planning stages of the project where decision making is most intense, such as when detailed design and estimating, planning, and scheduling are well underway. They are less suited to the inception/conceptualization stage of the project, where detail will be lacking and only ideas, often quite vague, are accessible.

Typical among such tools are: methods statements, WBSs, Gantt (bar) charts, and CPNs. These are each more fully discussed in Chapter 7. Generally, the greater the level of project decomposition revealed in these tools, the easier risk identification will be, but the later this process will have to be done.

For the most part, the primary purpose of activity-related techniques is to arrange what project activities must be carried out and the order of execution. They generally do not explain why the activities are necessary, how they should be undertaken, and what resources will be required. These issues are important to project risk management, and hence to risk identification, so it must be recognised that none of the techniques in this category constitute a completely sufficient tool for identifying project risks.

6.4.2 Analytical Techniques

Care is needed with this group of techniques. Any good search engine in an internet browser will unearth at least 50 tools claiming to represent effective ways of conducting 'risk analysis'. By the time this book is published, the count will probably exceed 100; such is the rate of development in this field. The great majority are based upon network analysis theory, which essentially aims to disaggregate causes and effects in networks and systems. In this regard, Figure 5.1 in Chapter 5 (Project Risk Contexts and Drivers) might be described as an analytical technique in that it distinguishes a project system from other systems operating in contexts external to the project itself. Figure 5.1, however, does not take us very far down the path of project risk identification, other than raising our awareness of such systems.

Almost without exception, the techniques popularly labelled as 'risk analysis' applications do not themselves directly identify risk and components of risk, nor do they quantify risk factors. They act as visually formatted tools that can help to *communicate* risk information and risk management decisions across and between organisations. 'Bow tie' and 'fish bone' diagrams are typical examples of such tools. Their diagrammatic structure has to be populated with information that must largely be derived from brainstorming. We will return to these techniques in Chapter 19 (Communicating Risk) to consider their communication effectiveness. The point to note here is that, for risk identification purposes, they represent a new tool that participants in a risk management workshop must become familiar with; whereas our preference is to adapt techniques that are already familiar from other project management applications.

Truly analytical techniques are usually 'data hungry' in a quantitative sense and are more appropriately used in the risk analysis stage for quantitative assessment (Figure 4.2, Stage C1). None were shown in Figures 6.1–6.3, and few are encountered in construction project management.

These techniques may be either inductive or deductive: tracing consequences flowing from a risk event, or auditing an event from its consequence path(s). Application timing is quite limited in terms of their use for project risk identification, and more often than not the structure of the technique must be altered to suit the process, hence their general lack of proactive use in project risk management (other than the last three shown in the list for this category in Table 6.1).

Analytical techniques include decision tree analysis (DTA), event tree analysis (ETA), fault tree analysis (FTA), failure mode and effects criticality analysis (FMECA), hazard and operability studies (HAZOPS), and safety hazard analysis (SHA). Chapter 7 explores them more fully.

6.4.3   Associated Representative Techniques

The associated representative group of risk identification techniques is also quite limited in terms of application timing and available techniques. The purpose of using them in risk management is mainly to raise risk awareness on the part of the relevant project stakeholder. Risk identification occurs through recognising project risks in aspects associated with the primary purpose of each tool.

Typical tools and resources in this category that can be used for risk identification include: visual media such as project sketches, drawings, and plans; performance-related documents such as project specifications; financial information such as budgets, estimates, cost plans, and quotations; contractual information such as contract conditions and subcontract agreements; and more general resources such as market analyses and economic forecasts.

Also in this category may be found official reports such as the findings of inquiries into accidents and disasters. Not always freely available, and probably least often used in project management, their usefulness in project risk identification is usually limited to confirming the presence of particular types of risks, or the circumstances that are likely to give rise to them. Nevertheless, such reports can lead a project stakeholder organisation to review its policies, practices, and processes with a view to avoiding particular threat risks and may act as a catalyst to identify particular types of risks on particular types of projects.

Some techniques rely on structured formats developed from previous experience or as part of deliberately framed policy implementation. Examples include checklists and risk type classifications. An ORR, if it exists and is arranged as an easily searchable database, is a primary resource for risk identification. The ORR is discussed more fully in Chapter 11. Company procedural manuals may also raise risk awareness. Knowledge of the planned stages for a project can act as a simple memory jogger for risk identification.

Some tools are models that attempt to assess the riskiness of projects and require user-input measurement values for project attributes or factors known by previous experience to be associated with significant project risk. An example based upon an information technology (IT) project will be shown in Chapter 7 (Table 7.12).

Checklists are probably the most commonly encountered tool in risk management and can be highly effective in capturing risks for projects which are familiar to the stakeholder and involve comparatively few unique features on each project. We will consider the limitations of checklists in Chapter 7. Their primary use is to act as a reminder for tasks or issues that should not be overlooked in undertaking project activities, or that can be checked off when completed. The origins of checklists are peculiar to each organisation. They usually develop organically from instances where something was either forgotten or not properly carried out on previous projects.

Risk classification guides (e.g. what weather risks do we face on this project?) such as those proposed in Chapter 2 are also useful in risk identification, but are not closely aligned to the project decision making process. As will be shown in this chapter, they are probably best used in combination with activity-related techniques.

6.4.4 Functional Value Technique

The functional value technique is quite restricted for application in project risk identification since it is primarily part of the investigative process of value engineering (VE; also known as value management [VM]). This process interrogates the required function and economy of project design at various system or component levels by asking the following questions:

- What is this system/component?
- What is it required to do (its function)?
- What does it cost?
- What is its functional value (worth to the project client)?
- What else would fulfil the function?
- How much does that cost?
- What is the functional value of that alternative?

During the process of speculating about alternative solutions and assessing their functional performance, cost, and value, the risks associated with each alternative proposal can be canvassed and explored, particularly those that may impose constraints on the actual implementation of an alternative solution.

Obviously, if VE/VM is not undertaken on a project, then this technique will not be used for risk identification. Furthermore, since the risks identified are those solely associated with specific alternative design solutions, even where VE/VM has been undertaken its use in project risk management is quite limited. On three occasions, VE/VM intervention was used in Case Study C (an aid-funded project). While each one dealt implicitly with specific project uncertainties, no explicit consideration was given to risk management.

6.4.5 Matrices

Techniques in the matrices type of risk identification mainly use combinations of techniques in a tabular matrix format. This can be a powerful approach, especially when risk classification is part of the matrix.

In Chapter 7, we will explore combination techniques that progressively employ additional project features, but care must be taken with this approach. It should not become so extensive that workshop participants are faced with too much detail and lose confidence in ever being able to complete the risk identification process.

6.4.6 Simulation and Visualisation Techniques

Although now prolific in project design processes, the use of simulation and visualisation techniques is still quite rare in project risk identification. Most of these techniques harness graphics application software properties and the power of computers to create virtual reality or augmented virtual reality representations of physical projects. Two-dimensional and three-dimensional visual perspectives may be enhanced by dynamic 'project tour' capabilities which can act as powerful mental triggers for risk identification. Such techniques can be particularly useful for hazard recognition in exploring project safety risks.

Storyboarding techniques are based upon processes used in the film and TV production industry. Individual scene layouts and actions are either planned and represented visually in a progressive series of pictures, often hand-drawn, or described in a similar way in text. This simple 'walk-through' approach can be adapted for different types of projects, and is frequently used in other contexts to explore the 'customer experience' before a new commercial procedure or policy is deployed. While rarely found in project risk identification, the power of storyboarding lies in its cognitive stimulation capacity to represent or describe actual situations and the activities surrounding them.

6.4.7 Speculative Techniques

Risk identification techniques in the speculation category include scenario testing and stress testing. Their speculation characteristics set these tools apart from simulation and visualisation techniques.

Scenario testing is a deductive technique for identifying project risks that really stands alone. Although similar to storyboarding in some ways, it differs in that greater imagination is demanded as the speculative cognitive stimulation involved does not arise from pre-planned project activities. Scenario testing also requires users to focus first on hypothetical external contexts for a project before relating those situations to the internal context.

Stress testing is a management technique, derived from engineering principles, that explores the strength and resilience of organisational systems. It may be used inductively or deductively. When using stress testing in project risk identification, the focus of the workshop participants is directed towards the key project decisions made by the project stakeholder organisation. The application of this technique is therefore heavily dependent on the timing of those decisions.

6.4.8 Structural and Management Tools

Structural or management tools that can be used to identify project risks include: organisational structures (for both the project and the stakeholder), governance and operational procedures (for the procurement and operational phases of the project if the latter is relevant to the stakeholder), and specific project management frameworks and applications. Using any of these to identify project risks essentially entails exploring the relevant material, be it structure, governance mechanism, procedure, framework, or application, by auditing and interrogating the inherent decision points – usually concentrating on the major decisions affecting the project – and asking what could affect a satisfactory outcome from that decision.

Organisational structures, for projects and project stakeholders, were considered in Chapter 3. In risk identification mode, workshop participants may usefully question not only how the organisations are structured in that way but also why.

A similar approach can be adopted for governance mechanisms and operational management procedures. Here the questions concern the hierarchy of governance in the relevant organisation, how decision making power is devolved throughout it, and how project-related decisions are authenticated and approved, and then again asking what uncertainties could affect a satisfactory outcome.

Project management guidelines and frameworks (such as PMI 2013), or their computerised applications, can be used as proxy checklists for identifying project risks. They represent typical processes for managing projects, usually in a linear fashion, and this means that they offer guidance about the progressive sequence of activities required to manage the procurement of a project. Since the activities relate closely to the decisions that initiate them, the now familiar interrogation approach can be employed to identify the associated risks. One disadvantage with these techniques is that they rarely, if ever, embrace the operational or disposal phases of projects. Thus, their use for risk identification is limited to the procurement phase. Another potential weakness is that project decision making is not made explicit, but must be deduced or traced from each process and subprocess description. On the other hand, most frameworks offer guidance about project risk management so that, unlike other techniques, you may get to identify the risks associated with risk management itself!

6.5 Summary

In this chapter, following the essential first stage of context establishment, we have commenced the 'narrowing' process of identifying the risks that a project stakeholder may face.

The ideal vehicle for this is at least one, but preferably several, risk management workshops timed to coincide with the stages of a project where decision making abounds, since auditing these decisions should expose the risks associated with them. We have referred to this as a process of targeted discovery.

Workshop timing will thus depend largely on the project processes adopted by individual stakeholder organisations.

Techniques are available to stimulate and assist with risk identification. For the most part, they consist of management techniques adapted from their original and primary purposes. The benefit of this is that workshop participants are likely to be familiar with them and comfortable with applying them.

The various techniques can be categorised in terms of their underlying approach, and suggested categories involve methods that may be activity-related, analytical, associative representations, categorical or list-driven, functional value-related, matrix combinations, simulation or visualisation, speculative, or related to structural or management purposes.

Brainstorming is the ingredient common to all risk identification techniques, and this must be carefully managed, not only in terms of encouraging all workshop participants to contribute but also in terms of the fatigue that cognition and intellectual effort can generate.

Speculative techniques demand the greatest stretches of imagination and thus present the greatest danger of steering project risk identification away from its intended course. Workshop leaders/facilitators need to be aware of this and keep the proceedings on track.

If there is an overarching principle for the process of risk identification, it is this: follow the timing and flow of project decision making.

In Chapter 7, we continue with project risk identification by examining some selected techniques and considering them in greater detail with illustrative examples.

Reference

PMI (2013). *A Guide to the Project Management Book of Knowledge*, 5e. Project Management Institute Newtown Square, PA: PMI. ISBN: 9781935589679.

7

Project Risk Identification Tools

7.1   Introduction

In Chapter 6, we looked at the approach to the risk identification stage in the project risk management cycle (Chapter 4, Figure 4.2). Brainstorming, used in small workshop environments at appropriate times in the project life cycle, is the basic technique advocated but, with few exceptions, should be structured rather than unstructured. Structuring this activity is best achieved by interrogating various attributes of the project and asking what could happen to affect the achievement of the objectives relating to it. Several techniques, which mainly derive from other aspects of project management, are available. These tools can be adapted to explore project attributes from a risk perspective. A typology of the techniques was presented, with outline explanations, in Chapter 6. In this chapter, we examine some of these tools in greater detail and use examples to show how they can be used to guide the brainstorming processes for workshop participants. No fully elaborated descriptions of their primary purposes will be given; rather, they are treated as multipurpose tools being used in a particular way. Our treatment follows the same sequence presented for the typology displayed in Table 6.1, although not all of those types are explored here.

In our experience, project risk identification can be quite arduous and is likely to be ineffective unless undertaken with serious intent. The tools described here are available to help with risk brainstorming. It is just a matter of choosing those that are most appropriate for the project (and project stages) that you are dealing with and adapting them where necessary.

For several reasons, precise risk statements are an important ingredient in effective project risk management. They are introduced in this chapter, but the topic will emerge again in other chapters, notably Chapter 11 (Project Risk Knowledge Management) and Chapter 19 (Communicating Risk).

Similarly, risk mapping plays a useful role in project risk management. It is discussed here but also arises again in Chapter 16 (Strategic Risk Management).

While this may appear to be a fragmented approach, it is deliberately intended to raise project risk awareness among readers in an incremental manner that is more easily understandable and which will reinforce risk learning.

Managing Project Risks, First Edition. Peter J Edwards, Paulo Vaz Serra and Michael Edwards.
© 2020 John Wiley & Sons Ltd. Published 2020 by John Wiley & Sons Ltd.

7.2 Activity-Related Tools

In this section, we present examples of project risk identification through the adaptation of three different activity-related management tools: work breakdown structures (WBSs), bar charts, and critical path networks (CPNs).

The power of activity-related project management tools lies in their close proximity to explaining what has to be done on a project (i.e. the tasks and the decision making that relates to them).

7.2.1 Work Breakdown Structures

Typically used in the construction industry, a WBS comprises a list of required activities conventionally displayed as a sequenced list.

Table 7.1 shows a WBS presented at a sub-activity level for a main activity (a repetitive casting cycle for reinforced concrete floor slabs in a multistorey building). Many main activities are required to construct the whole building, and each will involve several sub-activities, so a complete WBS for the whole construction project is a lengthy and comprehensive document. For example, construction of the reinforced concrete structural frame of the building will itself comprise main activities for building the load-bearing foundations at the lowest level, the columns needed to transfer the slab loads at each floor level, the structural walls surrounding lift shafts, the floor slabs and their associated beams, access and escape staircases, balustrade walls, and other integral concrete features.

Table 7.1 Typical activities for reinforced concrete floor slab casting cycle for a multistorey building.

Project activity
Reinforced concrete floor slab casting cycle
Assemble formwork materials
Setting out
Erect formwork
Rough-in services
Position block-outs
Assemble steel bar/mesh reinforcement
Fix reinforcement
Mix/order concrete (arrange pump?)
Pour concrete
Take test cubes
Cure concrete
Strip formwork
Clean and reassemble forms for next cycle
Take test cores

In practice, a complete WBS is too unwieldy to use for many construction projects. For project management purposes, only the relevant sections pertaining to current activities will be progressively released to site managers for the purpose of progress monitoring.

The level of sub-activity representation in the WBS is determined by the amount of task distinction needed. For the example shown in Table 7.1, it could be assumed that the structural design is for a monolithic flat concrete slab with integral edge beams. If the design were for a composite slab (e.g. with some precast components such as hollow blocks laid between precast ribs bearing on in-situ reinforced concrete supporting beams) or for a vaulted slab structure requiring coffer-dam forming inserts, then the sub-activity descriptions might be substantially different.

For risk identification, the WBS stimulates the cognitive capacity of workshop participants to address the question: What could happen concerning this activity (or sub-activity) that could lead to uncertainty about achieving a satisfactory outcome for it?

Clearly, at the main activity level the question might be asked once, whereas iteration is required for each sub-activity level.

For the activities shown in the Table 7.1 WBS example, typical brainstormed answers to the question might include:

1) Weather issues
2) Material supply issues
3) Material specification issues
4) Labour supply issues
5) Labour quality issues
6) Labour productivity issues.

Depending on the amount of decomposition revealed in the project WBS, a workshop-based risk identification process will almost certainly become quite protracted when this tool is used, especially at sub-activity levels. While risks similar to those identified at the main activity level for the project will recur (e.g. weather risks), others specific to the sub-activity will be found. Thus, for 'Position block-outs' (the insertion of temporary forms to create openings in the finished slab), there might be a specific risk of inaccuracy in positioning them; while for the 'strip formwork' sub-activity (the process of removing the temporary forms used to impart the required finished shape to the concrete), there might be particular risks associated with damaging the finished concrete during the removal process, or with damaging the formwork material such that it cannot be reused on a subsequent casting cycle.

It is thus easy to appreciate how the number of risks identified can multiply rapidly as the WBS tool is used at greater levels of detail. Deciding upon what WBS activity levels to use for risk identification is largely a matter of experience in project risk management, but it may also be influenced by strategic risk management policies in the project stakeholder organisation or by the need to comply with regulatory requirements such as those associated with occupational health and safety hazard identification.

It should also be noted that, in our view, using two or three-word descriptors for identified project risks is not sufficient, and each risk should be stated more precisely. The identifiers used for the example are kept short here not only in the interests of brevity, but also because this is what actually happens in risk identification workshops for projects in real life. However, if left as abbreviated descriptors, identified risks can become vulnerable to errors in communication within the stakeholder organisation and beyond. The issue of risk statements is addressed more fully later in this chapter

(Section 7.9), but preparing precise statements does require more time. Risk workshop planning should recognise and accommodate this.

In terms of our understanding about the constituents of projects (Chapter 3), the WBS approach to risk identification as suggested by Table 7.1 is also quite limited in two other ways. First and foremost, it depends upon the pre-existence of the WBS itself, at least at the level of the main project activities. That often precludes using a WBS-based tool for risk management in the earliest conceptual stages of a project, in the sense that decision making has not yet reached an appropriate point where a realistic WBS can be prepared. Secondly, it addresses only the task elements of projects. In the list above, of (1) to (6), only (1) is task-related; (2) to (5) are concerned with resources, (3) might also affect technologies, and (6) might be influenced by project organisation. To incorporate other project elements into the project risk identification process (and this is necessary as they too involve decisions), a *resourced* WBS, capable of representing project technologies, resources, and organisation, is desirable. This is shown in Table 7.2 as an expanded version of the WBS for the reinforced concrete floor slab casting cycle example.

While there is a substantial amount of repetition evident in the cell contents for the technology, resources, and organisation columns in Table 7.2, they are capable of stimulating the identification of unique risks for each sub-activity. The extent of risk interrogation thus expands commensurately, and the number of risks identified grows exponentially. Take the 'Assemble reinforcement' and 'Fix reinforcement' activities, for example. While similar technologies and resources are needed, the two activities are probably located in different physical locations on site: the first undertaken continuously in a dedicated assembly area, and the second iteratively in the unique final destinations for the assembled reinforcement components. Crane risks associated with the assembly activity are likely to differ in some respects from those incurred in fixing reinforcement in place. Similarly, in the sub-activity 'Take test cores' in Table 7.2, identified risks might now include issues associated with the coring activity itself, the drilling equipment and drill size, the availability of suitable power, the competency of the drillers, the precise timing and location of each core extraction, the experience and knowledge of the concrete specialist, and the consistency and reliability of the laboratory testing processes. There would also be risks associated with remediating substantial portions of the concrete work should any cores fail to meet the lower limit of the specified compressive strength.

We should also point out that, for the example depicted in Table 7.1 and Table 7.2, it is assumed that all activities are preceded by the necessary purchasing, payment, and delivery requirements for materials and the requisitioning of labour, plant, and equipment.

A WBS can be a powerful tool for risk identification, and it is amenable to iterative use as project decision making increases and information builds up. When developed to the resource and organisation levels of project constituents, a WBS will at least show *who* will be needed for the project. It does not normally address questions involving the time and cost objectives of projects. For these, other project management tools are available for the purposes of risk identification. Bar charts, for example, address time factors.

Table 7.2 Resourced slab casting cycle activity schedule.

Activity	Technology	Resources	Organisation
CONCRETE FLOOR SLAB CASTING CYCLE			HQ management Site management
Assemble formwork materials	Transport Cranage	Props, timber, deck plates, fixings Labourers Plant operators	Site management
Setting out	Survey	Surveyor	Site management
Erect formwork	Cranage Power tools	Props, timber, deck plates, fixings Formworkers Labourers Plant operators	Site management
Rough-in services	Cranage Power tools	Pipes, ducts, conduits, special fittings Formworkers Artisans Labourers Plant operators	Services engineers Site management
Position block-outs	Cranage Power tools	Timber, polystyrene shapes, custom inserts Formworkers Artisans Labourers Plant operators	Services engineers Site management
Assemble steel reinforcement	Cranage Power tools	Steel bar, mesh, tie wires, spacers Steelworkers Labourers Plant operators	Site management
Fix reinforcement	Cranage Power tools	Steel bar, mesh, tie wires, spacers Steelworkers Labourers Plant operators	Site management
Pour concrete	Concrete pumps Spreaders Tampers/vibrators	Ready-mix concrete Concretors Labourers Plant operators	Site management Concrete specialist
Take test cubes	Test cube pots	Ready-mix concrete Cube pots Concretors	Site management Concrete specialist Laboratory

(Continued)

Table 7.2 (Continued)

Activity	Technology	Resources	Organisation
Cure concrete	Water hoses Heating/cooling	Surface additives Labourers	Site management Concrete specialist
Strip formwork	Cranage Power tools	Formworkers Labourers Plant operators	Site management
Clean and reassemble forms for next cycle	Cranage Power tools	Formworkers Labourers Plant operators	Site management
Take test cores	Core drilling		Site management Concrete specialist Laboratory

7.2.2 Bar Charts

Bar charts align project activities against a time axis. Several figures in previous chapters allude to this (see Figures 4.3–4.6 and Figures 6.1–6.3), but a bar chart uses a specific format and time measure for this. Depending upon the nature of the project, the measure may be in minutes, hours, days, weeks, or years.

Table 7.3 represents a bar chart for the information technology (IT) project described in Chapter 4 (Figures 4.5 and 4.6). The table incorporates the main activities identified

Table 7.3 Bar chart for product personalisation IT project.

Project activity	Week 1	Week 2	Week 3	Week 4	Week 5
Idea generation	▓				
Concept development and budget	▓				
IT executive go/no-go	▓				
Brand business case		▓			
Submission to group board		▓			
Planning and resource allocation		▓			
Designer briefing			▓		
Test proof of concept			▓		
Arrange third-party inputs			▓		
Design, develop, and test back-end system				▓	
Design, develop, and test front-end system				▓	
Write up test cases				▓	
Test, modify, and retest systems				▓	
System deployment					▓
Evaluate and fine-tune					▓

for the IT project, and a duration is assigned to each one. No sub-activities are listed, nor are there any identifiers for the technology, resources, and organisational requirements for the project. It is thus a simple project schedule.

Including all these additional factors in the schedule would be possible but not necessarily desirable, as it would result in a very complex and overly detailed format for a management tool that essentially is intended for monitoring and controlling project time in the procurement phase and partly into the operational phase. Table 7.3 already represents time in relatively small units (i.e. days). Introducing sub-activities into this chart might necessitate breaking time allocations down into hours or even minutes. For some types of projects, that might be desirable and possible. For most projects, it is neither.

Using the IT project bar chart example for project risk identification takes the process just one step beyond that achieved by simply using the list of activities alone. For each activity, the additional question is: What uncertainties could affect the satisfactory completion of this activity in the scheduled time? Note that for a threat risk, this would imply an event occurring that would lead to exceeding the planned duration, whereas faster completion might indicate the presence of an opportunity risk.

Since bar charts attempt a realistic portrayal of the time requirements for projects, they do provide a useful tool for project risk identification, but suffer the same limitations as WBSs in this regard. They also assume a sequentially progressive and distinct time relationship between activities, albeit possibly with some overlaps as Table 7.3 shows. Because of the need to maintain simplicity for project management purposes, bar charts do not fully represent *dependencies* between project activities; yet in such relationships, risks are likely to arise. To explore these in a project time dimension, CPN *analyses* are needed.

7.2.3 Critical Path Networks

CPN analysis explores the time relationships and dependencies between the discrete activities required to deliver a project. It is almost exclusively reserved for the procurement phase (since this is usually where the greatest time criticality is found). For the operational phase, CPN either is not necessary (if time is not a critical operating factor) or is replaced by process flow control management. For a project disposal phase, CPN may be used if the time factor and disposal activity interrelationships are important enough to warrant it, but this is rarely the case.

For this technique, a simple domestic project is used as an example. You have decided to renovate a room in your house. We appreciate that for much of your house maintenance (that is, if you have that responsibility), you do not normally create detailed task lists – even if you admit to having a list-driven personality. The example also serves to demonstrate how we have become a project-driven society!

Table 7.4 lists 19 activities associated with this project. Some tasks that could have been listed separately have been combined here (where they do not involve interdependencies) in order to reduce the size of the array for Figure 7.1.

Note that this example has earlier activity antecedents. Prior agreement about the need for the renovation project must be obtained if other parties are involved and a suitable calendar opportunity negotiated. The 'go/no-go' decision alone may present unique risks in terms of attitude towards the proposal, and the amount of active support promised for it, shown by others. Existing equipment that may have been unused for a long time must be found and inspected, and old cans of paint opened and checked for freshness. Early booking is probably needed for the services of cleaning specialists that will be required after physical completion of the renovation but before furniture is reinstalled. Decisions about paint colours have to be made before any new paint is purchased. The list goes on!

Table 7.4 Task list for room renovation project.

Ref	Task/Activity
1	Renovation proposal (scope & dates)
2	Go decision & plan
3	Plan activities; check equipment & paint stock
4	Purchase new equipment & paint
5	De-clutter room & furniture etc.
6	Lay drop sheets; clean all surfaces
7	Repair, fill & sand; masking tape windows, frames erc.
8	First coat ceilings & wardrobe (3 hr drying time)
9	First coat walls (4 hr drying time)
10	First coat woodwork (6 hrs drying time)
11	Second coat ceilings (2 hr drying time)
12	First colour coat walls (4 hr drying time)
13	Finish coat walls (4 hr drying time)
14	Finish coat woodwork (6 hr drying time)
15	Clean & refix ceiling lampshades; clean & refix hardware
16	Remove masking tape; clean window glass; refix blind; clean & store equipment; seal & store cans; dispose waste
17	Steam clean carpet (5 hrs minimum drying time)
18	Re-install furniture & hang pictures
19	Final inspection & approval

Given this activity list, and the availability of a suitable computer application (such as Microsoft Project™), a CPN can be prepared. In order to do this, start times and durations for each activity must be determined, together with the precedence logic. The CPN diagram for the renovation project is shown as a screenshot in Figure 7.1.

For each activity, the basic risk identification approach for CPN is the same iterative questioning as that posed in the bar chart example: What uncertainties could affect the satisfactory completion of this activity in the scheduled time?

Now, however, the question must be extended: What uncertainties could affect the satisfactory completion of this activity in the scheduled time, *and* what activity relationship dependencies exist that could give rise to additional uncertainty?

An obvious dependency in the example is the time required for each coat of paint to dry before further coats can be applied to the same surface. Another is the need for wall paint to be dry before wood trims can be painted. Although this is a completely indoor project, cold and damp weather can affect drying times, not only for paint but also for newly cleaned carpets; thus, a weather risk arises. Other risk events might require remedial work to be carried out, thus also causing activity durations to be extended.

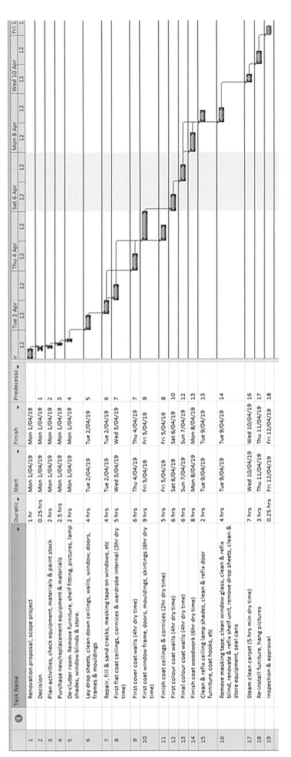

Figure 7.1 Screenshot (MS Project) of critical path network (CPN) example.

Activity-related project management techniques are powerful tools in risk identification as they reflect essential project decision making and its outcomes. However, they may not be available at the point where project risk identification essentially needs to commence. While they can provide effective stimulation to the cognitive processes of risk brainstorming, the greater the disaggregation of project activities, the more protracted the risk identification process will become. Project risk managers need to be aware of this when planning and conducting workshops.

7.3 Analytical Tools

The use of analytical techniques for project risk identification stems largely from the fields of decision science, manufacturing, and occupational health and safety.

From decision science, we find decision tree analysis (DTA), event tree analysis (ETA), and fault tree analysis (FTA). Manufacturing and processing industries have yielded tools such as failure mode and effects criticality analysis (FMECA; from the automotive manufacturing industry) and HAZOPS (hazard and operability studies; from the petrochemical manufacturing industry). Safety hazard analysis (SHA) derives from the burgeoning field of occupational health and safety and the efforts of dedicated government or quasi-government organisations such as the Health and Safety Executive (HSE) in the United Kingdom.

Space does not permit us to explore the whole range of analytical tools available to apply in project risk identification, and in the remainder of this section we examine only those mentioned here. An important consideration is that the primary purpose of most of these tools is in risk analysis, that is, the magnitude assessment of risk factors such as likelihood of occurrence, consequences, and relative severity (for threat risks). In their assessment role, the techniques can be applied quantitatively, and their representations can be susceptible to the calculation of mathematically predictable outcomes. We will return to this aspect briefly in Chapter 8 (Project Risk Analysis and Evaluation).

7.3.1 Decision Tree Analysis

DTA identifies logical consecutive decision nodes and explores their possible outcomes. It associates objectives with the decision making required to fulfil them. These two factors thus bring DTA into the orbit of project risk management. The risk identification questions are:

- What is the objective?
- What are the decision alternatives associated with the objective?
- What events can happen to affect each decision alternative?
- What are the possible consequences of those events?

Clearly, the greater the number of objectives to be considered, the greater the number of decisions to be examined. Similarly, the more decision alternatives there are available for each decision, the greater the likely range and number of events that could affect them, and hence the greater the range of possible outcomes. DTA is definitely not a risk management tool for the faint-hearted, and is rarely within the capacity of non-specialists to use in practice. In decision science, DTA is often harnessed with utility theory to inform decision making by calculating the worth of each outcome to the decision maker, with a view to optimising decisions. This introduces further complications, and the quantitative assessment it entails logically belongs in the two following stages of the project risk management cycle: analysing risks and evaluating risks (see Figure 4.2: Stage C_1 – Analyse risks; and Stage C_2 – Evaluate risks). In this section, we provide a simplified example of DTA, excluding quantitative assessment, as a vehicle for discussing its practicality for project risk identification.

For our hypothetical example, Director Mary Jane, who works in Melbourne, must attend an important project meeting in Sydney in two days' time. She must decide between alternative methods of travel. The choices available to her are: (i) fly with a commercial airline service, (ii) drive her car, or (iii) take the Very Fast Train (VFT) service. For reasons of diagrammatic simplicity, we have excluded other options such as chartering an executive jet plane, hiring a chauffeur-driven car, or using video-conferencing to avoid travel. The initial decision node alternatives are thus: fly, drive, or catch the train. Each has a two-option second decision node: take an early or a later flight; make an early or a later driving start; catch an early or later train. For each of these, events may happen to bring about three possible outcomes: early arrival in Sydney, arrival on time, or delayed arrival. These potential outcomes will have consequences for Mary Jane's attendance at the meeting. Figure 7.2 sets up the situation diagrammatically.

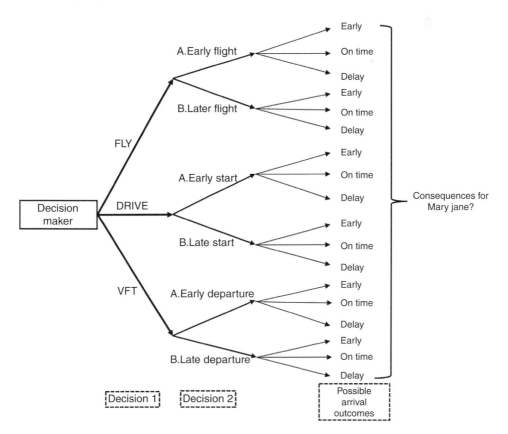

Figure 7.2 Decision tree analysis (DTA) example.

The example is deliberately removed from a *project* context (although actually not that far) in order to keep matters as simple as possible and avoid the undue complication of trying to replicate project specifics.

Mary Jane's objective is to get to a meeting on time. For various reasons (some perhaps relating to personal preferences), she has to decide between three travel alternatives. Each has two departure time options from Melbourne. Once the travel alternative and departure options have been decided, events may conspire to affect her arrival time in Sydney. For example, early-morning congestion at Sydney's Mascot International Airport (as arriving flights try to land as soon as possible after the overnight curfew) often affects departure times for domestic flights heading to Sydney from other Australian cities. Early-start departure by car from Melbourne may avoid city morning traffic jams but then encounter delays on the highway due to lane closures for overnight roadworks. An early train (the VFT is still just a hope for the future, by the way) may entail bus replacements for sections of the line where essential track works are in progress. Other events may similarly affect late-departure travel options. The outcomes of all such events not only will affect Mary Jane's success in getting to the meeting on time, but also may influence her attitude towards, and contribution to, the meeting itself: arrival delays may irritate her, longer journey durations may affect her diurnal rhythms, and leaving the previous day and staying in a Sydney hotel overnight may not be to her taste. Ignoring these consequences, what does this example tell us about the application of DTA in project risk identification?

The first point to make is that DTA can be used as soon as objectives are known (and understood) and decision making commences. It thus has earlier and broader application than activity-related techniques, where more detailed information (flowing from project decisions already made) must be available in order to stimulate risk identification effort. Secondly, decision making is seldom a single-issue dichotomy. Multiple options (the solution space), and the need to make related sub-decisions and consider their implications, complicate the process exponentially and thus the efficient capacity to identify risks. Thirdly, few decisions are free from personal preferences and bias, and, as we noted in Chapter 2, this also applies to recognition and perceptions of risk. In essence, the risk identification questions to be asked in the DTA example are: What uncertainties underlie the initial travel decision options? This is followed by: What uncertain events could happen (for each travel and departure option) to delay arrival in Sydney?

Our view is that DTA can be helpful in identifying risks in the early stages of a project when objectives have been established and the first key decisions are being made. It is less easily applied later on, when it is likely to be overwhelmed by the sheer number of decisions already made and the amount of information generated by them.

We will return to this example in Chapter 8 (Project Risk Analysis and Evaluation) to explore the issues of assigning values of worth to the outcomes.

7.3.2 Event Tree Analysis

ETA inductively traces a sequence of possible consequences flowing from an event. It is essentially the downstream adjunct to the FTA technique described in Section 7.3.3. Its application to project risk identification is thus limited to the possible flow-on events and outcomes rather than their original causes (the source events). Figure 7.3 shows an example that could derive from the operational phase of a maritime ferry service project. Here, the nodes are dichotomous yes/no possibilities for flow-on events, each with its associated uncertainty. The potential outcomes are presented in escalating levels of seriousness ending with disastrous possibilities of passenger/crew fatalities and the loss of the ferry vessel.

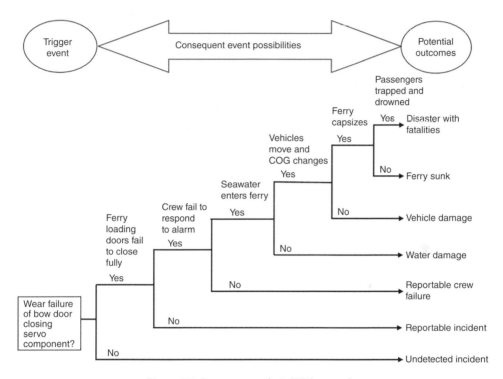

Figure 7.3 Event tree analysis (ETA) example.

As with DTA, this technique is used more as an assessment tool in decision science, but then the purpose would be to determine the probability of occurrence for each possible outcome by calculation from the combined probabilities of occurrence for each preceding event node.

For risk identification, ETA requires the primary 'trigger' event to be established first, and then the question becomes: What flow-on events can occur as a result of this one? Since other means must be used to identify the primary risk event, the use of ETA is somewhat limited in risk identification, but, by following chains of events, it is helpful in preparing full risk statements and, like DTA, in exploring quantitative aspects of risk assessment.

7.3.3 Fault Tree Analysis

FTA employs a top-down deductive process by starting with a risk impact. This is the primary, but not necessarily the final, consequence in a downstream continuation of outcomes such as that depicted for the ferry door failure in Figure 7.3. FTA then traces a logical sequence of possible causal or contributory factors to the primary consequence.

Like DTA and ETA, FTA tends to be used more in risk analysis than in the risk identification process. In practice, it is often used in engineering design to inform the necessary extent and nature of manufacturing quality control, component tolerances, built-in redundancy mechanisms, and operational maintenance procedures.

Figure 7.4 shows the tree logic diagram to illustrate possible prior causal events for the failure of an auto-deployment emergency evacuation chute for a commercial jetliner aircraft. By assigning a probability of occurrence to each fault node, and combining them in each of the logic chains, the probability of a particular node being the cause of the chute failure can be established.

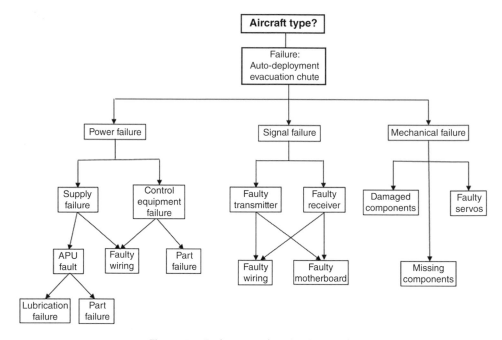

Figure 7.4 Fault tree analysis (FTA) example.

From the diagram, it can be deduced that only three primary contributors to chute failure are possible. Each of these may have two, or three, sub-causes beneath it, and so on. In fact, the tree diagram shown here does not necessarily arrive at terminal causal factors since each of those shown at the bottom of Figure 7.4 could have its own origins (e.g. through temperature extremes, rodent attack, or accidental physical impact).

For risk identification purposes, FTA is effectively best used as a stimulator to trace the antecedents of a risk event that has already been identified in some other way. Given this limitation, it can be applied at any appropriate stage in the project cycle. FTA can also help to distinguish between what should be regarded as a primary risk event occurring within the project system boundary for the stakeholder, and what is a prior contributory event beyond that boundary and thus a risk more difficult to control (see Figure 5.1 in Chapter 5). In risk analysis, FTA will show where probability combination chains must be considered for risk events.

7.3.4 Failure Modes and Effects Criticality Analysis

The FMECA technique is a derivative of ETA and FTA, customised for application in the automotive manufacturing industry to predict and control the ramifications of vehicle component failure. Major automotive manufacturers rely on agreements with external suppliers for many vehicle components, and such suppliers often service several manufacturers. Different supply arrangements may exist for each vehicle type and model according to the country where it is manufactured or assembled. FMECA has been adopted as a means of establishing a universal approach to dealing with the downstream impacts of component failure. Such failure may be detected first by local vehicle sales dealers, following customer complaints or reported incidents, and is then notified to the manufacturer. The nature of the failure determines whether or not the FMECA procedure is fully implemented. If it is, the web-based system underlying FMECA is operationalised, and relevant component suppliers are usually given less than 24 hours in which they must respond with missing information or proposals for remedial action. The completed FMECA then informs a higher management decision about the extent of any vehicle recall required. Table 7.5 shows a small and abridged part of a typical FMECA form entry, which would additionally contain references to the vehicle model and series, the relevant factory identification numbers, and the identities of the vehicle manufacturer and component supplier. 'Sign-offs' would show areas of responsibility for treatment actions. Estimated costs, recall decisions, and notification actions are also likely to be incorporated into the FMECA form.

Table 7.5 Failure modes and effects criticality analysis (FMECA) example.

Item	Ref. no.	Description	Tolerance	Quality control	Fault mode	Failure probability	Failure effect	Treatment plan	Sign-offs
1	GT204678T	LF subframe assembly	Specif. SFLF550	1 : 500	Tube spot weld failure = SF collapse	0.0002	Serious/ Fatal	Full fillet weld. XR (1 : 300)	Des. Eng. Prod. QM. Cost.
2	Etc.								
3	Etc.								

FMECA is not primarily intended for project risk identification purposes. It really bears more resemblance to a complete risk management system (RMS), minus the identification stage, in that it incorporates analysis, response, monitoring, and control actions and also provides a knowledge capture tool. Any risk identification that does take place is actually reactive and post-event. Its usefulness in proactive project risk identification is thus limited to creating a means of highlighting situations where the decisions or actions of other project stakeholders can affect or contribute to the risks faced by a particular stakeholder organisation.

7.3.5 Hazard and Operability Studies

The HAZOPS technique is predominantly used in the petrochemical processing industries in the design phase of projects, to explore potential hazards arising in the operational phase, with a view to preventing their occurrence or containing their effects through appropriate redesign. This tool is highly interrogative, using a series of situationally conditional questions to explore each operational aspect of the facility design. Framing of the questions must first be done by industry experts, and the implications of the answers must also be subject to expert consideration. A truncated example is shown in Table 7.6, which deliberately avoids any attempt to include answers to the questions. Also omitted for reasons of space, the format of the tool should include provision for quantifying possible outcomes, proposing design changes, monitoring and controlling actions, estimating costs, and assigning responsibilities, thus again creating a mini-RMS.

Table 7.6 Hazard and operability studies (HAZOPS) example.

Item	Reference	Specification	Question	Guide	Situation	Result
1	Dwg. A/21/123	Section: A21	What if …	no …	flow occurs through Flask AF1?	aaa…
2	Dwg. B/78/456	Section: B05	What if …	more …	pressure than specified occurs at Valve BV6?	bbb…
3	Dwg. C/51/789	Section: C48	What if …	less …	temperature than specified is generated at Purifier CP2?	ccc…
4	Dwg. D/032/099	Section: D03	What if …	reverse …	flow occurs at Filter DFL12?	ddd…
5	Dwg. E/83/004	Section: ???	What if …	as well as …	XX and YY chemical wastes both flow in Channel EC9?	eee…
6 etc.	???...	???...	What if …	???...	???...	fff…
7 etc.	???...	???...	What if …	???...	???...	ggg…

Note that the tool format deals with the identification of a risk event over three columns (Question, Guide, and Situation) but does not require the assignment of a probability of occurrence to that event. If the answer to the question is sufficiently adverse, in the view of the expert interrogators, the system design must be amended regardless of the likelihood of occurrence for the given situation.

HAZOPS specifically addresses the operational phase of a project, but is conducted sufficiently early in the design stage of the procurement phase so that any necessary system design changes can be made with minimum disruption and cost. As will be seen in Section 7.7, some visualisation techniques can also be used to identify project operational phase risks. Their use is intrinsically part of a process of exploring whether or not the project design is capable of meeting the 'fitness for purpose' objectives established for a project. HAZOPS does the same thing but with a specific emphasis on operational hazards.

It is probably feasible to adapt this technique to expose and explore risks associated with the project procurement or disposal processes, but with the exception of 3D modelling (also discussed in Section 7.7), we know of no techniques that attempt to do this in a formal or semiformal manner. Any adaptation would entail expert framing of the design interrogation questions, probably for each type of project – a requirement that might not yield a positive benefit–cost ratio. Perhaps the best use of a customised HAZOPS approach, to explore risks associated with the procurement phase of a project, would be to consider overlapping activities and ask the question: What events might happen due to this activity overlap that could give rise to uncertainty in their outcomes?

7.3.6 Safety Hazard Analysis (SHA)

Over recent years, increasingly greater attention has been given to issues of health and safety in the workplace. For safety particularly, such attention has spread to other environments including the home, school, university, and places used for business, leisure, sport, entertainment, and other social and religious activities; as well as to external environments and modes of transport. Many countries have established authorities to monitor the observance of safety regulations, investigate the need for further regulation, and encourage safer behaviours among citizens. ISO 45001 (2018) is an international standard for promoting occupational health and safety, requiring a risk-based approach with decision maker/worker consultation and third-party verification processes.

As part of this drive for greater safety, organisations involved in almost every walk of life are required to conduct analyses that identify threat risks, to employees and to third parties, that are associated with the way they conduct their organisational activities or with the environments in which they are conducted. While focusing mainly upon physical safety, SHA does not exclude other health issues such as emotional distress and mental illness, and these are now included in the ISO 45001 (2018) standard.

SHA is therefore inextricably bound up with all aspects of projects, including project participants and all project phases (procurement, operational, and disposal). It is actually a risk management system in its own right, since it requires consideration and action according to every stage in the risk management cycle. In the United Kingdom, the government-sponsored HSE plays a vital regulatory and monitoring role, and for construction projects the UK's Construction and Design Management Regulations (HSE 2015) have assigned safety responsibilities to project designers as well as to contractors.

Such is the importance of safety that responsibility for managing it often devolves to a specific department within an organisation. How, then, should it be treated in the process of project risk management? Firstly, we will look at an example of SHA in practice, and then consider how it might fit into the wider processes of project risk identification.

Table 7.7 presents a typical template work SHA record sheet for a project. Many variations on this template will be found in practice, and the format is easily adapted to suit various industry, organisational, or safety authority requirements.

One observation from this example is that while preparation of the safety hazard record might be the work of one person, it is subject to review by another – presumably someone with appropriate knowledge and authority. In practice, initial preparation of the record is more often than not the product of a team effort.

Table 7.7 Safety hazard analysis (SHA) example.

Work safety hazard analysis record				Ref. no.
Project				**Variations (including dates and authority)**
Location:				
Client:				
Contractor:				
Subcontractor:				
Planned activity:				
Activity location(s):				
Confined spaces?				
Equipment and services requirements:				
Planned dates:				
Planned times:				
Supervisor(s):				
Work crew(s):				
	Hazard description	Severity (H/M/L)	Proposed treatment and control	
Hazard 1:				
Hazard 2:				
Hazard 3:				
Hazard 4:				
Hazard 5:				
WSHA Prepared by:		Date:	Signature:	
Reviewed by:		Date:	Signature:	

The example shows that attention goes directly to relevant work activities without first considering project objectives, the assumptions being that the planned activity is a necessary part of achieving them and that major methodological decisions have already been made. SHA therefore does not usually consider alternative ways of doing things beforehand, although the inclusion of a 'Variations' column acknowledges the possibility that the actual task processes might be changed later on.

Because of its detailed attention to work activities, SHA is unlikely to assist with more general risk identification in the early conceptual development stage of a project. However, knowledge gained (and preferably retrieved through suitably archived SHA records) from previous projects that involved similar activities will at least provide early warning of the nature of the safety risks likely to be encountered on the current project. For ensuing project design, planning, and scheduling processes, when project activities are more fully developed and known, SHA comes into its own and is directly applicable as a risk identification tool, at least during detailed project planning.

The SHA tool has been dealt with here as an analytical type of risk identification technique, rather than activity-related, simply on the grounds that it incorporates more analytical detail for project activities and the environments in which they are carried out. It also deals with hazard assessment and response treatment. In the example provided in Table 7.7, risk severity is accorded a simple high/medium/low scoring option range. This would have to be underpinned by more extensive label definitions established by the organisation (or even by the industry or an independent safety authority) that are recognised and understood across the relevant parts of the organisation. This is given more consideration in Chapter 8 (Project Risk Analysis and Evaluation). In some formats, evidence of direct communication with workers has to be included in the SHA. In some jurisdictions, prior consultation with workers about any hazard is also mandatory.

An important consideration arises with SHA. For 'hard' physical endeavours, such as construction projects, site safety has become an increasingly large component of project risk management. The introduction of ISO 45001 (2018) suggests that even greater emphasis will placed upon its importance in future. Because of this, a strong argument can be made for the complete separation of safety risk management from more general risk management in an organisation. While this would inevitably lead to substantial duplication of systems and more extensive administration, it would avoid the danger of attention to other risks becoming swamped by the needs of safety risk management. System separation would also support the development of safety specialists as a distinct professional discipline. This issue is considered more fully in Chapter 16 and again in Chapter 17, but it may be noted that in Case Study A (a PPP correctional facility), safety risks were not dealt with under a separate system by the main construction contractor in the private consortium of the PPP but were included in the project risk management system.

Each of the analytical tools shown here has some relevance, directly or indirectly, for project risk identification. Given the increasing influence of guidance frameworks such as ISO 45001 and regulatory environments such as the CDM Regulations (HSE 2015), their adaptation and application to all types of projects are inevitable.

Another important point to make here about analytical techniques is that the pre-existence and sufficiency of the records pertaining to them may be given evidentiary status in official enquiries such as those following major accidents and disasters. FMECA, HAZOPS, and SHA are particularly treated in this way and should thus be accorded careful consideration in terms of the ways in which they are formatted and used, and the manner in which records are maintained in the short, medium, and long-term life cycle of a project. Failure to do this, properly and consistently, could be construed as evidence of a lack of a duty of care or even as a contributory factor to threat risk events and their consequences. Commissioners in charge of formal enquiries have been known to be disdainful of evidence submissions exhibiting superficial 'tick the box' form filling.

7.4 Associated Representative Tools

Associated representative tools for risk identification derive from consideration of factors that directly or indirectly drive project decision making by representing project characteristics or attributes from a different perspective. They may be inherently related to the project internal or external contexts, to financial or economic considerations, to project processes, or to project design, or they may be simply based upon types of risks. These tools include: contextualisation, checklists, financially related and procedurally related techniques, design/cost tools, and risk-related aids. Other resources in this group, but not dealt with here, include project drawings and specifications, and contract and subcontract agreements and clauses. All of these can be interrogated from a project risk perspective as they become available.

7.4.1 Contextualisation

Project context establishment was identified in Chapter 4 (Figure 4.2), and discussed in Chapter 5, as an essential first stage in the project risk management cycle. There, its purpose was described as raising risk awareness through understanding how the internal and external settings for a project could influence the severity of the risks identified for the project (i.e. when project risks have been identified, then examine how the contexts might shape them).

Inevitably, risks will arise that are directly associated with the project context itself, whether internal or external, and reconsidering the context can be useful at some point during the project risk identification stage. The necessary question framing becomes: What events may happen in this particular context that could lead to uncertainties in project outcomes? This approach also recognises that the contexts themselves are rarely static and are likely to change over the project life cycle. Remember that they are often the 'drivers' and 'shapers' of risk.

7.4.2 Checklists

Lists are almost irreplaceable as a necessary adjunct to modern-day living. Many of us, either through pride, gratitude, or shame, admit to being 'list-driven' thus giving list making a psychological dimension. Furthermore, many of us still rely on paper-based lists, despite the plethora of modern and convenient electronic equivalents. It is as though the very act of putting pen to paper to create a list acts as a memory reinforcement in itself. The need for a 'list of lists' has been touted as the ultimate example of list dependency!

Most lists are shortcut methods of signifying a series of needs or planned future actions and events, generally for the short to medium-term future. At a personal level, they tend to deal with matters such as appointments, shopping requirements, gift and birthday card reminders, desired travel destination itineraries, and household repairs and maintenance. They rarely pertain to long-term issues, except for the checklists used to record information needs and process reminders for project-based endeavours. Professional consultants working on projects tend to rely on checklists as a simple tool for practice-focused risk management (as distinct from client-focused project risk management). Table 7.8 depicts a typical checklist that might be used by an architect commissioned to undertake design services on a new building project.

A list such as this might include several hundred entries. As a tool that has more often than not been developed through an accumulation of items over many years of experience, it may be randomly ordered and cover a wide range of issues with little or no attempt at classifying them. Each project consultancy builds up its own checklist, based upon its particular professional discipline. Other than in multidisciplinary or in-house design/engineering departments, checklists are rarely coordinated between consultants in different disciplines.

After five or 10 years, such checklists often cease to grow. They may have a large accretion of issues to be investigated, but thereafter are rarely subjected to rigorous updating or reordering. Checklist items are occasionally migrated into a more formal and more structured procedural manual in an organisation.

Inspection of the checklist in Table 7.8 shows that it also resembles a comprehensive contextualisation of issues that could directly or indirectly affect the project design. Most issues derive from a physical context; a few derive from a social context.

Table 7.8 Typical architect's preliminary checklist for a construction project.

Item no.	Checklist item	Known	Unknown
001	Project name		
002	Project client and contact details		
003	Client contact person		
004	Land title registry entry		
004	Land title description		
005	Survey points		
006	Boundaries		
007	Land area		
008	Ground levels		
009	Permitted land use(s)		
010	Existing development/building permit		
011	Permitted GFA		
012	Height restrictions		
013	Side spaces		
014	Setbacks		
015	Existing potable water services		
016	Existing sewer drainage		
017	Existing stormwater drainage		
018	Existing electrical supply		
019	Existing gas supply		
020	Covenants over land		
021	Servitudes over land		
022	Overhead cables, etc.		
023	Telecommunications services		
024	Adjoining owners' rights		
025	Flora/fauna preservation		
026	Heritage/conservation orders		
Etc.			

The derivation and nature of the issues are what make checklists useful for project risk identification, and the question to be asked is: What uncertainties are associated with this checklist item, and the resolution of it, that could affect design, delivery, or operational outcomes on this project?

While checklists are certainly effective in project risk identification, they can suffer from shortcomings such as currency, sufficiency, and project phase focus. If checklists are not regularly updated, they may gradually lose relevance for contemporary project practices, technologies, and contexts. Even where updating is carried out diligently, problems of sufficiency can arise since no checklist, especially if it involves an informal accumulation of items, will cover every conceivable project or contextual consideration. Furthermore, most project checklists are aimed at the procurement phase of projects, and particularly the design stage. Although they may be used for the pre-operational commissioning stage, such use rarely extends onward into the operational phase, or back into the early development stages of the project procurement phase.

7.4.3 Financially Related Tools

Financially related management tools that could be adapted for project risk identification include budgets, quotations, purchase orders, and economic and market analyses and cost forecasts. However, none of these are really comprehensive risk identification tools, even when they are available in a breakdown itemised format. They all represent summarised information to some degree, and they are each susceptible to uncertainty in some way. The guiding question then becomes: What aspects of this financial instrument could give rise to greater than anticipated uncertainties in project outcomes?

These tools can be used at any point in the project life cycle as they become available, but their limited scope is likely to reduce their effectiveness for project risk identification purposes.

7.4.4 Procedural Tools

Procedural tools for project risk identification, as part of the suite of associated representative techniques, explore project procedures rather than direct process-based activities. Suitable instruments include procedural manuals, project stage lists, and events schedules.

As noted in this chapter, procedural manuals are essentially a more formal version of checklists. They may also flow directly from company/organisational policies and include authentication and compliance (e.g. review and sign-off) requirements. The epitome of procedural manuals is probably seen in aircraft flight and maintenance manuals, where a clearly communicated and auditable trail of amendments must be maintained between aircraft manufacturer and operator, and rigorous compliance with the preflight manual is required of every pilot and aircraft engineer before take-off. Note that procedural manuals at this level are likely to include instructions and guidance for instances where the procedural outcome does not comply with the intended outcome. In commercial and military use, such guidance is usually extended to compulsory flight simulator sessions during which emergency scenarios can be presented – but that takes us into a different sphere of risk identification techniques!

Project stages and event project schedules frame information that usually associates distinct parts of a project with measures of time. The measures may be either roughly or precisely defined. For example, in a construction project the design-bid stage (see Chapter 4, Figure 4.3) may span several months or even a few years. On the other hand, for an IT project the development stage (see Chapter 4, Figure 4.6) may be limited to a few weeks (as shown in Table 7.3).

Event projects may present quite disjointed schedules, as the two following examples will show. One deals with the stages delineated for a special floral display to be mounted in a botanical garden setting. The other related event project is the opening ceremony for the new display.

Table 7.9 shows the project schedule prepared for the special display event to be held in the gardens. Note how it employs a 'D-day' countdown routine for time, and in two places bold font is used to emphasise the need for additional schedules to deal with two critical sub-projects.

Table 7.9 Schedule for botanic gardens display event.

Schedule	Duration	Display event activity	Notes
D-180	20 days	Preliminary theme design and approval	Confirm theme and objectives Sketches and 3D graphics
D-165	240 days	Cost estimates, cost accounting process	Include opening
D-160	30 days	Detailed display design and colour planning	Plans and augmented 3D graphics Sculpture and artwork commissions
D-150	210 days	Sponsor search and negotiation	1 major; multiple minor
D-145	15 days	Source and purchase plants, and arrange delivery schedule	Internal sources, external sources, quotes, purchase orders
D-130	84 days	Design, create, and outsource display furniture manufacture.	Detailed design, in-house construction, quotes, and purchase orders
D-125	5 days	**Design, plan, and commence publicity campaign**	**Separate schedule**
D-120	5 days	**Opening ceremony design and approval**	**Separate schedule**
D-40	10 days	Commence display furniture and fittings installation Planting in display areas	Delivery, transport, lifting
D-35	28 days	Select and transplant flowers etc. to display areas	
D-6	4 days	Install signage	Delivery, transport, lifting
D-2	2 days	Install opening ceremony signs and furniture	Delivery, transport, lifting
D-ZERO		**DISPLAY OPENING**	

(Continued)

Table 7.9 (Continued)

Schedule	Duration	Display event activity	Notes
D+1	1 day	Remove opening ceremony signs and furniture	Lifting, transport
D+1	60 days	Display maintenance	Replacements?
D+61	4 days	Dismantle display and flower disposal	Hospitals, etc.?
D+65	6 days	Restore garden areas and prepare for next display event	
D+72	½ day	Project debriefing	

The formatting for this table, and for Table 7.10, is a mixture of those used for Tables 7.2 and 7.3, although the time axis is not displayed graphically and the items are not augmented with resource requirements. The latter could be added quite easily with a landscape format. The whole garden display project spans more than 200 days, with the display itself open to the public for 60 days. Since most display events follow a similar pattern, and the botanical gardens organisation currently has a policy for holding at least two per year, this means that planning for the next display project will commence no later than about D-40 on the schedule for this one.

Note the confirmation of event theme and objectives at D-180. Also note that, as with most event projects, the schedule covers all three project phases: procurement, operation, and disposal. Typically, the Table 7.9 schedule includes two nested sub-projects: first the associated publicity campaign, and within that the opening ceremony.

Table 7.10 depicts the sub-(sub)-project schedule for the display opening ceremony. In this schedule, EPM refers to event project manager. Two have been appointed for the whole display event. EPM2 is likely to be subordinate to EPM1, and may well be a trainee project manager. In this organisation, project management is undertaken through an

Table 7.10 Botanic gardens display event: opening ceremony schedule.

Timing	Duration	Display event opening ceremony	Notes
D-120	5 days	Opening ceremony design and approval	Date, time, duration
D-120	18 days	Selection of opening dignitaries	Informal diary checks with selected invitees
D-100	1 day	Opening dignitary formal invitations	Check any protocols
D-95	30 days	VIP and guest list planning and approval	**About 1000 invitees**
D-90	90 days	Security planning and arrangements	Level according to VIP status
D-85	85 days	Parking and transport planning and arrangements	Official, VIP and guest cars, special transport

(Continued)

Table 7.10 (Continued)

Timing	Duration	Display event opening ceremony	Notes
D-80	70 days	Special needs arrangements and testing	Disability access, sign language facility, audio loops
D-70	65 days	Communications planning	
D-65	30 days	VIP and guest invitations and security passes	(25% acceptance)
D-60	60 days	Pre- and post-ceremony publicity planning	**Coordinate with event campaign team**
D-55	2 days	MC and staff team planning and pre-briefing	
D-50	50 days	Catering and refreshment planning and orders	Special dietary needs, drinks policy
D-45	21 days	Marshalling requirements and training	Temporary marshals
D-40	28 days	Staff training	
D-35	½ day	Media briefing, camera locations, and facilities	
D-32	5 days	Media and press passes	
D-30	28 days	Dignitary, VIP, and guest gifts	
D-6	2 days	Speech scripting and approvals	MC + drafts from dignitaries
D-4	4 days	Comms cables and equipment + testing	Safety?
D-1	½ day	Rehearsals	Stand-ins, reserves
D-ZERO 07.00–08.00		**Gardens closed to public**	
08.00–09.00	EPM1	**Marshal and brief security team**	
09.00–09.30	EPM2	**Final briefing: staff team**	
09.30–10.45	EPM2	**Inspect and check display and signage**	
10.45–11.00	EPM1	**Check catering and refreshments**	
11.00–11.45	Staff team	**Guest arrivals and shepherding**	EPM1 and 2 to troubleshoot
11.45–12.00	MC	**Dignitary and VIP arrivals and welcome**	
12.00–12.15	MC, Dig2	**Platform party and greeting**	
12.15–12.45	Dig1, MC	**Speech and opening**	
12.45–12.50		**Platform party leave for lunch marquee**	
12.50–13.00		**VIPs and guests transfer to lunch marquee**	
13.00–14.15	MC+	**Lunch + short speeches**	
14.15–14.30		**Dignitaries and guests depart**	

(Continued)

Table 7.10 (Continued)

Timing	Duration	Display event opening ceremony	Notes
14.30–15.00		Security and staff team stand-down	
15.00		Gardens reopen to public	
D + 1	09.00–10.30	Debriefing	

internal project management office (PMO) which deals not only with events but also with more permanent projects. Project managers for the latter are designated permanent project manager (PPM) as a distinguishing nomenclature, although this is done mainly for external recognition purposes. In practice, cross-overs do occur. A small internal team generally handles design processes for events projects and occasionally the initial conceptual design for permanent projects. Specialist design and engineering consultants are engaged where required for both types of project.

MC in Table 7.10 refers to a master of ceremonies, but this role is not gender-exclusive and is carried out in-house. Publicity is also dealt with in-house unless complex or sophisticated media are involved.

In all instances, procedural tools (and they are found on most projects) can be interrogated for project risk by iteratively posing the question: What event/situation could be associated with this procedure that could lead to uncertainties about a successful project outcome?

Later in this chapter, we will show how the risk identification power of such tools is enhanced by expanding them into matrix techniques.

7.4.5 Design/Cost Related Tools

Design/cost related tools could be subsumed into the financially related category of techniques, but we deal with them as a separate type here because of their very strong association (and causal relationship) with project design. Probably the most detailed and comprehensive example of a design/cost tool is the elemental cost planning (ECP) technique used in the construction industry in many parts of the world. The technique is based upon distinguishing a unique design element in a construction project and assigning a capital cost to it, either as a straightforward lump sum estimate or expressed as the product of a cost per unit of a predefined measure for that element. For example, the estimated cost of the external façade element for a proposed building might be calculated from the elevational area of the façade multiplied by a forecast of the cost of the installed façade material expressed as a $ rate per m^2. The estimated elemental cost for the façade is then translated into a cost per unit of a measure applicable to the whole building (e.g. per square metre of gross floor area [GFA]), and is also expressed as a percentage of the total project cost.

This technique is used in several ways in project cost planning. It may be applied to establish target costs (budgets) in the early stage of project procurement when little is known about the project design other than broad parameters such as permissible GFA and number of floors. Thus, an adjusted GFA rate per square metre from a known historic project may be applied to a proposed project to set a target cost for the new façade. Or it may be used to fine-tune early and cruder cost estimates as more detail is known about each element, and to explore the balanced distribution of element costs (using the percentage data) within the whole building in comparison to updated historic cost plans for similar types of buildings. Any element cost found to be substantially different from the historical 'norm' can then be investigated for design-related causes and either justified or the design amended to restore the 'typical' percentage distribution balance.

A design element is usually distinguished as a part of a building (or its setting) that always serves the same function, regardless of construction material or technology. As project design progresses, sub-elements (or components) are introduced for selected elements. Standardised lists of elements and sub-elements have been developed in various countries by professional associations serving the construction industry, beginning in the late 1960s in the United Kingdom. Their formats all differ in some degree. More recently, attempts have been made to create a uniform international list of building design elements.

Table 7.11 illustrates a typical basic elemental cost plan that is not aligned to any particular professional jurisdiction. For simplicity, we have omitted unit rate analyses, as they are not part of the risk identification purpose we are demonstrating. We qualify this omission later in the chapter.

Table 7.11 Typical list of design elements for a construction project.

No.	Design element	Cost estimate
DE01	Site preparation and demolition	
DE02	Bulk earthworks	
DE03	Substructure	
DE04	Ground floor	
DE05	Structural frame	
DE06	Suspended floors, stairs, and ramps	
DE07	Roof	
DE08	External façade	
DE09	Internal divisions	
DE10	External finishes	
DE11	Internal finishes	
DE12	Fittings and fixtures	
DE13	Plumbing installation	
DE14	Electrical installation	
DE15	Heating, ventilation, and air conditioning	
DE16	Lifts, escalators, and conveyors	
DE17	Fire safety systems	
DE18	Communication systems	

(Continued)

Table 7.11 (Continued)

No.	Design element	Cost estimate
DE20	Security systems	
DE21	Building automation systems (BAS)	
DE22	Other specialist systems and equipment	
DE23	External features	
DE24	Roads and paving	
DE25	External water reticulation	
DE26	External drainage reticulation	
DE27	Other external services	
DE28	Alteration work	
P01	Preliminaries and overhead allowances	
P02	Contingencies allowances	
P03	Profit margin	
	Total:	

In this example, Element DE08 (external façade) might have sub-elements for walling, windows, external doors, and curtain wall panels; while DE14 (electrical installation) could have sub-elements for single-phase and three-phase power distribution, illumination, solar panels, stand-by power generation, and so on.

While it is not within the purview of this book to discuss the pros and cons of ECP for construction projects, we would point out that the 'cost' assigned to any element or sub-element is actually a *price forecast* since the cost planner is forecasting what amount would be charged by a contractor bidding for and carrying out that work. Since it is a forecast, each 'price' will be subject to uncertainty and thus represent a risk, at least to the project client. While much of that uncertainty may be ascribed to market conditions for the project, some at least may attributable to the design decisions associated with the element (hence our earlier qualification for Table 7.11).

Using a design/cost tool such as an elemental cost plan, the iterative risk identification question becomes: What conditions/events can be associated with this design element (or sub-element) and its estimated cost that could lead to additional uncertainty in its price or suitability?

The ECP technique focuses on design-related decisions and issues and their financial impact on a project. It is useful in project risk management, as risk identification can commence as soon as a base cost plan is prepared, and continue as the ECP is developed in parallel with the project design process. Currently, this technique is used exclusively for the project procurement phase and mostly by professional consultants for a project. It may be possible to adapt it for other project phases; and it should be easily adapted and applied to other types of projects if significant and suitable design elements can be determined.

7.4.6 Risk Related Tools

Here, we present two techniques that can be used for project risk identification: risk categories and project risk attributes. Both are quite strongly related to checklists in the associated representative group of tools.

Generic and customised risk classifications were presented in Chapter 2 (Tables 2.2, 2.6, and 2.7). In either format, they can be used to spur risk identification by progressively asking the question: What events/conditions in this risk category could lead to uncertainty in achieving desired outcomes on this project? This approach can be applied at any time to any phase of the project.

A project risk attributes technique is more likely to be used at the earliest stage of a project to inform stakeholder decision making about whether or not to proceed further with a project. It differs from other techniques in this chapter in that it is not an adaptation of a tool used for other purposes. The project risk attributes technique essentially reviews and assesses the 'riskiness' of a project for that potential stakeholder. Because it is also an assessment tool, some level of risk analysis and evaluation is incorporated into the tool design. Table 7.12 shows part of a typical tool for reviewing the riskiness of an IT project to an independent commercial developer. The risk identification questions are implicit in the list of attributes to be reviewed, but would typically be framed as: What events/circumstances relating to this attribute could yield uncertainties that would affect our continuing involvement with this project?

Table 7.12 An IT project attributes checklist for assessing project riskiness.

Project details	Comment	L/M/H risk
Project identifier:		
Client:		
Invoices to:		
PROJECT SCOPE		
Brief requirements		
Estimated work hours		
Estimated duration (weeks)		
Staff requirements (internal)		
Consultant requirements (external)		
Time zone implications		
CLIENT		
Experience		
Familiarity with IT systems		
Reputation		
PROJECT ELEMENTS		
Existing systems involved		
System interruptions needed		
Data migration needed		
Data security constraints		
Client–customer interfaces		
Customer–client interfaces		
Payment system requirements		

(Continued)

Table 7.12 (Continued)

Project details	Comment	L/M/H risk
Payment security requirements		
Data management requirements		
DESIGN/PROCESS		
Staff availability		
Proportion new code required		
Specialist software inputs		
Special hardware requirements		
Segment testing constraints		
Full system test constraints		
DEPLOYMENT		
Roll-out timing and extent		
Client training		
Extent of follow-up services		

The illustration is only partial as the full review tool would also incorporate parameters for the low/medium/high risk assessments for each attribute. Thus, 'Estimated work hours' might be assigned estimating parameters such as: low = <200 hours, medium = <500 hours, and high = >500 hours. Other attributes might be qualitatively graded; for example, the client brief might be assessed as 'clear and comprehensive', 'clear but incomplete', or 'poorly defined', and others rated through percentage bands for their constituent importance or presence in the project. Rejection of the project as too risky is usually made on the basis of a count of 'high' project attribute risk assessments exceeding a 'hurdle' number for the organisation. Lower but still significant counts might attract a contingency premium to its bid for the project.

Clearly, application of this tool requires substantial knowledge and experience of dealing with IT projects as a commercial developer. However, with some research and adaptation, it could be used in an in-house PMO setting, or indeed for other types of projects. The earlier this type of tool is used, the higher the level of subjective estimation needed to operate it.

Among the case studies, Case Study F (an aquatic theme park project) probably comes closest to a construction project example of a risk attributes assessment. The issues addressed by Table F.2 present an indicator of the potential riskiness of a project to the civil engineering company.

Almost all of the associated representative types of risk identification tools relate primarily to the procurement phase of projects. Some can be adapted and used for disposal-phase risk identification; a few can be modified to explore risks in the project operating phase.

7.5   Matrix Tools

Matrix tools for project risk identification simply combine two or more different techniques into one application. An activity-related tool might be combined with an associated representative method (or the other way around). Most commonly, this is found in project stage/project activity combinations, but matrix risk identification tools are probably most powerful when risk categories are incorporated into them, particularly if the combination then becomes project activity/risk category, or even project stage/project activity/risk category. Given the power of modern computer spreadsheet applications, even more elaborate matrices can be attempted, such as project stage/WBS activity/ resources/time duration/risk category; but the tool is then likely to become unwieldy, especially in a risk identification workshop setting. The cost–benefit trade-off (effort versus outcome value) should always be kept in mind. The power in matrix tools lies in their ability to stimulate and guide the mental processes of workshop participants towards two or more perspectives, if not simultaneously then in rapid succession. The brainstorming demands are thus quite tiring, especially for projects with complex WBS. Workshop leaders should be aware of this and build in refreshment breaks into the workshop programme.

Table 7.13 presents a matrix risk identification tool based upon the WBS example shown in Table 7.2 and the generic risk category system developed in Chapter 2 (Table 2.2). This example is probably at the upper limit of user-friendliness in a workshop, and we remind readers that this is only a partial project depiction, in terms of project activities and risk categories.

Using the tool, workshop participants can choose to proceed either downwards and from right to left, or downwards and from left to right. Usually, the first option is chosen. The relevant iterative question for this example becomes: What events/conditions in this risk category are we likely to encounter for this project activity (or this technology, resource, or organisational aspect of the activity) that will lead to uncertainty in achieving desired outcomes?

The cell entries shown under each risk category in Table 7.13 refer to code labels for identified risks. These are discussed later in this chapter. 'Mapping' the risks in this way permits rapid visual assessment of the frequency of similar risks or where they tend to be clustered in terms of project activities.

Other activity-related or associated representative types of tools can be used in a similar way (by adding columns on the right-hand side to include types of risks): for example, those illustrated in Tables 7.3, 7.4, and 7.8–7.11. The selection of risk categories to be included can be customised to suit the nature of the projects usually undertaken by the stakeholder organisation. Weather risk, for example, is unlikely to be applicable to IT projects, although it would certainly be encountered on an urban broadband communications network cabling project.

7.6   Simulation and Visualisation Tools

As the description label implies, simulation and visualisation tools for risk identification rely on visual images, either graphically created or mentally imagined, for

Table 7.13 Resourced WBS/generic risk category matrix.

Activity	Technology	Resources	Organisation	Selected risk categories				
				Weather	Technical	Financial	Management	Other
CONCRETE FLOOR SLAB CASTING CYCLE			HQ management Site management			R23	R22	
Assemble formwork materials	Transport Cranage	Props, timber, deck plates, fixings Labourers Plant operators	Site management	R1	R2	R3	R4	
Setting out	Survey	Surveyor	Site management	R1	R5	R6	R7	
Erect formwork	Cranage Power tools	Props, timber, deck plates, fixings Formworkers Labourers Plant operators	Site management	R1	R8	R9	R10	
Rough-in services	Cranage Power tools	Pipes, ducts, conduits, special fittings Formworkers Artisans Labourers Plant operators	Services engineers Site management	R1	R11	R9	R11	

Position blockouts	Cranage Power tools	Timber, polystyrene shapes, custom inserts Formworkers Artisans Labourers Plant operators	Services engineers Site management	R1	R11	R9	R11
Assemble steel reinforcement	Cranage Power tools	Steel bar, mesh, tie wires, spacers Steelworkers Labourers Plant operators	Site management	R1	R13	R14	R15
Fix reinforcement	Cranage Power tools	Steel bar, mesh, tie wires, spacers Steelworkers Labourers Plant operators	Site management	R1	R13	R14	R15
Pour concrete	Concrete pumps Spreaders Tampers	Ready-mix concrete Concretors Labourers Plant operators	Site management Concrete specialist	R1	R16	R17	R18
Other						

their purpose. The former are probably harder to create but easier to use in practice. Generally, the more developed the project, the more comprehensive and realistic the simulation or visualisation can be. In this category, you may encounter storyboarding, static and dynamic 3D graphic visualisations, 'rich pictures', and 'experience' simulations. Disappointingly, perhaps, we can only describe them in text. Whichever tool is used, the risk identification question is: What event or situation could arise in real life in this depiction that could lead to uncertainty in the anticipated outcomes?

Storyboarding derives from the film industry and originally comprised hand-drawn pictures of each proposed scene for a movie project, usually depicted as progressive series of slides formatted as four or six on a single A4 size card. In appearance, they resemble the frames of a very jerky animation sequence.

The storyboards are used for several purposes including determination of actor positions, camera angles, props types and locations, and continuity checks. Thus, they serve the needs of not only film directors but also set designers and other film crew. The technique is still used today for film and television production, but card storyboards have largely been replaced by images stored on handheld electronic devices such as tablets, notebooks, and laptops.

Applying storyboarding techniques to project risk identification is clearly contingent upon having artistic expertise available to quickly render the visual properties of project 'scenes' in a sequential and appropriate manner. Since the storyboards can be developed incrementally, they can be used from quite early on in the project life cycle and can be used in all phases of a project.

3D graphic images for project development have long been used in hand-drawn sketch renderings often presented as an aid for client decision making. Later, they became an essential adjunct in computer-aided design (CAD) applications: first as a static display of project features, and then as dynamic animations capable of reproducing views of physical projects from a variety of perspectives and angles which could be layered to provide an 'onion peeling' effect of eliminating outer skins to reveal internal structures. In the construction industry, these techniques are known as 'virtual constructed environments' (VCEs). When the graphics application is harnessed to a suitable relational database, design considerations such as solar shadowing and construction processes such as structural frame erection can be simulated. More recently still, and again with access to a suitable relational database, it has become possible to simulate on-site plant and equipment movements, operating radii for cranes, proximity to existing overhead and underground services, and regulated safety hazard exclusion zone requirements. The latter capacity provides the clue to using simulation and visualisation techniques for project risk identification – it is simply a matter of applying them beyond the safety risk perspective to identify other types of risk, particularly those that might be associated with overlapping and interdependent process activities.

In some ways, computerised graphic simulations and dynamic visualisations of projects are the closest approach so far developed for automated risk identification, but for effective exploitation of this they require the existence of, and access to, suitable input data and must be accompanied by meaningful interpretation of the visual results.

'Rich pictures' are a visual form of mind-mapping techniques whereby concepts, issues, causes, and effects can be displayed in a visual diagram as text-boxes or symbols linked by line arrows or vectors in a reasoned and meaningful way. Used for project risk identification, the content of each rich picture is interrogated for uncertainties affecting project objectives.

The 'customer experience' may be used as a visual or non-visual simulation for events projects and IT application projects. It is interrogative and progressively asks what a typical customer might encounter at each point in the processes delivered by the completed project, and how that customer might respond.

The 'tour experience' is a similar tool used to graphically simulate a visual journey around physical facilities such as museum and exhibition projects. It is rarely interrogative but often accompanied by audio commentary. The ultimate tool of this nature is probably a physical scale model or a mock-up. Case Study D describes a project involving the construction of a mock-up for a proposed high-capacity metropolitan train. The mock-up was originally intended for realistic consideration of the aesthetic and practical acceptability of the new train design. It eventually became available for practice drills carried out by emergency rescue services personnel.

While a tool such as VCE is obviously relevant to the procurement phase of projects, and storyboarding can be adapted for most types of projects for this phase, rich picture and 'experience' simulation tools and models are most often aimed at the operational phase, and their early availability for risk identification purposes is likely to be quite restricted since they are expensive to create.

7.7 Speculation Tools

Speculation tools for project risk identification differ from simulation and visualisation tools in that they normally derive from hypothetical contexts far beyond that of the project under consideration – although they could emanate from the external context of the project itself, as described in Chapter 5. As we noted in Chapter 6 (Table 6.1), two types of speculation tools can be used: scenario testing and stress testing.

7.7.1 Scenario Testing

Since scenario testing does not rely on a requisite amount of prior information about a project being available, it can be used at almost any time, even at the very beginning of the project. At the other extreme, although other techniques may have already been used for project risk identification, it is worthwhile completing a risk identification workshop process with a scenario testing exercise. Our argument to support a double dose of scenario testing is that, initially, project risk workshop participants will have relatively low project risk awareness but higher project context awareness. Toward the completion of the workshop, participants will have high project risk awareness, high contextual awareness, and hopefully reasonable world awareness.

This technique requires workshop participants to adopt a replication perspective for a particular historical or contemporary event which has been the catalyst for significant change. From this background, one or more hypothetical scenarios may be imagined. For any chosen scenario, the question is posed: If there were to be a similar situation today (or if this type of event were to happen whilst undertaking the project), how would our involvement in the project be affected, and what uncertainties could arise that might affect project objectives?

The scenario might derive from a world, national, regional, or local situation. If time permits, the risk identification workshop should embrace several different scenarios.

Examples of world scenario backgrounds might emerge from the periods of economic depression in the 1930s, the political lead-up to the Second World War (or the Korean, Vietnam, or other more contemporary wars) as an example of global conflict, contemporary conflicts in the Middle East and elsewhere, the removal of the Berlin Wall, the breakdown of Soviet Russian hegemony in Eastern Europe, the release of Nelson Mandela from detention and the change to democratic government in South Africa in 1994, the Millennium and the Y2K computer functionality concerns that preceded it (remember those?), the September 2001 destruction of the New York World Trade Center and attack on the Pentagon in Washington, or the causes and ramifications of the 2008 Global Financial Crisis. Many more scenarios can be devised, limited only by the powers of imagination of workshop participants.

National or regional scenarios might include the introduction of a new tax system, the devolution of central government powers to states, policy formulation with respect to food and water security, the relocation of central public administration departments from cities to regional towns, the release of public land for development, or the adoption of public–private partnership procurement for the delivery and operation of particular public infrastructure and services.

Local scenarios could be changes in population density caused by rapid influx of large numbers of immigrant and refugee families, the transfer of health facilities from small rural towns to larger regional facilities, the decline or closure of traditional local industries and the privatisation of public transport services, energy generation and distribution, potable water reticulation, and renewal of aging sewer systems.

The benefit of scenario testing in a risk identification workshop in the *earliest* stage of a project is that it can help to more easily place the project context establishment process (Chapter 5) into a risk frame. It may also gently free workshop participants from overly rigid attitudes towards risk management. At its very best, scenario testing is capable of exposing projects that should have been avoided but perhaps were too enticing or over-optimistic in their expectations for that stakeholder!

Part of the purpose of *concluding* a risk identification process with scenario testing is to encourage participants to break away from identification approaches that may have become very narrowly focused. In the early stages of the process, developing a narrow focus is desirable, but this can become counterproductive if workshop participants simply become immersed in too much detail. By now, several hundred (or more) project risks may have been identified in an intensive manner in the workshop. Enjoying a refreshment break and then finishing this stage of risk management with a scenario testing technique give participants an intellectual stimulus that perhaps broadens (rather than elevates) their risk awareness and encourages them to think more creatively about risks. It acts as a 'winding down' strategy for the risk identification process.

7.7.2 Stress Testing

Used deductively, stress testing is a narrower version of scenario testing. Imaginary but potentially realistic situations are now juxtaposed against key project decisions. The question is then: What uncertain effects would that event or situation exert on the resilience of that decision? Risks to the project are thereby exposed.

Ideally, stress testing will focus on resilience associated with decisions that include project activity relationships involving dependencies, much like the CPN tool discussed in Section 7.2.3. However, while decision stress testing is comparatively straightforward in the inception, conceptual, and early planning processes for a project, it becomes progressively more difficult to use in the later part of project planning as more and more decisions are made. This is not to say that stress testing cannot be used for risk identification during this time; rather, it will require more time and effort.

Project decision stress testing could include exploration of decision resilience to situations such as: the imminence of, and proximity to, volcanic or earthquake activity; potential cyberattack including hacking, denial of service, and fake news; a severe acute respiratory system (SARS) outbreak; or new environmental protection regulations. There are many more. The chosen situations, unlike the 'blue skies' options available for scenario testing, should bear some relationship to the project type and circumstances.

7.8 Structural and Management Tools

Structural and management tools applicable to the process of risk identification include those relating to organisational governance and authority or functional structure, and to those used for management purposes. For project stakeholder organisations, we refer readers to the examples shown in Chapter 2 (Table 2.5) and Chapter 3 (Figures 3.11 and 3.12). From a project organisational structure perspective, the relevant examples are Figures 3.7 to 3.10 in Chapter 3. Given the appropriate and timely availability of these structural organograms, the iterative risk identification question is: What event or situation associated with this part of the organisation could lead to uncertainty in its effective contribution towards achieving project objectives?

By employing a structural tool for project risk identification, it should also be feasible to expose conflicts within or between organisations involved in a project. This aspect of risk identification can be applied from the earliest point of project involvement. It should also be part of the consultancy's risk management system for every professional consultant acting in an agency capacity for a project client.

Several dedicated project management tools are available, either as text-based guides (e.g. PMI 2013) or as computer-based applications (see Chapter 18). The contents and components of these tools can be used as prompts to identify project risks for a stakeholder by posing the question: What events or situations pertaining to this aspect of project management could lead to uncertainties in the desired management outcomes? By exploring risks associated with the project management process itself, this tool also becomes important in the professional consultancy's practice risk management armoury.

7.9   Risk Identification Statements

In most instances, project risks identified through brainstorming in a workshop setting are simply recorded informally (i.e. on small sticky notes) using one or two-word descriptions. Examples might include: 'bad weather', 'scaffolding accident', 'price increases', 'low productivity', 'data loss', 'compliance failure', and so on.

While this may be adequate for the purpose of the workshop, a more precise statement of each risk is needed for good project risk management. Precise statements help to ensure that effective risk communication takes place within the stakeholder organisation and between project stakeholders where this becomes necessary. More importantly, in most instances, a precise risk statement will inform the ensuing risk assessment process and may well implicitly contain the seed ideas for an effective risk response.

A good risk statement will contain information about a prevailing situation, a trigger event and its likelihood of occurrence, and the nature and uncertainty of the consequences likely to flow from it. Time dimensions and the risk category, according to a generic or customised classification (or both), will also be included, with the latter possibly at the end of the statement in parentheses.

The risk statement should also correct any inconsistencies apparent in the brief informal descriptors. Looking more carefully at the examples in the opening paragraph of this section, some are really consequences of prior events.

A precise risk statement will at some point, but preferably at the beginning, include the words 'There is a chance that ...' The single-sentence statement should contain references to:

- A situational source event
- The likelihood of occurrence for the event
- The consequences of the event
- Any uncertainty about those consequences
- Time of exposure to the event
- Any time constraints on the consequences
- The risk category and a risk identifier code.

A typical example, for a contractor on a construction project, might be:

'There is a **p** chance that, during the major concreting activity on the high-rise office building project, the price of cement will rise by more than **$x** per tonne, leading to additional costs over this period (**n+** months) of at least **$z** (Economic risk R17).'

Readers will now easily appreciate why a risk identification workshop is unlikely to deliver precise statements for each risk during its brainstorming activity! Nevertheless, the task should be carried out as soon as possible thereafter. Anyone familiar with the construction industry will appreciate not only the reality of this risk but also the implicit seeds (provided in the precise risk statement) for responding to it. Values for the bolded italic factors do not have to be inserted at this point, as that will be part of the following analysis and evaluation stage in the project risk management cycle. The precision of the statement is the important issue here.

In our view, good precise risk statements are crucial for effective project risk management, and we provide another example from a different project context:

> 'There is a **p** chance that, during the data migration process for the customer services IT project, hardware malfunction will occur, leading to loss or contamination of key data, and extra costs for data recovery action and project completion delay amounting to at least **y** weeks and additional costs up to **$x**; as well as reputational damage to the client and to the organisation (Technical risk R45).'

Additional points can be made for both examples. In the first, the possible cement price increase for the construction project is actually a consequence of a prior risk event beyond the project system boundary. This implies that risk prevention or avoidance will not be possible, and alternative mitigation treatment will have to be considered. For the IT project, hardware malfunction must also be due to some prior causal factor, but the project system boundary is not crossed. Further investigation for the possible causal event(s) for the cement price is probably not warranted for the first example; whereas for the hardware malfunction it might be worthwhile, in the following risk assessment stage, to explore potential causal factors through a technique such as FTA.

The risk identifier codes are useful in populating matrix cells such as those shown in Table 7.13. Alpha-code suffixes can denote the type of risk.

For the project risk register (Chapter 4, Figure 4.2, Stages A–E), a separate item record will be prepared for each identified risk. Table 7.14 provides an example of how such a record might be formatted and what it should contain.

Much of the content of the item record, and its relationship to the stakeholder's project risk register and organisational risk register, will be dealt with in later chapters. The assessment of the risk is discussed in Chapter 8, response options in Chapter 9, and monitoring and control procedures in Chapter 10.

Incorporation of risk item records into the organisation's risk registers involves documentary processes, knowledge management principles, and an appropriate system architecture. All these aspects are explored more fully in Chapter 11.

It is important to note that the risk item record probably represents a base level of what might be expected in terms of stakeholder organisational governance in contemporary management. It is not inconceivable that such records might be called for as evidence of appropriate risk management in a legal or official enquiry setting. This is why we advocate the preparation of separate risk item records, and then extracting information from them to populate risk registers. Any official enquiry is likely to focus upon the treatment of specific project risks associated with the focus of its investigation, and may well be less interested in the wider aspects of the risk management system used by the organisation. Keeping separate risk item records also allows them to be quickly arranged and resorted, if required, in sequences different from that maintained in a more comprehensive register.

Table 7.14 Project risk item record.

PROJECT:	
Project Phase:	
Project Element:	
Risk Workshop Date:	
RISK ID REF.	
Risk Category:	
Risk Description:	There is a chance that….
RISK ASSESSMENT	
Likelihood:	RARE / UNLIKELY / POSSIBLE / LIKELY / ALMOST CERTAIN Reasons:
Impact:	INSIGNIFICANT / MINOR / MODERATE / MAJOR / CATASTROPHIC Reasons:
(Quantitative Analysis?)	
Severity:	NEGLIGIBLE / LOW / MEDIUM / HIGH / EXTREME
RESPONSE OPTIONS	RETAIN / REDUCE & RETAIN / TRANSFER / AVOID / (COMBINATION?)
Reasons & Procedure:	
Post-treatment reassessment:	Likelihood: Impact: Severity:
MONITORING & CONTROL PROCEDURES:	Describe:
Responsibility:	Assigned to:
Frequency:	Describe:
Emergency Action:	Describe:
Debriefing:	Process:
Sign-off:	
Date:	

Precise risk statements are a vital precursor to good risk assessment.

7.10 Summary

This chapter reinforces our assertions about the importance of risk identification in systematic project risk management. It is a chapter that you may want to return to again and again, as your project risk management experience grows and you seek to use additional techniques.

Among the large array of tools and techniques available, some derived from decision science and others from general management, organisational management, and project management, examples have been taken from the typology devised in Chapter 6, showing how they can be adapted to aid the identification of project risks.

We have explored techniques including activity-related tools, analytical techniques, associated representative methods, matrix combination approaches, simulation and visualisation tools, speculation techniques, and others based upon structural aspects of organisations and their governance, and from project management applications.

Two activity-related tools (bar charts and CPNs) could be moved to the analytical techniques group. While each has quantitative analytical characteristics (time and associated task dependency), they are retained in the activity-related category because of their reliance upon 'task identification' as the primary driving factor for their original purposes.

In almost every instance, some adaptation is needed to extend the original management purpose of a technique and make it suitable to use for risk identification. The effort is worthwhile, as the benefit of familiarity to users can then be exploited.

Matrix approaches, where two or more tools are combined in a matrix format, are powerful in project risk identification, particularly when risk categories are included in the combination. Most organisations could develop this approach. The effort involved should not be too difficult or time-consuming.

None of the techniques discussed here will guarantee capture of every project risk for a stakeholder organisation. Inevitably, some will be missed. However, diligent application (and practice) will improve the risk identification process, especially if multiple methods are used. Experimentation with different techniques, starting with those that are familiar and using old projects as vehicles for learning, is an excellent means of finding out which tools can be most easily adapted, which can be most conveniently applied, and which are most effective for project risk identification.

With the possible exception of 3D dynamic graphic visualisation and simulation modelling, where appropriate relational database capacity permits automatic risk flagging, all the techniques rely on brainstorming to fuel the risk discovery process. This mental process, and personal willingness to use it, also requires practice and some level of guidance and control to be fully effective. It should be structured or semi-structured rather than completely freewheeling. Workshop leaders must take seriously the intra- and interpersonal issues associated with brainstorming, and be prepared to deal with any barriers arising during workshop sessions.

Not all risk identification techniques and tools are suitable for every stage, or every phase, of a project. We have tried to provide some guidance about this, but again practice will be the best indicator of application timing and appropriateness.

All techniques employ an interrogative approach. For each of those dealt with in this chapter, we have suggested suitable question formats for users. Some analytical techniques may require the assistance of experts or specialists.

We have recommended a workshop environment for project risk identification, conducted within the relevant stakeholder organisation and attended by a small number of appropriately qualified and experienced participants. Single-person risk identification is never advisable, and multi-organisation project workshops are unlikely to yield the essential risk discovery that each organisation needs. Careful planning and management of the workshop should be considered beforehand.

The immediate outcomes of a project risk identification workshop are likely to be lists of inadequately framed risk descriptors. As soon as possible, these should be transformed into precise risk statements that will not only provide information to populate the project risk register, but also contain implicit guidance for the ensuing tasks of assessment and treatment response in the project risk management cycle.

Effective project risk identification provides the essential foundation for risk analysis, which is discussed in Chapter 8.

References

Health and Safety Executive (HSE) (2015). *Construction Design and Management Regulations*. London, UK: Health and Safety Executive (HSE). ISBN: 9780717666263.

International Organisation for Standardisation (ISO) (2018). Occupational health and safety management systems (ISO 45001). In: *International Organisation for Standardisation*. Geneva.

PMI (2013). *A Guide to the Project Management Bok of Knowledge*, 5e. Project Management Institute, Newtown Square, PA: PMI: ISBN: 9781935589679.

8

Project Risk Analysis and Evaluation

8.1 Introduction

Once project risks have been identified for a project stakeholder, they should be analysed in terms of their magnitude and their severity (for threat risks) should be gauged in some way. This takes us to Stages C_1 and C_2 in the project risk management cycle (see Figure 4.2).

Risk analysis may be undertaken quantitatively or qualitatively. Sometimes this is also described as 'objective' or 'subjective' assessment, but this interpretation is somewhat misleading. It assumes that all risk data are either objective (factual information) or subjective (perceptions and opinions of people). While this may be so, the two types of data are not necessarily mutually exclusive in terms of the approach to their use in risk analysis. In Case Study F (an aquatic theme park project), for example, the civil engineering contractor uses objectively derived historic weather data to quantitatively explore weather-related risks on its projects. Also, subjective data may find their way into quantitative analysis, and objective data are not excluded from qualitative assessment. It is important, however, to know when this happens, in order to appreciate the level of care required in such situations and to have some appreciation of the confidence which can be placed in the analytical inputs and outputs.

In risk analysis, essentially we are looking at recognising, understanding, and dealing with uncertainty. The uncertainty may relate to the likelihood of occurrence of a risk event; the nature, magnitude, and occasion of its consequences; or all of these.

For most organisations, qualitative risk analysis will be carried out first, as a rapid means of triaging threat risks. Indeed, for many projects this is all that is required. No further analysis takes place simply because the organisation recognises that some proactive form of risk treatment is inevitably required for particular risks and that extending the analysis further is unlikely to yield additional knowledge that will change that situation.

Ideally, qualitative risk analysis should take place immediately after risk identification has been completed, preferably in the same risk management workshop with the same knowledgeable participants, while the project contexts and project risks are still fresh in their minds. If this is not possible, another workshop should be held soon after the first and be led by the risk manager. Here the total number of participants can be reduced to three or four (but all should have attended the first workshop). Our earlier assertion still holds: project risk management is not the preserve of a single person within an organisation.

Managing Project Risks, First Edition. Peter J Edwards, Paulo Vaz Serra and Michael Edwards.
© 2020 John Wiley & Sons Ltd. Published 2020 by John Wiley & Sons Ltd.

In our experience, it is also unwise to analyse each risk as it is identified in its simple 'two or three word' descriptor form. Attempting this requires workshop participants to constantly switch between identification mode and analysis mode, and this is rarely effective as it interrupts the risk identification 'mind-set' that will have almost certainly developed in the workshop. More precise risk statements (as proposed in Section 7.9) provide a better guide as to what factors must be analysed for each project risk.

Precise risk statements help considerably in risk analysis. Besides properly framing each risk, the actual process of formulating the individual risk statements also stimulates the intellectual analytical and judgement capabilities of the brain. In fact, writing good risk statements is where the previous identification mind-set begins to change to a risk analysis track. Simply writing 'There is a "p" chance that ...' encourages the brain to simultaneously think about what that chance might be. Listing the consequences of the risk event stimulates thinking about the potential influence of factors associated with the internal and external contexts of the project. For many projects, before risk analysis commences, it may be worthwhile revisiting the risk statements to confirm their sufficiency and reminding workshop participants of the project contextualisation that has been done.

Two outcomes from the project risk identification stage, risk classification and the results of prior project contextualisation, can help where the number of risks to be dealt with is extensive. For some types of projects, this may amount to several hundred separate risks, and assessment thereby becomes arduous and onerous.

It may be possible to first consider risks in their *category groups*. Thus, for generic natural category weather risks, the likelihood of occurrence and damaging impacts of extreme weather conditions can be explored as general factors first, then modified for particularly vulnerable stages of a project, and finally adjusted and applied to individual risks. Other generic risk categories, and indeed other customised risk classification systems (see Chapter 2), can be used in a similar way.

Establishing the project contexts (see Chapter 5) will have revealed particular project circumstances. Each of these can be examined for its influence, and the results then applied to the relevant risks. External contexts will have been framed in terms of their physical, technical, economic, and social effects. These too can be explored generally and then applied to relevant project risks.

When embarking upon analysis, a stakeholder organisation should also investigate what prior risk allocation arrangements already exist for the project (see Table 4.2). There is little point in exploring a risk further if it has already been allocated to another party, through a mechanism such as a special clause in a contract agreement, unless there is a residual risk that your organisation must bear. For example, a subcontractor normally used to submitting tenders on a fixed price basis would usually identify the chance of high economic inflation as a threat risk factor on a project with a duration extending beyond the short term. However, if the head contract included clauses agreeing to share cost fluctuations between client and main contractor, with flow-on provisions for subcontractors, this would mean that the consequences of the inflation risk would be borne partly by the subcontractor and partly by the main contractor, who in turn would share the increased cost with the client. A contract agreement such as this has effectively allocated the economic inflation risk on a shared basis, and the subcontractor would need to reflect this in its analysis of the risk.

In theory, of course, all risks should be allocated to the party best able to control them. Practice, however, especially in the guise of contracts, has a habit of mocking theory!

In this chapter, we first consider qualitative threat risk analysis and explore linguistic measures of likelihood and impact. Similar qualitative measures are suggested for evaluating threat risk severity. (Note: Qualitative approaches for assessing opportunity risks are discussed in Chapter 15.)

A brief discussion of quantitative risk analysis is then presented, in line with the premise made in Chapter 1 about avoiding heavily mathematical treatments in this book.

The chapter concludes with discussion about the value of mapping project risks.

8.2 Qualitative Analysis

Qualitative approaches to risk analysis reflect the perceptions, opinions, and value mechanisms of the assessors. Much of the analysis deals in some way with the treatment of uncertainty, and this aspect of risk was explored in Chapter 2.

There will be uncertainty associated with the values of nearly all the input variables from which the estimates are derived for the component factors of risk. Any one of a range of values might actually occur for a variable.

Qualitative analysis usually is regarded as subjective because it relies almost entirely upon human judgement. That judgement will be influenced by experience and by personal biases that may lead to errors.

Errors in human judgement include those arising from biases such as:

- Anchoring (wrongly sticking with a first estimate even though it is inappropriate for the context) – a 'representativeness' heuristic bias.
- Selective recall (remembering only certain facts or incidents and ignoring other evidence) – an 'availability' heuristic
- Base rate fallacies (e.g. assuming that a particular trend line will continue at the same rate into the future) – also an 'availability' heuristic
- Over-pessimism
- Over-optimism
- Overconfidence in the assessment
- Inappropriate search for patterns in data
- Inappropriate framing of the situation.

Note that several of these biases are similar to those listed as barriers to good decision making in Chapter 3.

The advantage of subjective analysis is that it is generally quick to apply and simple to understand. The disadvantage is that errors in judgement are often difficult to detect and eradicate. Here, it is not so much that practice will make perfect, but more that periodic recalibration of assessment criteria, particularly at an intra-organisational level, would help to minimise any accretion of subjective biases. Exploration of the existence and nature of judgement bias in an organisation also helps.

The outcome of qualitative risk analysis should be a list of project threat risks that the stakeholder organisation faces, categorised according to type and grouped or ranked in terms of the level of severity they represent. The analytical data can be used to populate the stakeholder's project risk register (PRR).

The process involves applying estimating measures to factors such as the likelihood of occurrence of the identified risk event and the magnitude of its consequences. The two measures can then brought into association to determine a level of threat severity for each risk.

In qualitative risk analysis, these measures are most often *linguistic*, using words as descriptors for value or quantity estimates.

The most basic linguistic assessment measure is: low, medium, and high. Table 8.1 illustrates this three-point scale applied to the likelihood of occurrence and impact of a risk event.

Table 8.1 Three-point risk assessment measures.

High			
Medium			
Low			
↑ Likelihood Impact →	Low	Medium	High

The very simplicity of the three-point assessment approach can lead to problems in application. Firstly, the potential range of interpretations for each interval of the measure is wide. How low is 'low'? How high is 'high'? What does 'medium' mean? Establishing interval definitions (a strategic task for senior management in an organisation) is difficult, especially for impact measures where different types of consequences may ensue. Secondly, as will be seen later in this chapter, the three-point assessment measures flow through to corresponding three-point measures of risk severity, and this may not provide sufficient distinction between risks to allow confidence in their ranking or priority: the scale is too coarse. Thirdly, the simple structure of the three-point scale infers an erroneous assumption that the intervals are equidistant, with 'medium' as a midpoint. More explicit interval definitions may reveal that this is not the case, but that fact may not be fully appreciated by project risk workshop participants working rapidly under pressure to deal with many risks.

We advocate the adoption of more extended linguistic measures for qualitative project risk analysis. Sections 8.3 and 8.4 explore these more fully for likelihood and impact.

8.3   Assessing Likelihood

If three-point linguistic risk measures are too coarse, how many intervals should be incorporated into the scales? We have encountered as many as seven and nine, but in our view *five* intervals best serve the purpose. This opinion is supported by the suggestions offered by the old risk management handbook AS/NZS HB 436 (2004). As noted in Chapter 4, these documents have since been superseded by the international standard ISO 31000 (2018) and the handbook SA/SNZ HB 89 (2013).

Customised, linguistically based, ordinal-scale measures for risk assessment can be developed by any organisation to suit its particular purposes, but some principles of scale construction should be observed:

- The word descriptors for scale intervals should be precise and distinctive (and this is where *seven* and *nine*-point scales usually fail, because the descriptors become too blurred and difficult to separate).
- Descriptors should preferably be single-word interval labels, avoiding qualifiers such as 'very', 'highly', and so on, as they add little or nothing to clarity.
- The descriptors and underlying definitions should be confirmed in terms of strategic risk management policy established by senior management.
- The descriptors and definitions should be appropriate to the internal context of the organisation and the activities which it undertakes.
- The scale progression of the risk assessment measure must be logically progressive (but not necessarily with equidistant intervals).
- The scale interval descriptors and definitions should be meaningful and uniformly understood across the organisation (i.e. not just in particular parts of it).

Table 8.2 shows a generic five-point measure for the likelihood of occurrence for a project risk event applicable to all types of projects, organisations, and industries.

The ordinal descriptor labels and their expanded definition should be easily grasped, not only by project-based staff but also by managers and employees across the organisation. The same measure can thus be applied to project risks and to other risks faced by the organisation.

Note that in the first column, the two extremes cannot be labelled as 'Never' and 'Certain'. The uncertainty represented by likelihood of occurrence is mathematically expressed by probability values greater than 0 and less than 1. Since 'Never' must equal 0 and 'Certain' must equal 1, no uncertainty is present, and therefore no risk arises.

Table 8.2 complies with the principle of scale progression whereby '1. Rare' represents the lowest expected frequency and '5. Almost certain' the highest, while each of the intervening labels escalates the frequency in a logical manner. The underlying interval

Table 8.2 Five-point measures of likelihood.

Probability/likelihood scale interval label (AS/NZS 4360)	Defining description (AS/NZS 4360; AS/NZS HB 436)	Possible project frequency basis
5. Almost certain	Expected to occur in most circumstances	75 or more out of 100 projects
4. Likely	Will probably occur	Fewer than 75/100 projects
3. Possible	Might occur at some time	Fewer than 50/100 projects
2. Unlikely	Could occur at some time	Fewer than 10/100 projects
1. Rare	May occur in exceptional circumstances	Fewer than 5/100 projects

Source: Adapted from AS/NZS HB436 (2004).

definitions follow the same logic, although the linguistic distinction between '2. Unlikely' and '3. Possible' is quite subtle and reinforces the need to ensure uniform understanding across the organisation.

The third column in the table is offered as a more quantitative, project-based suggestion of likelihood, with intervals corresponding to the linguistic descriptors and definitions. Mathematically, the probability extremes shown are <0.05 and ≥0.75. This column reinforces our earlier caution that scale intervals are not necessarily equal. However, to apply this ratio scale implies that the organisation has sufficient objective evidence, gathered from a sufficient number of similar (but not necessarily identical) projects, to construct a reliable assessment scale. A further caution might be that such a scale requires periodic recalibration to reflect changes in organisational practice.

Table 8.3 presents a more customised version of the generic measure shown in Table 8.2. While adopting the same basic interval descriptor label scale, this organisation has developed its own underlying definitions.

Here, the probability definitions must relate to the likelihood of occurrence of the risk events (although this is not explicitly stated) rather than the frequency of projects in which they might be encountered.

Table 8.3 Alternative five-point measures of likelihood.

Probability/likelihood	Description
5. Almost certain	• >75% probability *or* • occurrence is imminent *or* • could occur within 'days to weeks'
4. Likely	• ≤75% probability *or* • on the balance of probability will occur *or* • could occur within 'weeks to months'
3. Possible	• ≤50% probability *or* • may occur but a distinct probability it will not *or* • could occur within 'months to years'
2. Unlikely	• ≤10% probability *or* • may occur but not anticipated *or* • could occur in 'years to decades'
1. Rare	• ≤5% probability *or* • occurrence requires exceptional circumstances *or* • very unlikely, even in the long-term future *or* • only occurs as a '100 year event'

While we do not recommend adoption of this particular alternative to replace the generic measure, it does raise the important aspect of time. In most mathematical expressions of probability, time is usually implied or made explicit in the probability itself, for example 'a 50% chance of heavy rain showers occurring over the next 24 hours' or 'a 20% chance of surviving this surgical operation beyond five years'.

Traditionally, projects (particularly in the construction industry) have been regarded as relatively short-term undertakings and viewed largely from the perspective of the delivery or procurement phase. In those terms, time was not seen as a major consideration for assessing the likelihood of occurrence for project risk events and uncertainties, and could be almost ignored. Our exploration of projects in Chapter 3, which introduced ultra-short and ultra-long durations for different types of projects, forces us to consider time more carefully. In adopting generic or developing customised measures of likelihood for project risks, therefore, project stakeholder organisations should pay careful attention to the ramifications of time (with its associated uncertainty) and the risks they face.

Incidentally, a problem arising with Table 8.3 (which has been taken from a real-life project stakeholder organisation's risk management system) is that in part it adopts an 'imminence to the project' concept of time (e.g. levels 4 and 5) which may be misleading.

8.4 Assessing Impacts

In exploring assessment measures for the impacts of risk events, we remind readers that our focus will be upon threat risks, and the consequences will thus be negatively framed.

For three-point qualitative measures of impact (low, medium, and high), the same limitations apply as those pointed out in Section 8.2 – possibly to a greater degree, since while the likelihood of occurrence of a risk event is a single concept, there may be multiple consequences of different types. In turn, multiple consequences entail greater attention to ensuring that each is suitably defined, described, and understood.

Table 8.4 illustrates five-point risk impact measures (again adapted from the Australian/New Zealand standards). While measures 1 to 4 on this table could apply to threat and opportunity risks, the level 5 'Catastrophic' label clearly anchors this measure to threat risks. Two impacts of risk events are embraced by this measure: financial and personal harm. If the type of project is such that no risk event could result in personal harm, then that impact measure would not be applicable. It is tempting to place information technology (IT) projects in this category, but more careful consideration might suggest that the harm descriptions should be modified rather than abandoned.

Table 8.4 Five-point measures of impact.

Impact/consequence (AS/NZS 4360)	$Loss	Harm
5. Catastrophic	Huge. Greater than $d (or financial collapse of organisation)	Death occurs
4. Major	Major. Less than $d	Serious injuries; hospitalisation; long-term disability
3. Moderate	High. Less than $c	Medical treatment needed
2. Minor	Medium. Less than $b	First aid treatment required
1. Insignificant	Low. Less than $a	No injuries

Source: Adapted from AS/NZS 4360 (1999) and AS/NZS HB436 (2004).

Some IT projects, like any others, may be subject to impact uncertainties associated with personal stress and anxiety.

As with Table 8.2, the progressive logic of the impact levels in Table 8.4 is clear, but an important point to note is that, unlike the measure of likelihood, the levels of impact assessed for any risk could differ in terms of the type of impact. For example, a safety risk impact might be assessed as moderate in terms of personal harm, but minor in terms of financial cost to the organisation. A further assessment principle is thus necessary: *where multiple consequences can arise for any risk, the highest impact category rating should be assigned to the whole risk.*

Table 8.5 shows how an organisation might expand its risk impact measures. Six types of impact measure are shown in this table. For some organisations, there could be several more, but caution should be taken when attempting to create more extensive measures in this way. Too many different types of impact measures will directly affect the capacity of the multiple measure to be meaningfully understood across the organisation. Applying the impact measures in a project risk management workshop will take more time (and probably lead to more argument among participants!). The value of precise risk statements is demonstrated again here, as they will determine what risk consequences *should* be assessed as distinct from those that *could* be assessed. Other types of risk event consequences are considered in Chapter 16 (Strategic Risk Management).

As with the measures of likelihood, it is important to remember that the rating levels for impact, while they are ordinal, do not signify equidistant intervals on the scale and therefore do not permit any quantitative arithmetic to be performed. Numeric labels of 1 to 5 must be treated with caution. These measures also require periodic recalibration, particularly as an organisation grows and its perceptions of financial and reputational impacts change.

Qualitative analysis of the uncertainties associated with the likelihood of occurrence and the consequences of project risks is followed (or even simultaneously accompanied) by evaluation of their severity to the project stakeholder organisation, as indicated by Stage C_2 in the project risk management cycle (Chapter 4, Figure 4.2).

8.5   Evaluating Risk Severity

Great care is necessary in devising measures for evaluating threat risk severity, particularly where qualitative measures have been used for assessing likelihood and impact.

In *quantitative* risk analysis, risk severity is a straightforward matter of calculating the product of the probability and impact values for each risk, as each is represented by data values that are continuous in type. The calculation is therefore mathematically valid, and reliable relative comparisons can be made between evaluated risks.

No such validity attaches to *qualitative* analysis, and any capacity for relative comparison therefore suffers. The tables presented in this section cannot be regarded as *mathematical* representations of risk severity. The nominally ordinal linguistic measures for likelihood and impact are set alongside each other simply for the purpose of assigning ratings from an ordinal scale of severity. That scale must be developed by senior management as part of a strategic organisational and project risk management policy. It is usually aimed at informing risk response decision making.

Table 8.5 Multiple-impact risk assessment measures.

Rating	Financial	Safety	Environmental	Social	Legal	Reputational
5. Catastrophic	Loss greater than $d (= financial collapse)	Death occurs	Irreversible long-term environmental harm National/international impacts	Adverse effects on whole organisation	Major litigation/ prosecution likely	Medium/ long-term adverse media inquiry
4. Major	Loss less than $d	Severe irreversible disability; off work more than six months	Measurable environmental harm Medium/long-term recovery Required to inform government authority Local/regional impacts	Adverse effects on sub-organisation	Significant litigation involving lawyers and senior counsel	Medium-term adverse media attention
3. Moderate	Loss less than $c	Significant medical treatment; long-term recovery or off work less than six months	Medium-term recovery Some local effect	Can be managed within organisation	Breach of regulation with internal investigation and report to authority Prosecution and heavy fine possible	Short-term adverse national press attention
2. Minor	Loss less than $b	Medium-term largely reversible disability or off work less than one month	Short-term recovery Minimal local effect	Can be managed within sub-organisation	Minor non-compliance or breach of regulation Small fine possible	Short-term adverse local press attention
1. Insignificant	Loss less than $a	First aid or minor medical treatment; not required to be off work	Little or no impact	Entirely a project issue; no further action needed	No legal breach	Little damage

Table 8.6 shows a risk severity measure that might accompany the three-point low/medium/high measures for assessing likelihood and impact measures. The severity measure is also a three-point scale with low, medium, and high interval labels. The distribution of the three severity markers to each of the nine severity rating possibilities is the responsibility of senior management in the organisation. They are shown in Table 8.6 as different grey-scale shadings, getting darker as the severity increases, but in practice coloured 'traffic light' cell shading could be used (e.g. low = green, medium = amber, and high = red).

Table 8.6 Three-point measure of risk severity.

	High	**R7, R9**	**R5**	**R6** *High*
Likelihood	**Medium**	**R4, R8, R13 R18, R19, R20, R21**	**R12, R14** *Medium*	**R2, R10, R16**
	Low	**R11** *Low*	**R1, R3, R22**	**R15, R17**
	/ *Severity*	**Low**	**Medium**	**High**
		Impact		

As with the other three-point tables in this chapter, the distinction between severity intervals is not sufficiently nuanced, and only one cell is allocated to low-severity risks in this measure. Extending the severity range helps to address that shortcoming. The extended range could include any number of intervals, influenced only by the sophistication of the risk response, monitoring, and control guidance sought by the stakeholder organisation and its capacity to devise suitable interval descriptors. As with the likelihood and impact measures, we have found a five-point scale for risk severity to be the most suitable. Table 8.7 illustrates this approach.

Here, too, the grey cell shading and hatching can be replaced by colour shading. Note how the expanded capacity created by the five-point measures permits more nuanced

Table 8.7 Five-point measure of risk severity.

Likelihood →	Almost certain					
	Likely					
	Possible					
	Unlikely					
	Rare					
		Insignificant	Minor	Moderate	Major	Catastrophic
		Impact ↑				
		Severity key ↓				
		Negligible	Low	Medium	High	Extreme

Source: Adapted from AS/NZS HB436 (2004).

severity evaluation allocations, capable of reflecting the strategic risk management policies of the organisation. As Table 8.7 shows, the risk severity distribution among the 25 possible cells does not have to be symmetric. It is perhaps disappointing to see low, medium, and high reappearing among the risk severity interval labels, but we invite you to stretch your command of the English language to suggest alternatives!

Practice and experience will improve familiarity with, and rapid practical application of, qualitative project risk assessment and evaluation using the measures suggested here. The approach should suffice in most circumstances.

The project risk item record example (Chapter 7, Table 7.14) presents a format for entering data from qualitative risk analysis and severity evaluations. Table 4.1 (Chapter 4) provides a spreadsheet-based PRR template for ongoing risk management purposes. In the spreadsheet, the column cells provided for risk severity can incorporate suitable 'if', 'and', 'or', and 'then' algebraic logic to determine the appropriate severity from the likelihood and impact ratings for each risk recorded in the two previous columns. This is discussed more fully in Chapter 18 (Computer Applications). Such logic should be capable of accommodating asymmetric severity distributions.

8.6   Quantitative Analysis

While we have claimed that qualitative analysis is usually sufficient in project risk management, we do acknowledge that, for some project risks, more sophisticated quantitative analysis may be necessary. This means that further investigation is warranted for the most severe risks revealed through earlier qualitative assessment. More often than not, formal exploration of the uncertainties associated with those risks is the required focus, and tools are available to help with this task.

In Case Study A (a public–private partnership [PPP] correctional facility), the main contractor's qualitative risk assessment was deemed sufficient to allow risk response decisions to be made unless the risk parameters were such (high uncertainty in probability or impact, high level of severity) that prudence suggested further quantitative analysis. The results then informed risk decision making outcomes, such as calculating a contingency amount to include in the bid, extending the search for a risk avoidance option, or even declining an invitation to bid for the project. On this project, the ten highest priority (i.e. most severe) threat risks for the main contractor were reassessed quantitatively.

As noted in this chapter, since quantitative risk analysis involves mathematical analysis, risk severity can be calculated as the product of likelihood and impact and used for valid relative comparisons between risks. This is not possible with qualitative risk analysis. We cannot claim that an 'Extreme' project risk is x times more severe than a 'High' risk, only that in our judgement one is more (or less) severe than the other for our organisation on this project.

However, the benefits of valid risk severity comparisons derived from quantitative approaches do not come without cost. Several issues must be considered and dealt with, including:

- Availability and sufficiency of essential input data
- Nature, accuracy, and reliability of input data
- Reliability of the analytical and modelling techniques involved
- Competent application of techniques
- Nature, accuracy, and reliability of output data
- Competent interpretation of outcomes
- Time availability
- Costs of analysis.

Quantitative approaches to investigating uncertainty in project risks generally use mathematically based statistical analyses of historic data (updated in some way if necessary) and may involve simulation. At a simple level, the analysis might comprise sensitivity testing of the critical variable values: changing the value of one variable by one incremental step at a time and observing the effect on the outcome. The limitation of sensitivity testing is that it is not suitable for exploring the effect of multiple changes in several variables at the same time.

At a more sophisticated level, Bayesian analysis, fuzzy sets, Monte Carlo simulation, and other techniques may be appropriate. However, we have stated that this book was not intended to be a mathematical treatise on probability. Many texts are available that deal comprehensively with Monte Carlo and other forms of simulation and statistical analysis, and several computer software applications for such analyses are commercially available, either as independent applications or as add-on modules for popular spreadsheet, project scheduling, and project management tools.

A strong caveat is that all such analytical and simulation tools are vulnerable to the issues noted here.

If sufficient and appropriate data are not available, or cannot be collected, in order to service the input requirements of assessment tools, then the reliability of output information from such analyses is diminished and confidence in ensuing judgement and risk decision making may be compromised.

The nature and quality of the data required for quantitative risk analysis add to this dilemma. It cannot be assumed that all data will be objective; some will represent the individual or collectively gathered subjective opinions of people. This exposes such data to the uncertainties, errors, and biases of qualitative assessment noted in Section 8.2. The sufficiency and quality of historical risk data for projects rarely match that for data collected by the insurance and manufacturing industries. Sample sizes are often too small, and the data are not sufficiently homogeneous.

High levels of uncertainty associated with input data will be reflected by greater uncertainty in output data, and this too has to be acknowledged and dealt with in decision making.

The convenient availability and often 'tick-box' user-friendliness of commercially available software applications can be a dangerous trap if users are not fully competent in understanding and using them, and in being able to interpret outcomes appropriately. Technical understanding of the tool itself is obviously important, but so too are grasps of the underlying theory, the concepts of risk, and familiarity with the particular target project risk contexts.

Quantitative analysis at this level cannot be rushed simply to meet the time pressures associated with contemporary projects. It is also likely to be expensive, particularly if domain experts and specialists have to be hired and briefed to carry out the analytical work.

To demonstrate some of these issues, we return to the decision tree analysis (DTA), event tree analysis (ETA), and fault tree analysis (FTA) examples presented in Chapter 7. In each case, we have inserted data (admittedly hypothetical) that permits quantitative analysis of the three situations.

Figure 8.1 is expanded from the example of a trip to Sydney shown in Chapter 7 (Figure 7.2).

Probabilities for the likelihood of occurrence and worth values are added to each of the travel outcome nodes. The three arrival probabilities for each travel option must sum to 1.0, as no other possibilities are entertained. The expected utility value is then calculated for each arrival node as the product of probability and worth. Some comment is necessary for each of these. Firstly, however, we must admit that, for the sake of diagrammatic simplicity, the model is not complete. For each of the travel alternatives options shown in Figure 8.1, there are two decision nodes: whether to take the early departure option or the later departure option. Each of these then generates three arrival possibilities: early, on time, or delayed. However, between the departure options and arrival outcomes, we should insert another set of possibilities associated with departure. Will that travel mode depart on time, or will it be delayed? For the 'Drive' travel mode, three departure possibilities actually exist for the 'Early start' and 'Late start' options. For each of these options, Mary Jane (our DTA 'victim') might be able to leave earlier than planned, she might be unavoidably delayed and have to start her journey later, or she might leave at the planned time. For the other travel options, there is only a remote possibility that the flight or train will actually leave earlier than the scheduled time. For these two options, we have also omitted any further complications associated with Mary Jane's journey to the airport or railway station!

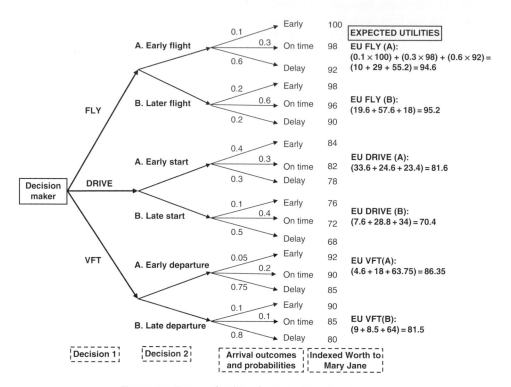

Figure 8.1 Expected utilities for DTA of travel outcomes.

For the expanded example, we have noted that the data insertions are hypothetical. How would actual data values be derived, and are the values objectively or subjectively determined?

Comprehensive airline and railway schedule compliance data are almost certainly compiled by airline and railway companies, but also almost certainly inaccessible to the general public. In order to assign the relevant probability values, Mary Jane will have to either exercise personal judgement or rely on the experience and judgement of her travel agent. She might conduct a small survey among colleagues and staff in the organisation, but would thereby encounter problems associated with sample size. Without access to authoritative objective data, Mary Jane's input data values will be largely subjective and vulnerable to all the errors and biases we have identified. When we turn to the estimates of worth, these are likely to be almost entirely subjective estimates as they pertain only to Mary Jane's personal preferences, albeit that they may be conditioned by her previous travel experiences and made with the best possible intent.

The DTA outcomes in Figure 8.1 suggest that Mary Jane's best travel option (with the highest expected utility) is to book on the Sydney flight leaving Melbourne later in the day. Her worst option is to drive to Sydney with a late departure. This is what the quantitative analytical model tells us. In real life, of course, Mary Jane's travel decision would be most heavily influenced by how much she hates flying, driving, or travelling by train, or getting up early!

Note also that the DTA example models only one consequence – uncertain arrival time in Sydney. Attempting to use DTA to quantitatively assess uncertainties associated with multiple risk impacts, such as those shown in Table 8.5, becomes highly complicated.

Next, we examine the ETA example of the potential disaster flowing from the closing failure of ferry loading doors. Figure 7.3 from Chapter 7 has also been augmented by adding probability values to the nodes. Figure 8.2 displays the results.

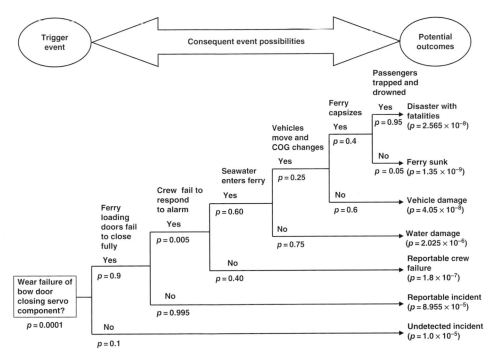

Figure 8.2 Outcome probabilities for ETA of ferry vehicle loading door incident.

After assigning a probability of occurrence to the door failure as the initiating event, each of the dichotomous 'Yes/No' consequence tree nodes of the diagram is assigned a probability, and the probability for a possible outcome is calculated from the product of all the previous probabilities that can be traced back from that branch of the tree. As with the DTA example, the assigned probabilities are hypothetical, and the same questions arise about these data.

The model for this example suggests that the chance of a ferry disaster with loss of human life and vessel, due to wear failure occurring in a component of a loading door servo-motor, is about 3 in 100 million ($0.001 \times 0.9 \times 0.005 \times 0.6 \times 0.25 \times 0.4 \times 0.95 = 0.000\,000\,025\,65$; or 2.565×10^{-8}). This probability assumes a 'domino' sequence in that the doors do not close fully, the crew then fail to respond to an automatic alarm, seawater enters the ferry, the vehicles on-board move and the vessel's centre of gravity shifts, the vessel capsizes and sinks, and passengers are trapped and drowned.

Note that there is a smaller chance (slightly more than 1 in 1 billion) that the same sequence could occur without resulting in any loss of life.

Design engineers might be able to assign an objective probability to the servo-motor component failure, but almost certainly a pessimistic subjective probability must be assigned to the next link (the loading doors do not close fully). Some measure of crew incompetence might be objectively assessed through reference to employee profiles and training records in the human resource department of the ferry company. Prevailing sea conditions and the likelihood of seawater entering the ship might be gauged from meteorological forecasts, tide tables, and analysis of the ship design; but the extent to which these will produce objective data is questionable. Vehicle movement and changes in the ship's centre of gravity could be calculated from first principles, as could the likelihood of the vessel capsizing in such circumstances. The likelihood of passengers being trapped and thereby drowning might be assessed from historical ferry disaster incidents, but almost certainly a pessimistic subjective interpretation would be given to the length of time the ferry might remain floating, the depth of the sea at this location, and the proximity of effective response help.

As noted in Chapter 7 for this example, the ETA does not explore any further upstream causal investigation of the reasons for the component failure. We can claim, however, that for analysis of circumstances such as these, the assignment of probability values (as measures of associated uncertainty) will derive from objective and subjective sources, and objective data are likely to be subjectively conditioned.

Our third example takes the FTA diagram (Figure 7.4) for the deployment failure of the aircraft emergency evacuation chute and assigns probability values to likelihood of occurrence for the various nodes on the fault tree. Figure 8.3 depicts these values (again hypothetical).

As with the ETA, sub-causal probabilities in this FTA example are multiplied in the nodal branches of the three 'trunk' possible main causes of chute failure (power failure, signal failure, and mechanical failure). Thus, the chance that chute deployment failure is due to a lubrication failure in an auxiliary power unit (APU) is about 3% ($0.3 \times 0.65 \times 0.45 \times 0.3 = 0.026\,325$). How are the node probabilities determined? Objective probability data would have to be gathered from prolonged testing under controlled conditions, possibly augmented by evidence from whatever real-life incident analyses are available. For mechanical components, failure estimates might also be derived from complex design engineering and safety factor analysis. However, these data are only objective because the context involves a highly engineered aircraft system. Other contexts might reveal a considerable reliance on subjectively determined probabilities.

The three examples (DTA, ETA, and FTA) all suggest that context plays a large part in determining the extent to which quantitative risk data are objectively or subjectively derived and how this might influence our confidence in using them for quantitative analysis of project risks. All three examples incorporate probability assessments. DTA reflects different values for decision node outcomes in terms of their worth to the company director, and these could be transformed into financial values (albeit of doubtful precision). The ETA example presents possible outcomes in ascending levels of severity, while the FTA example shows one outcome (chute deployment failure). Neither example places any financial values on the outcomes, although this could be attempted.

Expected monetary value (EMV) is an analytical technique that deals directly with financial outcomes of risk events. It is a financial version of expected utility (EU) or may provide an additional level of analysis for that tool. EMV is also known as 'expected mean value', as the next example will show.

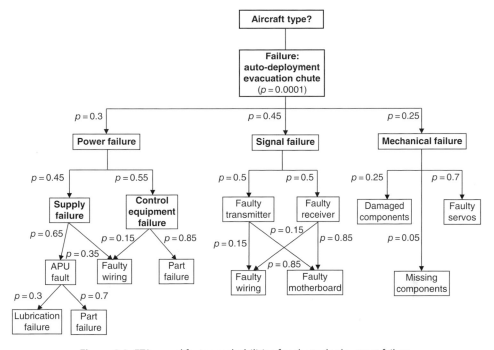

Figure 8.3 FTA causal factor probabilities for chute deployment failure.

For this example, we assume a business situation where a newly formed company is exploring a project to purchase and operate a franchise opportunity. The price of the franchise is $250 000. Before entering into a franchise agreement, however, the company examines the business case and realises that in the first year a trading loss is likely to occur due to the necessary start-up costs. The exact amount of the loss is not known with certainty, but the company has sought advice from the national association of business franchisees. This association has collected extensive data about franchise performance during the first three years of operation. Table 8.8 shows the information provided by the association for potential first-year losses on a franchise costing up to $300 000.

Table 8.8 Franchise first-year trading loss EMV.

Expected trading loss in first year of franchise	Associated probability (p)	$Loss \times p$
≤$25 000	0.55	$13 750
≤$15 000	0.25	$3 750
≤$10 000	0.15	$1 500
≤$5 000	0.05	$250
	EMV =	$19 250

From this information, the company can calculate an EMV of loss for the first year of operation of the franchise. The calculation is shown in the right-hand column of the table, where each possible loss is multiplied by its likelihood of occurrence. The EMV is the sum of the calculated losses.

The EMV suggests that the company should prudently plan for a trading loss of about $20 000 in the first year if it purchases the franchise, but 'prudence' in this context suggests an attitude towards risk that is discussed more fully in Chapter 9 (Risk Response and Treatment Options).

While the analysis in the example now deals directly with uncertainty associated with the financial impact of first-year trading operations for the franchise, the expected loss is not explained other than ascribing it to start-up costs. There might well be additional explanations (insufficient customers, bad debts, etc.) that constitute the prior risk events. In this sense, the example is too simplistic in terms of risk identification. Furthermore, the quality of the information provided by the franchisee association is not known in terms of: the sample size that the data are drawn from, representativeness (type of franchise), and 'shelf life' (contemporaneity). Inadequacies in any of these will degrade the data quality and affect organisational confidence in the EMV.

In the construction industry, EMV can be used for quantitative exploration of several project threat risks, as long as the caveat about data sufficiency and reliability is observed. The technique is used to calculate a monetary amount that should be included in the project bid price, or on more direct means of mitigating the risk. Case Studies A (a PPP correctional facility) and B (an aquatic theme park) indicate use of this risk strategy, but its application is limited by the data requirements. Dealing with weather risk is perhaps the most frequent application of EMV. The use of cranes on construction sites is, among other factors, determined by weather conditions and particularly by wind speeds and wind gusting. Access to meteorological records will indicate the probability and number of days in any season where extreme winds are likely. Costing records will indicate the average hourly cost of operating a crane on site and the costs of lost productivity per day for different types of projects. The seasonal flow of site activities for the target project will also be considered. A contractor can then bring data for all these factors into the EMV analysis to determine an amount that should be priced into its bid. The analysis will indicate minimum, maximum, and mean values, and, depending upon its risk appetite and assessment of the tender market, the contractor can choose which amount to add to its bid.

This section is not intended in any way to criticise the use of quantitative risk analysis. When used correctly, this approach is a valuable contributor to effective project risk management. As with qualitative risk assessment, our purpose has been to highlight the issues that must be resolved, and important among these is the danger of using values for uncertainty that are 'pseudo-objective' – they have the appearance of quantitative objective data but are actually subjectively derived – and then treating them as though they are completely objective data. A guiding principle to quantitative risk assessment is 'caveat emptor' ('Let the buyer beware') for both the analytical techniques and the data that service them.

While qualitative risk analysis is generally faster than quantitative analysis and can be carried out rapidly by a small experienced team, it should not be undertaken as a 'once-over lightly' exercise just to satisfy the compliance requirements of a project stakeholder organisation's risk management system and policies. Nor should it be just a means of avoiding the greater rigour of quantitative analysis. Good, appropriate, and thoughtful risk analysis, and evaluation of risk severity, substantially informs organisational decision making about how risk should be treated, monitored, and controlled.

Particular care should be taken with what are often referred to as 'long tail' threat risks. This occurs where the probability or likelihood of occurrence of the risk event is extremely low, but the period of exposure to the consequences is very long and the impact magnitude is very high. Such risks do not fit comfortably into qualitative risk assessment models and should be flagged and accorded individual attention. We will return to this dilemma in Chapter 9.

Information arising from project risk analysis and evaluation is used to populate the stakeholder organisation's project risk item record and thence to its PRR and organisational risk register (ORR) systems. A good organisational project risk management system will provide access links (e.g. on a dedicated cloud-based platform) to any quantitative data and analytical processes.

The four examples (DTA, ETA, FTA, and EMV) demonstrate that quantitative risk analysis is not for the faint-hearted! In our view, EMV is likely to be the most practical tool for project risk management, and then only if the user has sufficient confidence in the financial input data.

A useful adjunct to qualitative or quantitative risk assessment is that project risks can then be mapped.

8.7 Risk Mapping

Risk mapping is a very under-utilised tool in project risk management, yet it can yield substantial benefits, not only for the project at hand but also strategically. 'Clusters' of risk are quickly revealed, and the purpose of mapping is to discover where abnormally large or unusual clusters exist. Follow-up comparison with the risk maps of previous projects can indicate signs of emerging risk trends.

Perhaps the simplest approach is to map project risks according to severity level.

Table 8.6 showed three-point risk likelihood, impact, and severity measures. The coded risks (which are amplified in individual risk schedule items) can be mapped onto the matrix according to the severity assigned to each one. The matrix shows that the highest preponderance of threat risks occurs under 'Medium' severity. This information is useful as it could be compared with severity matrices from previous similar projects. If the comparison is found to be unfavourable, the risk item records may provide explanations, and the organisation might need to consider changes in its project management or methods of undertaking projects. If the number of 'High' risks mapped is excessive, the organisation might even want to consider abandoning that project. The larger matrix of Table 8.7 could be used in a similar way.

Another approach is to map risks according to category type. This can be done in a simple manner by recoding risk identifiers according to their risk category: for example, if Risk R1 is a weather-related risk, it may be recoded as R1-W, and so on. Should the cluster of weather risks prove substantially and unusually large for the project, thought can then be given to a cluster mitigation response, such as using an innovative temporary canopy for a construction project.

Ideally, this approach should be combined in a matrix format with some of the risk identification tools dealt with in Chapter 7, such as the 'listing' types represented by Tables 7.8–7.11. The ultimate combination matrix format is probably that shown in Table 7.13.

Mapping according to risk type category explores risk source events. While mapping risks in terms of likelihood of occurrence of the risk event is unlikely to provide much useful benefit, risk impact mapping can prove to be very useful. For instance, mapping risks that have large environmental consequences (see Table 8.5) may give insight into the environmental 'footprint' of the project and trigger consideration of alternative project technologies.

A unique form of comparative risk mapping that uses the graphical charting capacity of modern spreadsheet software applications is found in 'spider charts'. Table 8.9 shows hypothetical severity assessments for 10 different threat risks that reoccur across three project alternatives.

Table 8.9 Comparative project risk severity assessments.

	Risk 1	Risk 2	Risk 3	Risk 4	Risk 5	Risk 6	Risk 7	Risk 8	Risk 9	Risk 10
Project A	100	90	80	70	60	50	40	30	20	10
Project B	85	85	60	80	40	65	30	70	45	50
Project C	60	95	110	40	75	35	60	50	60	30

Considerable care is needed for this type of risk mapping. While the continuous data available from quantitative risk analysis are suitable for charting, ordinal data from qualitative analysis cannot be used for this purpose unless the ordinal scales are arranged so that ascending step progressions between scale intervals are equal and similar scales are used for all risks and all projects. Inevitably, some form of objective indexation should be applied to the risk severity values, so as to form a common relative basis for comparisons. In this example, the 10 risks for Project A are indexed from the severest of the 10 (A1 = 100), and the similar risks for the alternative projects (B and C) are then rated in comparison to A1.

Figure 8.4 displays the ensuing spider chart for the projects and their risks. The visual risk 'footprints' for the three projects suggest that Project B and Project C are each probably 'riskier' than Project A, but overall differences might be more difficult to detect between B and C. Risk C3 looks abnormal and may warrant further examination for this project alternative. Other project risk comparisons deserve similar attention.

Charting comparisons are not limited to alternative projects. They can be prepared for different parts of a project where similar risks (but with differing severities) tend to arise.

Experimentation with various charting formats, and with data required for them, could yield valuable risk knowledge for a project stakeholder organisation. Our trials suggest that up to 20 project risks can be charted in this format. Beyond that, the chart becomes too cluttered and illegible, but we leave it to the creative ingenuity of our readers to explore this technique further!

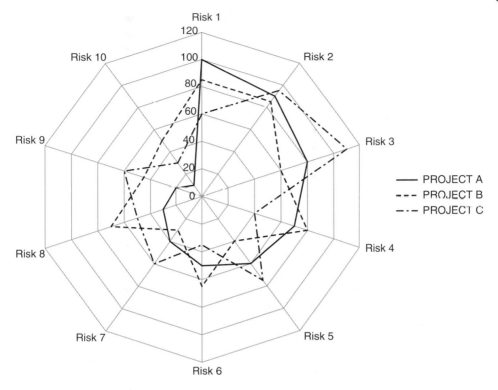

Figure 8.4 Risk severity spider chart.

8.8 Summary

The outcome of the risk analysis and evaluation processes should be a list of *all* the identified threat risks that the stakeholder faces on the project. Each risk is analysed, either qualitatively or quantitatively (or both), in terms of the uncertainty associated with the likelihood of occurrence for the risk event and the consequences that will flow from it. From these analyses, threat risks are then evaluated in terms of their severity to the project stakeholder organisation that faces them. For qualitatively analysed risks, subjective linguistic interval labels are applied to an acceptable meas-ure. For quantitatively analysed risks, objective mathematical values of severity can be calculated.

If subjective linguistic measures are adopted for qualitative risk analysis and evalua-tion, it is important to remember that it is your 'model' measuring your risks, and you must be comfortable using it as part of your risk management system. Establishing the measures and defining intervals are the responsibility of senior management in an organisation and should be done through careful consideration and with an acceptable logical underpinning. In the contemporary world of projects, always be mindful that your organisation might be called upon, in an official enquiry, to explain and justify its risk assessment methods.

Similarly, 'long tail' risks will almost certainly warrant separate attention.

Risk identification is often claimed to be more important than risk analysis for the purposes of project risk management, since you cannot proactively manage unidentified risks and it may be obvious that some risks must be addressed regardless of their likelihood and impact characteristics. Despite this, risk analysis is always worthwhile as, like the earlier stages in the project risk management cycle, it raises the level of risk awareness among project participants.

The uniqueness necessarily associated with any organisation's approach to assessing its risks supports our ongoing argument that a universal project risk management system (i.e. one shared by all project stakeholders) is impractical.

Finally, the usefulness of risk mapping should not be neglected. This activity does not necessarily have to be carried out in parallel with risk analysis, nor at the same time, but such mapping may provide further risk insights for a project stakeholder organisation, and the results can influence decision making about risk treatment options and responses. This is the topic of Chapter 9.

References

Australian Standard/New Zealand Standard (AS/NZS) (2004). *Risk Management Guidelines – Companion to AS/NZS 4360* (HB 436). Homebush, NSW: Standards Australia.

International Organisation for Standardisation (ISO) (2018). *Risk Management: Principles and Guidelines* (ISO 31000). Geneva: International Organisation for Standardisation (ISO).

Standards Australia/Standards New Zealand (SA/SNZ) (2013). *Risk Management – Guidelines on Risk Assessment Techniques* (SA/SNZ HB 89). Homebush, NSW: Standards Australia.

9

Risk Response and Treatment Options

9.1   Introduction

Up to this point in the book, we have first focused upon the theoretical and conceptual underpinnings of risk, and then upon preparatory requirements and activities for the early stages of proactive project risk management. Now, as a project itself moves forward from the planning and design stage towards an actual delivery or implementation process, so too does exploration of the nature and magnitude of perceived project risks turn towards deciding what should be done about them.

We have claimed that risk is a psychosocial construct – it is experienced by people as individuals and as societies. This means that each of us will have thoughts and feelings about risk, and especially about the risks we face in our own lives. Our risk attitudes (and by extension those of organisations and societies) will influence our decisions about risk. Fundamentally, this is a matter for behavioural psychology, and we can only dip briefly into this vast field.

By definition, project risk management seeks to deal with the effect of uncertainties upon project objectives, and particularly those associated with decision making. However, it is important to appreciate that uncertainties associated with risks will also be found in the treatment decisions made about them. Responding to risks is thus a risky business!

Any response to, and treatment of, a project risk is likely to incur costs to the project stakeholder organisation at some point. For all risks, whether they represent threats or opportunities for the risk taker, 'trade-off' situations will arise. The cost of treating a risk must be balanced against the financial magnitude of any consequences of the risk event. How much should be spent to mitigate the loss arising from theft of materials from a construction site? How much would it be worth spending to fully exploit the potential benefit of an opportunity risk? It is not our intention to examine these 'trade-offs' in detail in this chapter, but we will endeavour to point out where decision uncertainties that could influence them may lie in risk treatment.

There are basically four types of response to threat risks: avoidance, transfer, reduce and retain residual risk, and retention. Responses to opportunity risks are obviously different and are discussed in Chapter 15. Given these four types of response, actual threat risk treatment options may be as plentiful as the capacity of project stakeholders to devise them. This stage in the project risk management cycle (Figure 4.2, Stage D)

Managing Project Risks, First Edition. Peter J Edwards, Paulo Vaz Serra and Michael Edwards.
© 2020 John Wiley & Sons Ltd. Published 2020 by John Wiley & Sons Ltd.

can be quite creative, certainly for those who enjoy problem solving! We offer some treatment principles as a guide.

The strategic risk management policies of an organisation may provide direction as to which project risk responses are preferred and should be adopted.

Once risk responses and treatment options have been explored, those involving some form of treatment should be re-analysed on a post-treatment basis, hence the dotted feedback arrow shown in Figure 4.2. The reassessment not only acts as an indicator of whether or not the treatment is likely to be effective, but also determines if the revised severity of the treated risk now meets the risk policy (hurdle) requirements of the project stakeholder.

Risk response and treatment decisions implicitly signal a point of commitment to a project for the stakeholder organisation. This is not to say that the stakeholder can no longer 'walk away' from the project, but that the decisions represent an intention to expend real time, effort, and resources on managing project uncertainties. The organisation should now feel able to go ahead with its involvement in the project, confident in the knowledge that it has considered as many risk eventualities as possible and is prepared to deal with them should they arise.

9.2 Risk Attitudes and Appetites

Project risk response and treatment options are influenced by the risk profile or risk attitude of the decision maker. These attitudes and profiles are culturally, socially, and psychologically conditioned. This conditioning applies to individuals and organisations and, to some extent, societies.

In organisations, decision making is carried out by people on behalf of the organisation. Where this power is legitimately and expressly authorised and exercised, the risk profile of the organisation will match those of its key decision makers (and vice versa). Problems in this regard (e.g. an organisation becoming overexposed to project threat risks) can, and often do, arise from unauthorised assertion or misuse of decision making power. Responding to this 'risk about risks' means that every project stakeholder should clearly demarcate the lines of authority and limits of decision power in its organisational structure, and ensure effective communication, understanding, and acceptance of these lines and limits across the organisation. This may be easier said than done, and we refer readers forward to some of the dilemmas of organisation culture discussed in Chapter 13.

Our knowledge about individual risk attitudes derives largely from behavioural psychology and gaming theory. Suffice to say here that research tells us that in theory, people exposed to decisions about the uncertain outcomes of alternative choices may be risk averse, risk seeking, or risk neutral.

We emphasise 'in theory' as, in our view, people are rarely risk neutral in making real-life decisions. Yet neutrality is claimed in the experiments conducted to elicit risk attitude. The experimental techniques tend to offer participants limited and uncertain choices about outcomes of proposed wagers (hence the association with gaming theory). Other approaches adopt a different frame using contexts associated with utility theory. In most trials, participants are asked to respond iteratively to similar questions but with different probability values for the likelihood of occurrence (including certainty in some instances) and different outcome magnitudes. Individuals who opt for a higher outcome that has a lower probability of occurrence are said to be risk seeking (willing or even eager to accept the gamble). Those who opt for a lower outcome but with a higher (or even certain) likelihood of occurrence are said to be risk averse (conservative in their decisions). Participants exhibiting indifference to the choices (either throughout or at some point in the trial) are deemed to be 'risk neutral'. Our contention, however, is that people do not normally 'sit on the fence' in this way. They are more likely just to switch from one state to the other as the circumstances (probability and outcome) change and they feel comfortable (or uncomfortable) at that point. This is not really a state of neutral indifference, but is more akin to ambivalence or even disinterest. Neutrality in this context is rather like saying that a chair is neither comfortable nor uncomfortable. It is one or other, but we do acknowledge that the degree of comfort associated with the chair might vary according to the perceptions of the sitter.

In the same way, while risk attitude experiments may show shifts from risk seeking to risk aversion occurring at different points for different people, we cannot say that, at a particular point, Person A is twice as risk seeking (or risk averse) as Person B. We can only use terms like 'more' or 'less' without asserting the absolute degree of either.

Two other problems arise with behavioural psychology experiments to determine individual and group risk attitudes. One is the representativeness of the participant sample, usually a convenience sample drawn from cohorts of university students studying psychology courses. We rarely know if such tests have been controlled for variables such as age, timing of administration in the university calendar, the stress levels of participants, and 'maverick' responses from bored or disgruntled students. Another problem is controlling for reality in the probability value variations and particularly in the magnitude of the monetary amounts at risk. Taking either of these factors too far from the reality expectations of trial participants could lead to responses that signify indifference rather than genuine choice. We are beginning to learn much about risk attitudes and appetites, but there is still much more that we do not yet know.

To complicate matters further, risk profiles are not necessarily consistent across different risks, nor over time or different contexts. While this may be generally more applicable to individuals than to organisations, the latter are by no means excluded in this regard. We do know that people in professional disciplines tend to be risk averse in their professional lives, but they may be risk seeking, or just continue to be risk averse, in their private lives. An engineer may practice a cautious and conservative approach to a structural design, but be anxious to visit the local casino on Friday evening. An accountant may be prudent with clients' accounts and with his own financial affairs, but prone to engage in extreme sports. A surgeon may be extremely risk averse at the hospital operating table, be untroubled by pricking her thumb on a rose thorn during weekend gardening, and be looking forward eagerly to a day out at a prominent horse racing event!

Organisations may show similar variability, but be less mercurial and more long-term in their attitude shifts. Changes in governance, structure, and personnel can influence organisational risk profiles, especially in terms of appetite for risk. Adverse project 'experience' may sometimes lead to greater unwillingness to accept threat risks or exploit opportunities. Conversely, 'familiarity' may lead to greater willingness to take on more risks, especially for similar types of projects. Spectacular success almost inevitably leads to increased risk appetite. Organisations may become increasingly comfortable with some types of risk than with others. Keywords in this risk aversion/risk seeking dichotomy and dilemma are 'progress' and 'change', and organisational attitudes to either will influence profile shifts.

An important consideration for individuals and organisations in project risk management lies in decision *heuristics*. These are the 'rules of thumb' that we all tend to rely upon in our decision making. They are often developed over years of experience and, more often than not, bring real value in application. However, they are not infallible, and two common heuristic biases should be treated with caution: availability and representativeness (Kahneman *et al.* 1982).

The 'availability bias' occurs when undue influence is exerted by an experience that is personal or quite close in time. Thus, someone who has been involved in, or witness to, a 'falling from heights' accident will tend to assign a greater but unwarranted probability to an identified safety risk on a new project. The bias acts to skew judgement in either direction, but for threat risks an optimistic availability heuristic has more serious implications.

A 'representative bias' skews risk decision making when undue reliance is placed upon historic information without first checking if it is still appropriate. For example, a project contractor might have data about the delivery performance of three different suppliers on previous projects: Supplier A has delivered on time for three out of five projects, Supplier B's on-time record is six out of eight records, while Supplier C was late on two out of four projects. Before awarding the next supply contract to B, however, it would be prudent to investigate if any change has taken place in the business situations of the three competitors. Perhaps the efficient warehouse manager in Supplier B has left the company after prolonged dissatisfaction with pay and conditions, and has now joined Supplier C?

These biases are insidious because they are associated with common, easily applied heuristics that tend to save us much time in our decision making.

Slovic (2000, 2010) has long explored perceptions of risk and is a seminal researcher in this area. Latterly he has focused upon an *affect* (feelings) heuristic in risk decision making, whereby risk judgement is influenced by our positive and negative feelings about each risk (see also Chapter 2, Section 2.6). An emotional perspective, such as the one this heuristic introduces, may be difficult to accept in the largely unemotional climate we tend to prefer for most projects, but careful thought suggests that it is still valid. We often tend to 'feel good' or 'feel bad' about particular project risks, and these feelings are real. Slovic (2010) notes that the feelings may or may not be accompanied by complete awareness of the situation, hence our discussion in the value of the systematic cycle of risk management in raising awareness of project risks. He also cautions that we should integrate those feelings with technical analysis, not only to support or reject them but also to make our risk decision making and communication more effective. We would also advocate the converse: separating our feelings about risk from technical analysis in order to better understand the nature of the risk and to avoid the 'black box' trap that can accompany quantitative risk analysis. To critics who might claim that the *affect* heuristic cannot apply to project stakeholder organisations, we point to our argument earlier in this chapter that the risk decision making of organisations reflects the risk profiles of the relevant decision makers in it.

At the highest management level, an organisation should periodically make 'temperature checks' of its risk appetite and profile. The preliminary question might be: *Has anything taken place, or is anything taking place, in this organisation that could influence the way in which we respond to, or treat, the project risks we are likely to face?* The check could be extended to deal with heuristics.

Ideally, answers to this exercise should then be tested against known risk parameters from previous projects: *If we were to face this risk again on a future project, would our response and treatment decisions then be any different?*

The first question could be posed as part of the project contextualising process described in Chapter 5, but the periodic 'temperature check' ensures that senior management become involved.

Similar questions could be asked of themselves by individuals in an organisation, and sole practitioners should also interrogate their risk profiles on a regular basis.

9.3   Existing Risk Controls

Before proceeding too far with deciding how to respond to project threat risks, the risk management team should consider what risk controls may already be in place to deal with them. The second part of the project risk register template (Table 4.2) alludes to this.

Existing risk controls may be found in many areas of an organisation, although they are not necessarily identified as such. 'Bow-tie' analysis (See Chapter 18, Sub-section 18.2.1) is a diagrammatic tool popular in engineering industries to explore control systems and their effect on risk events and impacts.

A project stakeholder organisation with a mature approach to risk management will have recourse to the knowledge captured from previous projects (see Chapter 11). The archived risk registers for those projects (or the organisational risk register) will provide substantial information and guidance about existing controls.

Policy documents, procedural manuals, and clauses in contract agreements are rich sources for further information. Usually these are located in appropriate departments in the organisation to deal with matters such as:

- Finance
- Occupational health and safety
- Quality management
- Information and computer technology management
- Human resource management
- Environmental management
- Customer and public relations
- Sales and marketing
- Legal affairs.

However, while these sources should satisfy most of the guidance needs for treating and responding to project threat risks, they are not infallible. Nor are they always sufficient.

Using the greater risk awareness they have developed for the project at hand, workshop team members should consider what gaps there might be in existing controls and what additional treatment might be necessary or desirable. The nature of the risks identified for the new project might justify different responses and different treatments from those employed on previous projects. Organisational policies and procedures may have changed. New forms of contract may be in place. Regulations may have been amended.

Reconsidering existing risk controls and identifying gaps are themselves important contributors to raising project risk awareness. They also help to reduce some of the uncertainty associated with risk decision making.

9.4 Risk Response Options

As we noted in Section 9.1, four types of response to project threat risks are encountered: avoidance, transfer, reduce and retain residual risk, and retention. Combinations of transfer and reduction are possible. Risks are rarely eliminated. Even where this is possible, they are almost invariably replaced by other risks and, should it become necessary to revert to the former project situation, the original risks will inevitably return.

9.4.1 Risk Avoidance

Avoiding a risk means deliberately taking another course of action so that it cannot arise in the new circumstances.

From a project perspective, the ultimate form of risk avoidance is not to proceed further with the project. However, abandoning a project is seldom a response to one project risk factor in isolation. Such a decision should only be made after the more important risk issues have been reconsidered and their overall influence on the project assessed.

For example, an investment project opportunity is not usually rejected simply because it exhibits a potential rate of return which is not commensurate with the investor's criterion rate, but rather because sufficient improvement in performance cannot readily be achieved in the major factors that contribute to the rate of return calculation. This is typically the case with a commercial building development project. The internal rate of return (IRR) for such projects is often used as a measure of the riskiness of the project that facilitates comparison with other investment options. IRR is a form of discounted cash flow (DCF) modelling and, for the development project, is calculated from actual or forecast cash flows occurring on the project. Negative cash flows derive from factors such as:

- Land costs
- Land transfer costs
- Land treatment and/or demolition costs
- Legal fees
- Loan finance costs
- Costs associated with planning permissions and building permits
- Design costs
- Procurement costs (tender costs)
- Construction progress payments
- Variation costs
- Completion, commissioning, and opening costs
- Refinancing costs
- Marketing and leasing costs

- Tenancy incentives costs
- Utilities costs
- Rates and taxes
- Loan repayments
- Repair and maintenance costs
- Equipment replacement costs
- Refurbishment costs
- Sale or disposal costs (including demolition and land remediation costs if necessary)
- Loan redemption costs.

The list is not exhaustive.

Some of these items will be represented by one-off payments; others will occur periodically (regularly or irregularly) through the whole or parts of the life-cycle horizon envisaged for the investment, from inception to disposal or demolition and redevelopment. This may be anything from 15 to 50 years and beyond.

On the other side of the financial feasibility model, positive cash flows will include income from tenant rental payments and other sources such as fees for advertising space granted in or outside the completed building and rental from cellular communications companies wishing to erect aerials on the roof.

Essentially, the DCF model attempts to reflect the effect of time on each of the net annual cash flows occurring throughout the project life cycle. For IRR, iterative trial and error calculations determine the rate that discounts all net annual cash flows to a zero sum, and this rate is claimed to represent the true yield on the investment.

The IRR for a commercial building development is most sensitive to income flows, operating costs over the investment period, and the discount rate selected for modelling the IRR. The capital costs of land and construction, while representing large amounts, are diminished in significance by the effects of discounting and the relatively far longer period of income generation.

Few of the investment model factors are known with certainty, and the uncertainties associated with them, or at least with those that have significant influence on the model, render each of them 'risk-laden'. Each therefore warrants deeper investigation and risk assessment before the IRR alone is used as the risk avoidance basis for abandoning a project.

IRR itself is not a completely fool-proof technique, and abnormal results can occur under certain conditions. The topic of IRR, however, is beyond the remit of this book, and we recommend that readers interested in this area of financial risk management should seek more appropriate texts.

In exploring avoidance as a response to particular risks, a stakeholder organisation should always bear in mind that this response might thereby increase the severity of other risks, introduce new risks, or result in lost opportunity. An example serves to illustrate this:

A construction project, such as a multi-storey residential apartment building, has been designed with a basement parking level. However, the site has a history of occasional flooding following periods of exceptionally heavy rain. The risk is the likelihood that a heavy rainstorm will occur and flood the basement, causing damage to the vehicles parked there, damage to electrical and other services at that level, and inconvenience to the residents.

In this case, a risk avoidance treatment might be to redesign the building to omit the basement level and replace it with ground level and above-ground level parking. The risk is thus avoided, since there is no longer a basement to flood. Instead, however, the risk of flooding occurring at ground level is now increased – although the impact of this might not be as serious as that for the basement. Furthermore, such redesign will inevitably alter the external visual effect of the façades of the completed building. Changes to the entrance design and parking access routes will also have to be made. The original ground floor apartments might now lose any individual garden areas. All these changes might be enough to adversely affect the selling or letting market opportunities for the residential units (ugly car parking is now visible for one or more levels above the ground). Finally, of course, the cost of making the design changes (in terms of the design costs, construction costs, and any approval or delay costs incurred) has to be weighed against the potential impact costs of the risk that the project owners intended to avoid.

Risk avoidance is rarely a straightforward or easy response option. Because it is highly effective in the right circumstances, however, avoidance and its treatment should always be considered first among the possible options available to a stakeholder. Depending upon its perceived effectiveness, risk management workshop time might then be spent in creatively exploring what other responses and treatments are available.

9.4.2 Risk Transfer

In electing to transfer risk, a project organisation is seeking to shift the potential burden of a particular risk to another party or parties who may be directly or indirectly involved with the project.

This is a common response in project situations where stakeholder supply chains or networks are easily distinguished, often under contractual agreements. Attempts will be made to transfer risks progressively along the supply chain or to the more distant extremities of the network. For example, a project client will transfer risks to the head contractor, who in turn will transfer them to subcontractors or suppliers.

The mechanisms typically used to transfer risks in such situations include, inter alia, head contract agreement clauses, subcontracts, and supplier agreements.

Risk transfer often forms the distinguishing characteristic between many alternative forms of project or service procurement and delivery, such as joint ventures, public–private partnerships, franchises, and agencies. In the construction industry, it is noticeable in building procurement systems such as design and build (D & B), design-build-finance-operate (DBFO), build-own-operate (BOO), build-own-operate-transfer (BOOT), and many others. Common to most of these is the intention to transfer risk away from the client stakeholder and towards the contractor.

Rarely in such 'cascading' transfer arrangements is any strong observance found of the principle of allocating a risk to the party best able to manage it. To mix metaphors, a party at the end of some contractual supply chains is often the one that is also at the bottom of the food chain – highly vulnerable to threat risks and without sufficient capacity to deal with them. Nevertheless, explicit risk transfers through contract agreements are a legitimate risk response.

Insurance is a transfer mechanism for risks that are insurable such as theft, injury, damage to property or equipment, and some third-party liabilities. This introduces another stakeholder to the project, albeit indirectly. Performance bonds and payment guarantees are used in a similar way to deal with the impacts of the threat risk of default by parties in the execution of project contracts and agreements.

Two things should be noted about risk transfer: it is usually expensive, and it is rarely 100% effective.

Every risk transfer has a cost. This is reflected in the price the transferee charges for accepting the risk, such as the premium payable to an insurance company, the fee set by a bonding agent, or the commission charged by a bank when offering a guarantee. Other than in client/contractor/subcontractor chains, the transferee usually has greater power to determine the conditions for transfer.

The direct costs of risk transfer may be obvious. Indirect costs can occur when, for example, a risk is passed to a project subcontractor who does not manage it effectively and is then affected by the impact of a risk event. If the subcontractor survives the risk impact, but cannot recoup its losses from the main contractor, it will attempt to do this on other future projects. If the subcontractor does not survive the risk impact, the losses are eventually borne by the relevant industry and finally by society itself.

Few risk transfer mechanisms completely discharge the transferor from all responsibility associated with risks. One cannot insure possessions against theft, neglect to take any precautions against theft occurring, and then still expect to receive full recompense from the insurer if something valuable is stolen. In a similar way, a client stakeholder might transfer the risks of accident and injury involving project workers or third parties, or non-compliance with environmental regulations, to the project contractor. However, should a serious accident occur, or if a major 'green' issue arises on the project, the client stakeholder will inevitably become involved, simply by reason of its association with the project. While the direct impacts of any subsequent risk incident may be borne by the contractor, the client organisation is unlikely to escape other reputational risk impacts, such as critical media attention. It might also find itself (even in other areas of its activities) the target of unwelcome public protest demonstrations. Contemporary views of corporate ethics and corporate social responsibility mean that organisations cannot absolve themselves completely from such risks simply by transferring them to other parties. Transfer of liability for the consequences of risk events may be nominally achieved, but accountability often remains in some degree, not least in terms of reputational damage.

Case Study C (an aid-funded civic project for an island nation) provides an interesting twist on risk transfer by insurance. Most professional consultants in the construction industry deal with their professional liability risks by taking out professional indemnity (PI) insurance. Other professions (e.g. medicine and law) adopt a similar approach. It is usually regarded as a necessary business expense. The professional consultant in Case Study C takes a different view, informed by two perceptions. The first is that PI insurance is highly expensive and that policies contain far too many exceptions. The second perception relates to his actual role in aid-funded projects. He is appointed as a senior advisor to the donor government, and his role is exactly described by the title. He also advises the recipient government. His is not a decision making role per se, but one of guarding each government's interests through the timely provision of advice. In essence, this role aligns more with the Latin tag '*Quis custodiet ipsos custodes*?': Who is to guard the guardians? It relates to the decision making and supervisory activities of other project stakeholders, notably the design and project management team and the contractor in this case. As far as possible, the senior advisor seeks to facilitate the recipient government's interests in the project, while also protecting the donor government from poor decision making and project outcomes that might reflect badly on its regional interests.

The relevance of this case study here is that insurance is not always an effective risk transfer mechanism. This was recognised by the donor government, which was able to distinguish between the implications of discrete design and supervision decision making (which could directly affect project success in the short and long terms) and those of advice relating to its interests. PI insurance may deal with the former, but not fully with the latter. In this case, the donor government seeks to lessen the uncertainty associated with project decision making by acquiring additional professional advice, but does not further insist upon PI insurance to manage the risk of possibly receiving mistaken advice. At some point, the transfer chain has to end!

9.4.3 Risk Reduction and Retention

No risk should be avoided, transferred, or retained without first checking to see if it is possible to first reduce it to a more acceptable level and then retain the residual risk.

For anyone with a professional interest in risk management, reducing risk (risk mitigation) is probably the most absorbing area of involvement. The stakeholder is deliberately attempting to lower the risk severity in some way. All dimensions of the risk should be examined, since it may be possible to reduce the probability of occurrence, the impact consequences, or the duration of exposure to the risk. Combinations of any of these may be contemplated in some risk circumstances. The nature of the risk will influence which of the risk factors can be mitigated, and the context will influence the possible extent of mitigation. It is here that the true value of precise risk statements is observed, since it is likely that the statement itself will provide clues about possible mitigation treatments.

The risk reduction exploration process also allows the value of a risk classification system to be exploited. Whenever a new project risk is encountered, it is possible that an effective treatment can be found among more familiar risks (i.e. from the organisational risk register) in the same category. Risk mapping will also reveal where clusters of similar risks lie. Treatments that are uneconomic for a single risk may become worthwhile for a cluster of similar risks.

Inevitably, the process of exploring risk reduction requires a return to the analytical processes of risk assessment. Care must be taken in doing this. There is now a natural inclination to try to replace earlier subjective qualitative analysis with techniques that are more quantitative. Provided adequate and sufficiently reliable data are available or readily accessible, this is acceptable. If better data are not available, then it is better to continue with a qualitative approach since by now the subjective judgements involved should be better informed about each risk.

Risk reduction exploration may continue iteratively until the point is reached where the residual amount of a risk is acceptable (in terms of a strategically determined level) and can be retained by the project stakeholder organisation. This means that more than one return to the risk analysis stage may be required.

9.4.4 Risk Retention

A decision to retain a threat risk without any mitigating action should only be made upon an informed basis, after appropriate investigation has been carried out and it is found that any treatment is unlikely to be effective or delivers a negative benefit–cost ratio.

Beyond this, of course, it must be conceded that there are likely to be risks unwittingly retained by the organisation because they have not been identified.

In some instances, it is possible for a stakeholder organisation to reward itself for retaining a risk. This presupposes that the risk identification and assessment processes of risk management have been implemented, at least to some extent.

Three examples are offered to facilitate better understanding of the 'risk/reward' approach to risk retention:

- An excavation subcontractor tendering for a construction project retains the risk of stormwater flooding the excavations but rewards itself by increasing the unit price rate for the work.
- A developer increases the 'hurdle' or required rate of financial return for a project as a reward for retaining a variety of risks, such as the economic risk of bringing the completed project to market in an unstable commercial climate.
- A financier increases its loan interest rate as a reward for retaining the risk of payment default on a loan.

In each situation, the risk itself may be unchanged in terms of probability, impact, and duration. The risk response option is actually closer to that of risk reduction and retention. In these cases, however, the reward is not contingent upon the risk event actually happening. The risk taker's reward is certain (if the project goes ahead) and thus, in terms of our earlier definition, technically these are no longer risks!

The 'risk and reward' approach, however, must be carefully balanced as it introduces its own additional risks. The excavation subcontractor might find that its bid is unsuccessful because competitors have submitted lower prices. The developer might find competing projects selling (or letting) at lower prices as their developers are satisfied with smaller returns. The financier might find its share of the loan finance market diminishes as borrowers seek alternative loan sources at lower rates. While the principle of risk retention, therefore, is that it should be done on an informed basis, the principle of seeking certain compensatory reward should always be on the basis that the stakeholder is not thereby exposed to other, more severe, risks.

9.4.5 Combination Responses to Risk

As noted in this chapter, combinations of risk transfer and risk reduction and retention are possible.

The most common example of combination risk responses is the transfer of risk through insurance, while at the same time retaining a share in the potential impact by accepting liability for a fixed excess sum in the insurance policy agreement.

A slight variation on the insurance example is where the risk transferee (in this case, the insurance company) requires the transferor to carry out mitigating action as well as imposing an excess liability. For burglary insurance, an insurance company might require the policy holder to fit special window and door locks, and install an alarm system. These precautions reduce the probability that unlawful entry will occur. Risk transfer and partly shared retention are therefore combined with risk reduction.

Another example can be found in the 'target cost' variation of the 'cost-plus' type of procurement system sometimes used for construction projects. In such projects, the risk impact of project cost overrun is sometimes contractually shared between client and contractor, as an incentive for the client to avoid making scope changes and for the contractor to work efficiently. Contract clauses will make the sharing arrangement explicit (e.g. 50/50 or some other proportional ratio). Coincidentally, similar clauses might deal with the sharing of cost *savings* on the project, but this probably falls into the category of opportunity risk – although the contractor's potential loss of anticipated profit on any cost saving might be regarded as a threat risk.

9.5   Risk Treatment Options

In seeking to reduce or mitigate a threat risk, the decision maker plans deliberate action to reduce one or more components of the risk and then retains the residual risk.

If it is decided that the response to a risk should involve some treatment of the risk (i.e. risk reduction), what might such treatment options comprise? Deliberate risk mitigation treatment aims to:

- Lower the probability (likelihood of occurrence) of the threat risk event.
- Lower the impact (consequence) of the event should it occur.
- Shorten the exposure period (duration) to the risk event or its consequences.
- Carry out some combination of any or all of these.

We offer some project examples:

A) A construction site manager instructs scaffold fixers to erect additional horizontal guard rails and higher toe boards (the low vertical boards along the external edges of scaffolding platforms) on scaffolding to upper floors, to reduce the chance of 'falling from height' accidents. This reduces the likelihood that something or someone will fall from the scaffolding. It is a treatment that may go beyond existing legal safety compliance requirements. However, if a 'falling' incident should occur (person or object), the consequences are not thereby reduced. Suitably strong catch nets strung from the scaffolding would be one way of reducing the impact of an accident, but note too that this alone would not minimise the chance of the accident happening. Note that none of the additional precautions addresses the root *causes* of scaffolding accidents. Better safety training and safety motivation for workers are the only things likely to do this.

B) An electrical engineer specifies more individual power circuits, and circuit breakers with lower kilovolt-amp (KvA) breaking capacities, to protect valuable equipment in a laboratory project. Here, the uncertainty associated with the potential damaging impact of a power surge or 'spike' event is reduced, together with the duration of the impact (the breakers will rupture more quickly). The probability of the trigger event is not reduced. Note that there will almost certainly be a causal event for the power surge, but this may lie beyond the project system boundary, and the power surge is therefore regarded as the trigger event.

C) For a theatre performance project, the director will appoint understudies for the leading roles. Doing this does not reduce the likelihood that a leading actor will become indisposed for a particular performance (although that may not be strictly true, since the motivating effects of having someone following closely in your foot-steps are well known in the performing arts!). The extra rehearsal payments to understudies, plus the additional performance payments if an understudy does have to step into the role, plus the costs of advising patrons and offering refunds to a few disgruntled ticket holders are the risk premiums (treatment costs) for averting the possibility of having to cancel performances completely and thereby losing signifi-cant ticket sales revenue.

D) In a book publishing project, the publisher might decide to increase the frequency and range of marketing activities during a given period for the book. In doing so, the publisher is trying to reduce not only the chance of a publishing failure, but more specifically the time during which the financial return on his or her investment will be at risk. The additional marketing and publicity activities are intended to generate more sales more quickly, so that the sales break-even point can be reached earlier. The trade-off is the extra cost of the additional marketing activities.

E) In another construction example, it is sometimes necessary to disturb or remove the foundations of an existing building, in order to deepen them or replace them with new foundations (perhaps for a new adjoining building). This operation is known as underpinning and is regarded as a risky process since it involves the tem-porary removal of structural support from the existing building. Where the existing building is situated on an adjacent property title with different ownership, the adjoining owner's entitlement to joint structural support may substantially increase the severity of the risk. A common risk reduction treatment in this case is to carry out the work in small sections at a time (each perhaps no greater than 1 or 2 metres in length), with each one executed in a strictly planned sequence so that no two adjoining sections in the linear length of the existing foundation are exposed or removed at the same time. This underpinning technique reduces the likelihood that a foundation collapse will occur, and the addition of temporary lateral support (such as angle shoring) will reduce that likelihood even further.

Admittedly, none of these examples demonstrates a risk response treatment that is particularly innovative, and we are sure that readers could do much better with exam-ples from their own project environments! Useful project risk management practice might be gained if you were to step outside of your own comfort zone with a few col-leagues and brainstorm alternative treatments for these five cases.

Perhaps the important learning from these examples is that the nature of the risk itself usually provides guidance for choosing the best treatment alternative. The mitigation treatment is intentionally planned and not haphazard or coincidental. The examples may also allow us to detect some principles for risk reduction.

9.6 Risk Mitigation Principles

A principle detectable from these examples is that reducing the *likelihood* of occurrence of a risk event generally means doing things better or doing them differently. Value management techniques can be useful in these investigations, since these are intended to explore alternative solutions to delivering required functions.

Attempts by stakeholder organisations to reduce the likelihood (chance/probability) of occurrence of a threat risk event on a project tend to focus upon implementing or increasing the effectiveness of other management techniques. Measures might include: more stringent safety precautions (occupational health and safety management), more frequent financial audits (financial management), more frequent stock checks (logistics management), more stringent or more frequent inspection of finished work (quality management), and regular or advanced training for staff (human resource management).

Reducing the *impact* of a risk event calls for a different approach. Risk impact reduction usually depends upon the prudent provision of extra resources, so that the potentially adverse consequences of a risk event are diluted in some way. The old financial investment adage about 'not keeping all your eggs in one basket' is a typical example of risk impact reduction. At one extreme, risk impact reduction precautions might amount to having additional sources of supply readily available for essential project materials or making forward purchasing arrangements to lessen the impact of price volatility. At the other extreme, money might be kept in reserve to pay off aggrieved people opposed to the project (yes, this is a form of risk impact reduction!).

The tools of risk impact reduction, therefore, are most often found in precautionary measures such as backup resources and recovery planning, contingency allocations, reserve funds, public relations expertise, and portfolio investment.

The topic of contingency sums deserves some discussion. In our understanding, these are sums of money set aside in project budgets and estimates in order to cover, at least in part, the financial impacts of events that cannot be foreseen in the earliest stages of a project. Prudent project stakeholders seek to ensure that the contingency amount is preserved for that purpose until the unforeseeable events are no longer possible. Some stakeholders reduce the contingency to match the decline in uncertainty as the project proceeds and information increases. Since the events are unforeseeable, so too are their consequences (they are thus 'unknown unknowns'), so an obvious question is: *How much contingency should be set aside?*

Our experience suggests that the determination of contingency is rarely done with any contextual rationalisation. More often than not, it is set as an arbitrary percentage of the project budget (perhaps 2.5–5% or more) on the basis that 'this is how we always do it'. This might be acceptable, especially if experience with previous projects provides sufficient justification, were it not also for our second concern: that, in practice, project contingency amounts are rarely expended on unforeseeable events as intended. They are most often used to pay for later changes to project scope or variations aimed at improving project quality. Cavalier regard for project contingency sums not only defeats their purpose, but also weakens effective risk management and encourages a culture of risk management indifference and ignorance.

Trying to reduce the period of exposure to a risk can be a complicated process, since it is almost always necessary to consider the exposure to the risk event separately from the exposure to its consequences. The tool most appropriate for the former is rescheduling, while effective instruments for the latter may be the implementation of shorter periods for guarantees and warranties offered to customers.

Rescheduling seeks to accelerate progress or at least reduce progress uncertainty. Introducing shorter warranty and guarantee periods attempts to minimise periods of liability.

For many natural risks (Tables 2.2 and 2.3), it may be difficult to reduce the likelihood of occurrence or duration of exposure. Planning a project to avoid adverse seasonal weather is one approach for weather risks. Possible treatments for natural risks are therefore more likely to be found by exploring the consequences more fully.

To sum up the principles of risk treatment: reducing the likelihood of occurrence usually means trying to do things better by enhancing or increasing management interventions, controls, techniques, and tools.

Reducing the adverse consequences of risks is often a matter of ensuring extra resources and backup facilities are available, using 'workaround' routines, or developing emergency procedures.

Reducing the duration of exposure for a risk may be achievable through rescheduling or increasing scheduling flexibility, or by reducing the duration of longer term liability commitments.

Some risk responses and treatment options may be predetermined, or at least shaped, by the strategic risk policies of a project stakeholder organisation. This is considered in Chapter 16 (Strategic Risk Management).

9.7 Strategic Use of ALARP ('As Low as Reasonably Practical')

While strategic risk management is discussed more fully in Chapter 16, it is appropriate to consider one way in which it might be used to influence organisational decision making about treating and responding to project threat risks.

If an organisation has adopted linguistic severity categories for risks, then it can assign preferred (strategic) response options to the categories.

Frequently, such strategies are based upon retaining (without treatment) as few risks as possible, and using avoidance or transfer mechanisms for as many extremely severe risks as possible. The first aim is achieved through the ALARP principle, whereby the severity point, beyond which other risk treatments will be considered more effective, is set *as low as* reasonably *practical*. At the other end of the scale, the 'Avoid' and 'Transfer' options are intended to embrace as many 'High' and 'Extreme' severity threat risks as possible. Figure 9.1 illustrates the strategies.

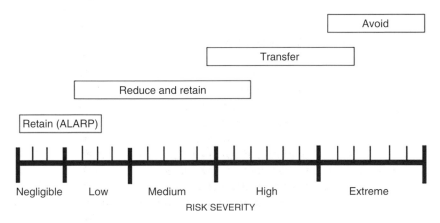

Figure 9.1 Strategic risk responses.

The number of intervals for each severity rating category corresponds to the number of cells for that group shown in Table 8.7.

The more risk averse the organisation, the lower it will set the risk severity boundaries for the risks it is prepared to simply retain. Setting these at the lowest level is not always feasible, hence the ALARP principle.

For a risk seeking organisation looking to exploit project opportunities, the converse AHAPM (*as high as possibly manageable*) principle and strategies might be applied.

9.8   Reassessment

Earlier in this chapter, we noted the need to re-analyse risks for which 'reduce and retain' treatments are under consideration. This is important because the criteria used in the pre-treatment analysis for these risks may have changed because of the particular treatments proposed for them.

Reassessment should be carried out before any treatment action is implemented, especially if the proposed treatment is likely to give rise to additional risks or will affect the responses and treatments for other risks already identified. In part, therefore, reassessment considers not only analysis and evaluation of the treated risk, but also any implications such treatment brings with it.

If reassessment does not demonstrate that the treated risk now falls within the organisation's strategic limits for retention, further additional treatment must be considered.

9.9   Recording Decisions

The proposed treatment and response option for each identified and analysed project risk signal an intention to manage it in some way. This is a decision and commitment point on the part of the project stakeholder. The decision should be entered into the project risk item record (Table 7.14) and thereafter into the project risk register (Tables 4.1 and 4.2). While information for the latter can be entered in abbreviated form if necessary, the item record entry should be as comprehensive and complete as possible, not least because of the potential evidentiary status of this document.

From Table 7.14, it will be seen that the response decision can be simply selected and circled on the risk item record, but this should not be allowed to become the only entry for this part of the risk item record. Any intended treatment should be fully described, including the particular aspects of the risk addressed by the treatment (e.g. 'Reduce probability by increasing frequency of pre-inspections from weekly to twice-weekly').

Organisational oversight of the sufficiency of risk decision records is a responsibility of senior management.

9.10   Summary

In our view, risk identification, analysis and evaluation, and response (Figure 4.2, Stages B, C_1, C_2, and D) are critical stages in effective and successful project risk management in an organisation. Ideally, all of these processes should be carried out in a project workshop environment by a small team that is conversant with the project and committed to the task.

Time spent on these stages is never wasted, and the benefits flow through heightened risk awareness and greater project risk knowledge. Overhasty or ad hoc approaches, shifting task responsibility to one individual or bypassing stages entirely, are all false economies that may leave a project stakeholder not only ignorant of the nature and extents of the risks it faces, but also inadvertently becoming risk seeking to a degree that it would not otherwise accept.

Organisations would do well to examine their risk attitudes and profiles on a regular basis, particularly when undergoing extensive organisational change. Existing risk control mechanisms in the organisation should also be checked for their efficacy in changing circumstances.

Responses to project risks should be considered in the light of feasible and available treatment alternatives, always bearing in mind that such treatments and responses may affect the severity of other risks or introduce new risks for the stakeholder.

Project risks should be reassessed after the treatment *proposal* and prior to the treatment *implementation* to ensure that they will then meet any strategic risk response policies of the organisation.

This stage essentially marks the completion of the proactive elements of project risk management. From this point, a project stakeholder's attention turns to monitoring and controlling risks during the delivery phase.

References

Kahneman, D., Slovic, P., and Tversky, A. (1982). *Judgement under Uncertainty: Heuristics and Biases*. New York: Cambridge University Press.

Slovic, P. (2000). *The Perception of Risk*. London: Earthscan Ltd.

Slovic, P. (2010). *The Feeling of Risk: New Perspectives on Risk Perception*. London: Earthscan Ltd.

10

Risk Monitoring and Control

10.1   Introduction

If project risks have been identified and assessed, and if decisions have been made about responding to them and treating them, then it makes sense to watch over them as the project continues to unfold. The project development stage, comprising planning, designing, and organising, now moves towards actual implementation of all the preliminary activities necessary to ensure that the project is delivered to its operational stage. To neglect or ignore project risks throughout this stage not only would be poor project risk management, negating all that has been done so far, but also would mean that any such management actually needed would then simply become reactive. It would be forced to deal with risk events only after they occur, rather than delivering the greater benefits of proactive preparedness.

The likelihood components of risks are now much closer in time for the project, yet the risks are still in the future, so monitoring risk activity should continue in a manner that provides the best means of attempting to control them. However, monitoring will also signal situations where particular risk uncertainties can no longer affect the achievement of project objectives for the project stakeholder. Those risks can then be 'closed off' in the project risk register.

Each project stakeholder must decide how to implement effective monitoring procedures and control measures for the risks it faces. Issues associated with this include: the assignment of responsibility, the type and frequency of monitoring required, suitable control measures, reporting methods, the identification and treatment of new risks, and the instigation of any necessary remedial or recovery planning and processes. While control measures will already be in place through the risk response treatments implemented as part of Stage D of the project risk management cycle, monitoring will indicate if more are needed.

The monitoring and control stage (Chapter 4, Figure 4.2, Stage E) should move seamlessly in parallel with project delivery activities. At appropriate points, it should also embrace Stage F in the cycle (Capture Project Risk Knowledge) as risk closure occurs.

Effective risk monitoring and control procedures not only sustain risk awareness among project personnel, but also heighten situation awareness generally, thus making staff more sensitive to project matters which may be of concern but are not directly related to project risks and their associated uncertainties. The processes may also reveal new risks for the project stakeholder.

Managing Project Risks, First Edition. Peter J Edwards, Paulo Vaz Serra and Michael Edwards.
© 2020 John Wiley & Sons Ltd. Published 2020 by John Wiley & Sons Ltd.

Risk monitoring and control will inevitably involve communication about project risks, whether within or beyond the project stakeholder organisation. Although the wider aspects of risk communication are dealt with in Chapter 19, some communication issues will be considered here where they are pertinent to Stages E and F in the project risk management cycle.

It is also inevitable that much of the discussion in this chapter will be general rather than project-specific. Given the huge range of project types and contexts, it would be impractical to attempt to cover all of them. Our approach will be to present general considerations and principles, and illustrate them with examples drawn from different project circumstances.

Two immediate reactions to this chapter by readers are likely to be that: (i) it is mainly all about common sense, and (ii) many aspects really involve strategic risk management. We would agree with both observations, but neither lessens the importance of the chapter and its content; and strategic issues will be considered more fully in Chapter 16.

10.2   Assigning Responsibility

Several considerations arise in assigning responsibility for monitoring and controlling project risks in a stakeholder organisation.

The first point is that different risks may require different monitoring procedures and control measures at different times by different people. In any project stakeholder organisation, it is rare for one person (or even one group of people) to bear responsibility for all the risk management activities of the organisation. As we have already seen in exploring risk response (Chapter 9), the treatment of risks may originate in areas such as financial management, quality management, sales management, safety management, and many others. Risk management support may be needed from any or all of these.

Secondly, it is also important to note that at least some of the people responsible for monitoring and controlling risks are likely to be different to those involved earlier in identifying and analysing those risks, and/or those involved in deciding upon risk treatments and response options. Organisational structures influence these personnel changes, as do employee churn rates.

Thirdly, assigning responsibility may also be influenced by the organisation's overall approach to risk management. Some aspects of this will be more fully discussed in Chapter 17 (Planning, Building, and Maturing a Project Risk Management System), but relevant questions here include:

- Are there separate risk management systems for each project and another for the organisation as a whole, or does an organisation-wide system embrace all project work?
- Has a separate safety management system been implemented in the organisation?
- Are other dedicated systems in place (e.g. for quality assurance or environmental management)?

If a separate project risk register is established for each project, then management responsibility should be assigned to project staff with appropriate ongoing decision making authority. This might be a small team headed by a project or contract manager. The team should be able to meet as easily and frequently as the project situation requires. The frequency of meetings may need to be changed to suit project progress and dynamics.

Where risk management is dealt with through a unified system implemented across the stakeholder organisation, then appropriate representatives from relevant departments in the organisation should constitute a team with assigned responsibility for managing project risks. Whether or not one team will be responsible for all projects, particular types of projects, or just a single project will depend upon the nature, number, and size of the projects undertaken by the organisation. In a large project-driven organisation handling perhaps 10 or 20 projects at a time, particularly one with in-house project management office capacity, it is not unusual to find several risk management teams deliberately established to include crossover multidisciplinary expertise.

Organisations that maintain separate departments for safety, quality, environmental, and other specialised areas of management may need to appoint small teams in each department. Each team is then assigned management responsibility for project risks associated with their specialism.

Fourthly, however project risk management responsibility is assigned, a representative of senior management at the highest level (e.g. a board member such as a vice president or projects director) should be appointed to provide global oversight of all project risk management, particularly during the monitoring and control process. This also accords with an overarching principle illustrated in Figure 10.1: the higher the risk severity, the higher should be the level of management assigned to deal with it.

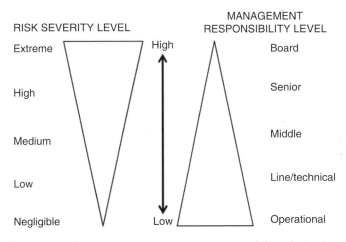

Figure 10.1 The risk severity–management responsibility relationship.

Operational and line/technical staff should rarely be assigned responsibility for managing risks beyond 'negligible' or 'low' severity, and then only on a temporary basis until higher level management capacity becomes available. The logic for this is not just an 'above my pay grade' argument but, as will be seen in this chapter, it is also on the basis that many risks at the lower levels of severity will be simply retained by the stakeholder and monitored by easily followed routines.

Finally, it is also important that risk monitoring and control activity during the project delivery stage must not be allowed to become a 'convenience' issue, taking place only when workers and senior management can find the time. Co-located meetings for risk monitoring and control management, attended by all team members, probably work best. However, modern communications technology allows other real-time alternatives to be used, particularly where live visual media resources can be included. Smartphones and other handheld devices are particularly helpful.

The importance of this issue is such that organisations should give careful consideration to the deliberations and decisions surrounding risk monitoring and control for their projects. Video and audio-recordings, as well as written minutes, are valuable and *auditable* information records to augment the project risk register (and eventually the organisational risk register). Should a responsible senior management representative be unable to attend a meeting or lack immediate access to its real-time communications technology equivalent, the stakeholder should implement appropriate sign-off procedures to ensure that the absentee has eventually seen and heard the recordings and agrees with all decisions made or has proposed amendments.

While this may seem somewhat 'over the top' to readers, the increasingly litigious nature of much of contemporary business practices makes the suggested procedure essential rather than simply a prudent management option.

The regular progress meetings arranged under many project management approaches can provide suitable occasions for monitoring risks. However, care is needed if such meetings are attended by other project stakeholders, since we have already argued that your project risks are not their project risks, nor is your project risk management system (PRMS) their PRMS. Instead, it is often better to follow on from the main project progress meeting with a separate and smaller risk management meeting for your organisation.

A cautionary reminder is also appropriate here for project participants working in an agency capacity for project clients, as the dual responsibility discussed in Chapter 4 (Section 4.1) surfaces again here. An architect, for example, commissioned by a client to provide professional services on a project must be mindful of not only the client's project risks but also the practice risks he or she faces professionally on the project.

When management responsibility is assigned, consideration can be given to how and when project risk monitoring should be undertaken.

10.3   Monitoring Procedures

The nature and frequency of project risk monitoring required will be influenced by the type of project and the individual characteristics of each risk. Logic suggests that monitoring intensity, in terms of both nature and frequency, should increase commensurately with risk severity. As noted, however, monitoring frequency will also be influenced by the dynamics of the project.

As a general principle, we would advocate monthly risk monitoring meetings for projects where the post-design implementation stage is longer than 12 months, particularly where the risk-related matters can be combined with, or follow, monthly progress meetings. For projects of shorter duration, fortnightly meetings may be more appropriate; while in some types of project, or periods where project activities are particularly risk-intense, weekly or even daily meetings may be needed. Events projects can present great difficulty for risk monitoring and control since the event itself may last only a few hours or a few days. In practice, this calls for more intense risk management during the lead-up to the event, and probably continuous monitoring of high and extreme-severity risks during the event itself.

Note that, apart from the exigencies of events projects, we are generally referring here to meetings (i.e. occasions attended by two or more people for the primary purpose of monitoring project risks). As will be seen in this chapter, for risks of relatively low severity, the monitoring may be arranged as part of other staff responsibilities. At the other extreme, continuous or semi-continuous automatic monitoring may be required in some project risk circumstances. For example, critical ambient temperatures may have to be checked at certain times, or other sensors installed to respond to smoke detection, air or hydrostatic pressure, sound or movement, security breaches, authorised access, and many other conditions. Closed-circuit TV (CCTV) serves a similar purpose. Automatic or semi-automatic monitoring of this nature may also incorporate control and response mechanisms such as fire sprinkler activation, emergency exit door release, lift shutdown, and so on. Even the autosave device on your computer acts as a monitoring and control system to minimise the data-loss consequences of the risk of computer failure.

Monitoring procedures are considered here in terms of the project threat risk severity evaluation developed in Chapter 8 (Table 8.7). The severity levels are negligible, low, medium, high, and extreme.

10.3.1   Negligible Risks

For negligible risks, minimum resources and staffing levels should be needed. It is likely that monitoring procedures for these risks can be combined with other project management activities or the work activities of line and operational staff. Reporting outcomes will probably be by exception, that is, only where observation indicates circumstances or situations happening (or likely to happen) beyond what could reasonably be expected to occur normally for such risks. Few data may be generated, and communication will be kept to a minimum. While 'tick box' methods can be used at this level, care must be taken to ensure that they are not abused.

10.3.2   Low Risks

At low levels of risk, monitoring and control activity will increase slightly in intensity compared with that adopted for negligible risks. Monitoring frequency might increase, but is still likely to be combined with other routine tasks, again with the objective of flagging deviations from the norm. Communication will be on a more formal basis, with performance data relating to each risk included on a systematic basis. 'Tick box' lists may still be used, but the parameter observation entries for each risk are likely to require enhancement. Management oversight will also increase in frequency, and possibly also in status.

10.3.3   Medium Risks

With medium-level project risks, monitoring procedures should take on a more distinctive and systematic character. Ideally, the frequency of monitoring will be clearly established for all the affected areas of the organisation, and will be formally incorporated into procedural policies and manuals. As part of their job descriptions, staff involved will spend at least some of their time directly in risk monitoring and control activity. Formal and specific data recording and reporting procedures will be implemented, using planned communication channels. Higher levels of management oversight are needed on a planned and frequent basis.

10.3.4 High Risks

High-severity threat risks generally require the appointment of dedicated and trained staff to undertake monitoring and control activity on a frequent, if not continuous, basis. Monitoring procedures must be consistent, repeatable, and easily auditable. Significant senior management oversight is necessary on a frequency close to being continuous. Reporting should be at least to strictly defined minimum standards, and fast communication channels should be made available, with additional alternatives. Complex data may be generated (some automatically), and sophisticated data analysis techniques (with appropriate database access) may be required. By way of illustration, commercial and military aircraft operation and nuclear power generation provide examples of high-risk monitoring and control norms that are not necessarily completely beyond the means of most project stakeholder organisations to implement.

10.3.5 Extreme Risks

With risks of extreme severity, the highest state of preparedness should exist in the organisation. Monitoring activity should be consistent and may have to be continuous. Senior management must be directly and continuously informed and involved. Reliable high-level monitoring data collection and analysis procedures will be implemented in conjunction with secure database access. Resilience and redundancy are likely to be features of monitoring and control equipment, whereby such systems will be capable of recovering from or bypassing damaged or nonfunctioning parts. Communication capacity will be extensive and flexible across multiple channels, allowing rapid and reliable alternative means for urgent communication between any part of the organisation, and with external emergency services if necessary.

Advances in computer and communications technology have led to greater use of monitoring devices for projects. Drones are increasingly used to inspect locations and situations where human access would previously have been difficult, if not impossible. Even confined internal spaces such as fuel tanks and boiler cylinders can be examined in this way, thus avoiding danger to human life. Laser equipment serves a variety of monitoring roles, including establishing precise setting-out lines and component alignment, and alerting for physical security breaches. Radio frequency identification tags permit monitoring of installations such as steel bar reinforcement placement after they have been covered over. The virtual 3D graphic 'tours' from the design development stage can be replaced by real-time video camera filming and time-interval photography to monitor progress, with outputs uploadable to secure internet sites for access almost anywhere in the world. Technology such as this is changing the ways in which we work, and stakeholder organisations can exploit it to enhance the effectiveness of their project risk management.

10.4   Control Measures

For mitigated risks (i.e. those where risk reduction, avoidance, or transfer actions have been implemented as part of the risk response treatment plans), it is important to check the efficacy of those measures in real time. Are the measures actually in place? Are they functioning correctly and sufficiently? Does more need to be done? These questions can be posed for any level of risk severity. They are particularly important for risk responses that originally included transfer or avoidance, since the effectiveness of these treatments may be more difficult to detect.

It is also important to ensure that the assessment parameters have not changed for risks that were originally dealt with on a straightforward 'retain' basis.

10.4.1   Negligible Risks

Measures to control negligible risks can usually be left to ad hoc actions by appropriate operational and line management staff. Depending upon the nature of the risks, however, basic and refresher training should be premeditated and implemented beforehand.

10.4.2   Low Risks

For low-severity risks, control measures are likely to be a mixture of ad hoc and pre-planned activities. Management oversight should match that for risk monitoring.

10.4.3   Medium Risks

Medium-severity threat risks require control measures that are almost entirely pre-planned. A 'chain of responsibility' approach should be adopted, ending with senior management at the higher end.

10.4.4   High Risks

For high-level risks, appropriate emergency response and recovery plans and procedures should be in place and rehearsed, especially for risks with serious consequences for human health and safety. The escalating levels of responsibility for controlling such risks must be known and clearly understood across the project stakeholder organisation, and should reach the executive board or partner level.

10.4.5   Extreme Risks

Control measures for extreme risks may involve people throughout the whole stakeholder organisation and beyond. Emergency response and recovery plans and procedures should be pre-planned, agreed through consultation with external agencies, formalised, rehearsed, practised on a randomly arranged and informed basis, and stress-tested. Backup personnel and other resources should be readily available. Adequate public relations measures should be in place and tested.

Much of what we have advocated here represents a 'common sense' approach to risk monitoring and control, showing how each escalates in intensity according to the severity level of each threat risk. We conclude this section with some practical examples of different types of projects and project circumstances:

1) For construction activities involving crane operations, it is usual to arrange safety exclusion zones where unauthorised entry or intrusion into the zone is not allowed. Risk monitoring may indicate that the operational environment originally foreseen has changed or is impacted differently in reality. A more stringent control measure may be required (e.g. expanding the exclusion zone, or narrowing the entry and intrusion authorisation parameters). At the same time, monitoring of seasonal weather patterns might show that high winds are occurring more often than anticipated. The construction manager could decide to check local weather radar forecasts and site anemometer readings more frequently, and reduce the time gap for instructing crane operations to cease. All these are post–risk monitoring control measures.

2) A public event is due to take place over a holiday weekend. Only a few days beforehand, it becomes known that another event will be occurring in the neighbourhood at about the same time. In the original project risk schedule, the risk of crowd control failure was mitigated by arranging for marshals to be in place during the event. The pre-event risk monitoring indicates that the number might not be sufficient now, so this project risk control measure is increased by 30%.

3) An online product shopping service has been designed with an external third-party payment gateway facility. In beta-test monitoring, the likelihood of customer payment transfer error is found to be higher than anticipated. The payment stakeholder has been asked to improve its system and, as an additional control measure, a further authorisation stage is to be inserted on the store side before customers are transferred to the payment processing stage.

4) For a tunnelling project, there is a risk that unexpected geological formations will be encountered that will lead to misalignment of the tunnel boring machine (TBM). The original risk treatment was to reduce the impact by setting the auto-alignment sensor parameters to smaller tolerances. Risk monitoring shows that the treatment has been too pessimistic, and geotechnical data show that the most difficult strata have now been passed. As a risk control measure, the sensor parameter tolerances are increased, permitting faster tunnel boring.

5) Monitoring the sales risk for a new book publishing project has shown that the rate of incoming purchase orders is less than anticipated, despite a widespread bookshop publicity campaign attended by the author. As a risk control measure, the marketing director decides to intensify the national TV and radio interview campaign opportunities for the author and send specimen copies to radio talk show hosts in regional areas. This will result in about a 2% overspend for the marketing budget but should improve sales orders.

6) The advance bookings for a new musical production, planned to open soon in a city theatre, have been slower than anticipated, thus threatening the 'angel' backers with more uncertain returns on their financial investment. The producer authorises the marketing manager to increase the TV advertising campaign as a risk control measure, counterbalancing the additional cost with a reduction in the probability of losses occurring on the production.

7) The project manager responsible for introducing a more efficient triage system in the Accident and Emergency (A & E) section of a hospital is monitoring patient waiting times and queuing congestion during the system changeover period. Uncertainty associated with the urgency and direction of treatment decision making suggests that the staff learning curve is steeper than anticipated. As a temporary control measure, the project manager asks the hospital director to substantially increase the number of staff shifts available in A & E over the next four weeks to reduce stress on the system and on staff. Micromanagement staff training measures will also be introduced on a temporary basis, along with opportunities for staff and patient feedback.

Whatever the levels of risk monitoring and control implemented on projects, some means of reporting and recording such activity, and any ensuing decisions, is usually required.

10.5   Reporting Processes

It seems rather obvious to state that the way in which the evidence and decisions of risk monitoring and control are reported and recorded will depend upon the nature of the risks and the risk consequences. We do so only because the means should be predetermined and not allowed to become a reactive post–risk event measure. Some reporting methods may involve external organisations. Others may form part of formal compliance requirements. All may require further risk decision making and action. Reporting processes may be automatic, semi-automatic, or manual. They may be continuous, intermittent, or periodic. Reporting media may be verbal, written, or audiovisual, and we have pointed to some examples in Sections 10.3 and 10.4.

Verbal risk reporting may be the most direct and realistic method of reporting risks, but it is also the least reliable and the least consistent. Given the many barriers to effective verbal communication, it should always be followed up as soon as possible by more formal reporting methods.

Written reports of risk monitoring and control activity can be either formal or informal. The latter might comprise abbreviated handwritten notes of observations and discussions and should be augmented soon after with more formal records. These might include minutes of meetings, project logs, and updated project risk item records, followed by updated versions of the project risk register itself.

Audiovisual reporting is a useful adjunct to verbal and written reports, and it provides a reliable means of transmitting project risk information to project participants who may be physically located far from the central hub of project activities. In most instances, audiovisual reports will have to be augmented, or at least backed up, by written reports.

Before the project risk monitoring and control stage commences, some communication issues are likely to require resolution.

Data processing and storage processes, and any associated security issues, must be decided and tested for reliability. The nature and level of any inter-organisational communication should be determined. Where compliance reporting is prescribed, the minimum requirements must be established and confirmed. Feedback confirmation should be obtained for any reporting undertaken beyond the project stakeholder organisation.

Finally, we reiterate our caution about the potential evidentiary status of project risk matters. Where there is even a low probability that this will be demanded, the structure, content, and treatment of risk monitoring and control reports must be carefully considered beforehand. While the most obvious situations for this will lie in risk events and consequences involving human health and safety, or in adverse environmental impacts, it should not be assumed that these are the only occasions when project evidence for judicial proceedings will be called for.

10.6   Dealing with New Risks

The processes of risk monitoring and control may reveal new risks (threats and opportunities) for the project stakeholder organisation, requiring a return to the earlier stages of the risk management cycle. This reiteration of the cycle may also be necessary for risks that have already been identified and treated, since it has already been noted that many risks are subject to changes in probability and impact over time.

New risks tend to emerge during the project implementation stage mainly because decision making, albeit on a relatively smaller scale, tends to increase as scope changes are contemplated or design amendments become necessary. Decisions now tend to be driven more by current circumstances rather than original plans. Project contexts become closer and clearer than at the outset. Uncertainties are probably less obscure and more specific. Finally, project participants should be more risk-aware and are also likely to be more risk-averse now. For example, construction contractors tend to be risk-seeking prior to bidding, but quickly become risk-averse after their bids have been successful.

Sadly, one area where new risks may emerge at this stage of the project cycle is in corruption. Corruption has been defined as 'the misuse of entrusted power for private gain' (UNDP 2008) and includes issues such as conflicts of interest and unethical conduct (Transparency International 2010).

Our experience in the construction industry shows that corruption risks may arise at any time on a project, sometimes at the project concept development stage but most frequently and most prolifically during the project implementation period.

At the early design and development stage of the project, corruption may arise in matters such as obtaining permits, finance, or for appointment as the preferred bidder. Later, during the implementation stage, as the number of project participants tends to increase substantially, corruption may be encountered (or suspected) between contractors, subcontractors, and suppliers, or with officials holding powers of approval over compliance. At the beginning of this stage, corruption may also occur at the opening of competitive tenders or with the tender award, most often through the complicit actions of public sector officials.

Corruption is itself risk-moderated. In a dynamic model of corruption (Edwards *et al.* 2017), the drivers for corruption are identified as:

- Pressure to engage in corruption (exerted by others through inducements, favours, leverage, threats, nepotism, or bribery)
- Propensity to be corrupt (through self-rationalisation, disdain for the rule of law, or distortion of moral compass)
- Potential for corruption (the capacity to exploit existing or create new opportunities to engage in corruption).

The model is dynamically risk-moderated through the perceived probability of evading detection or avoiding the consequences of detection.

Corruption is pervasive and can be found at all levels and in all areas. It is insidious, escalating in frequency and degree; and its effects are not only financially damaging to projects, but also demoralising to project organisations, and the industries they represent, as the effects surface.

In attempting to identify corruption risks during any of the stages of project risk management, the essential question to be answered is: What evidence of corrupt activity can be found on this project that will lead to uncertainty in achieving project objectives and damage to the organisation?

Since corruption risks are heavily influenced by organisational, local, regional, and even national cultures, they can be intractable and difficult to treat. For philosophical, ethical, and practical reasons, simple retention of corruption risk is not a feasible option as it signifies condonation, is wrong, and inevitably escalates if left untreated.

Mitigation through reduction is the only practical response to corruption risk, as avoidance is often impractical. In the short-term context of projects, effort should be made to lower the probability of evading detection or avoiding the consequences of detection. Randomising the frequency of monitoring process such as financial auditing is one approach. Careful observation of behavioural anomalies in the decision making of project participants is another.

In the longer term, the issues of corruption should be addressed through policy direction to change organisational cultures: encouraging, rewarding, and protecting whistleblowers; penalising corrupt behaviour; and refusing to become complicit in the corrupt activities of other project stakeholders. Interestingly, in terms of the generic approach to risk classification (Chapter 2, Table 2.2), justification could be made for placing corruption into any of three categories – political, legal, or sociological – hence the difficulty in dealing with it.

We have devoted considerable space to the risk of corruption, partly because of its unique nature and consequences, but also to emphasise our view of it as a threat risk, despite the view sometimes held that corruption (particularly if regarded as 'petty') is inevitable and opportunistic, and provides a justifiable means of 'oiling the wheels' of business. This view prefers a more euphemistic metaphor than 'greasing palms' and, to use another metaphor, represents the start of a very slippery slope!

Generally, new project risks are identified during the processes of monitoring and controlling risks that are already captured in the project risk schedule. Each new risk, together with those existing risks that are found to have changed circumstances, should be directed back through the iterative loop of the analysis, evaluation, treatment, and response stages of the project risk management cycle.

10.7 Disaster Planning and Recovery

Since even the best PRMS cannot prevent risk events from happening, project stakeholder organisations need to prepare and maintain plans to deal with the more severe risks they face, especially those where the consequences could be considered disastrous.

Many issues are involved here, some of which include:

- Assignment (or re-assignment) of responsibility for coordination and action
- Recruitment and training of emergency teams (and reserves for them)
- Acquiring/dedicating suitable and sufficient resources and equipment
- Defining key places or routes for access, exit, or congregation
- Defining key system control points
- Rehearsing general staff in emergency procedures
- Alerting of emergency services and external specialists
- Provision of adequate communication facilities
- Supplies, equipment, and spare parts logistics
- Psychological and counselling facilities for project staff (and affected members of the public)
- Public relations; the press and other news media
- Statutory reporting requirements
- Preservation and collection of evidence
- Alternative arrangements for project delivery or completion (designing deliberate resilience and redundancy measures into the project).

The exact nature of each of these issues will be unique to particular types of risks and to specific risk events. Some relate more to physical risk environments, but recovery and disaster planning should not be limited to physical risks alone. Contemporary organisations in many fields have had to give considerable thought to the consequences of data loss arising from information technology (IT) project systems failure or vulnerability to cyberattack and denial of service.

Addressing all these issues in greater depth is beyond the scope of this book. Indeed, they warrant a complete book. We seek here only to alert readers to this important aspect of project risk awareness. Also, while there is some resemblance between these issues and those relating to risk monitoring and control discussed earlier in this chapter, it should be borne in mind that the recovery and disaster planning process is not the same as risk monitoring and control. Different people may be involved at every level of the organisation. The reporting and decision making processes are likely to be different. Public relations requirements will certainly not be the same.

10.8   Capturing Project Risk Knowledge

Ignored even more often than the systematic monitoring and control of risks, the capturing of project risk knowledge is important for a stakeholder organisation in terms of being able to learn from its risk experiences. If an organisation lacks any formal means of collecting information, processing it, and placing it into the organisational memory (the organisational risk register), experiential risk knowledge is simply left to reside in the individual memories of people in the organisation. Whilst the latter happens anyway, regardless of the existence of a PRMS, failure to capture valuable information leaves the organisation vulnerable to the demise or departure of the personnel involved. Even if they remain active in the organisation, such people often move from one project to the next, sometimes in rapid succession. Their memories may grow dim over time, especially where negative aspects of former projects are concerned.

A good PRMS, therefore, will include the means to capture the risk 'stories' of the people involved in the project. Since the staff from the stakeholder organisation are also likely to have had interactions with personnel from other stakeholders during the progress of the project, valuable risk learning from other perspectives may be acquired.

In most cases, post-project debriefing is an appropriate method of collecting risk information. Sometimes it may be appropriate to do this periodically throughout the project, often when defined milestone stages have been achieved.

While the information collection process can be conducted informally, it is important to formalise the analysis and archival recording of the collected material in order to maximise its usefulness to the organisation. The archival format may be incorporated into the formal project risk register or dealt with as separate databases for each part of the organisation.

We will examine the processes of risk knowledge management more closely in Chapter 11.

10.9   Summary

Risks that are not monitored and controlled during the project implementation stage simply become risks that are not managed proactively but will have to be dealt with on a reactive basis should any eventuate.

Important aspects that should be considered and arranged beforehand by the project stakeholder organisation include: assignment of responsibility, monitoring frequency and processes, and control measures. As a general principle, each should be matched to the ascending order of threat risk severity. Modern computer and communications technology has substantially increased the means and quality of these processes for projects of all types.

Risk reporting also requires careful consideration in terms of method, content, and status. It is likely that new risks will emerge during risk monitoring and control, and the circumstances of previously identified project risks may have changed. These will require further analysis, evaluation, and response decisions.

The monitoring and control processes may also reveal the need for disaster and recovery planning for post-risk eventualities where the risks themselves have become much closer in time.

Finally, occasions should be found during the project implementation stage to capture risk 'stories' and knowledge. The importance of this activity, and how the risk knowledge should be dealt with, is discussed more fully in Chapter 11.

References

Edwards, P.J., Bowen, P.A., and Cattell, K.S. (2017). We can fix it: Corruption in the construction industry. In: *The Handbook of Business and Corruption: Cross-sectoral*

Experiences (ed. M.S. Asslaender and S. Hudson) Chapter 16. Bingley: Emerald Publishing Ltd.

Transparency International. (2010) Corruption and Public Procurement (Working Paper #5). Retrieved from http://issuu.com/transparencyinternational

United Nations Development Programme (UNDP). (2008) Primer on Corruption and Development. Retrieved from http://www.undp.org

11

Project Risk Knowledge Management

11.1 Introduction

When confronted with a new situation, people tend to look for similarities with previous situations they have experienced. They look for relationships between new and old, and, through a process of successive comparisons, they seek to understand and assess this new situation.

People also tell stories – to teach, to share, and to preserve and transmit knowledge through generations. Some stories may be invented, but will still allow the audience to imagine real situations from known details. These are the ancient traits and customs that make us human. For this chapter, they serve to illustrate how important knowledge management is to risk management.

Drucker (1993) predicts that knowledge will be the new basis of competition in a post-capitalist society, stating that: 'one thing we can predict: the greatest change will occur in knowledge – in its form and content, its meaning, its responsibility, and what it means to be an educated person'. Since 1999, Davenport and Prusak (2000) have argued for the conceptualisation of a firm as a collection of its knowledge, history, and culture, all of which need to be aligned with technology for effective and productive knowledge management. Fruchter and Demian (2005) claim that: 'knowledge is the only unlimited resource and is the essential element that can grow the more it is used'.

Paradoxically, within contemporary organisations, individuals do not always share knowledge naturally or easily. We live in an era where knowledge is equated with power or confers potential competitive market advantage. Historically, knowledge and know-how (technology) have been associated with social prestige, status, and various forms of power. Those holding power had exclusive access to knowledge (simply because they could read in an age of illiteracy), and those holding knowledge could leverage it into yet more power or financial gain. Secrecy was the key to gaining or retaining power. 'Confidentiality' and 'commercial in confidence' are now everyday synonyms for secrecy, but the rationale remains the same. Knowledge, as a rule, was only transferred through a careful selection process – from master to apprentice (through a long-term locked-in relationship), or to trusted allies. Nor was this selectivity only a feature of trades and artisanship; most of our professions began in the same way and continue in that manner today. Knowledge 'silos' persist.

There is thus a deep social history behind the tendency for individuals to avoid sharing knowledge and to retain what they perceive as a competitive advantage. True

Managing Project Risks, First Edition. Peter J Edwards, Paulo Vaz Serra and Michael Edwards.
© 2020 John Wiley & Sons Ltd. Published 2020 by John Wiley & Sons Ltd.

knowledge management seeks to reverse that, at least to a more acceptable and cooperative degree. At the organisational level, harnessing, sharing, and creating new knowledge are thought to deliver a competitive advantage better than more aggressive means. Additionally, since new knowledge is now created at an ever faster rate, maintaining secrecy about 'old' knowledge makes less and less sense.

Project risk management largely depends on being able to use existing information and knowledge to deal with future risks, whether they are threats or opportunities, and whether the risk management is proactive or reactive (see Chapters 2 and 4). The concepts of knowledge management are, in fact, naturally integrated into mature theories and systems used in formal risk management. Moreover, ISO 31000 (2018) highlights the iterative nature of risk management and gives emphasis to the importance of recording and reporting. The second edition of this standard recognises the role of new experiences, knowledge, and analysis in the whole risk assessment cycle and that this can be achieved using knowledge management systems (KMSs). Knowledge, when managed effectively, supports sound risk management, and where an organisation is a project stakeholder it improves project risk management. Effective knowledge management within an organisation directly influences project outcomes. When knowledge is proactively integrated with project risk management, the achievement of project objectives is made more secure. Knowledge is a counter to uncertainty.

The swift and widespread emergence of academic interest in knowledge management during the mid-1990s has established it as an independent field of its own. This chapter affords readers an opportunity to become familiar with knowledge management as a topic in its own right, and then to appreciate how important it is to organisational risk management and consequently to project risk management.

Many readers will have a preconceived or perhaps intuitive view of what knowledge is. Our first objective is to define knowledge, as distinct from raw data or information, and to consider types of knowledge. Then, we explore concepts of knowledge management and the knowledge creation cycle. Having provided this conceptual understanding of knowledge and knowledge management, we consider this in the context of project risk management systems (PRMSs) and their architecture.

All systems present challenges in terms of implementation. These are discussed with the aim of providing recommendations for KMS implementation.

While Chapter 19 of this book takes a more in-depth look at project risk communication, at the end of this chapter we try to show the benefits of integrating knowledge management alongside good communication practices and systems.

11.2   Knowledge Definitions and Types

The study of knowledge (epistemology) has been pursued since the era of the ancient Greek philosophers. Plato thought of 'knowledge' as 'justified true belief'. If we bypass later and more controversial claims about spiritual sources of knowledge as belief, a contemporary meaning is that knowledge is a cognitive understanding of something, obtained through theoretical and practical reasoning.

'Knowledge management' arrived much later on the scene and has been formally studied since the middle of the twentieth century. Several definitions have been put forward, and research has been undertaken to identify underlying principles and the best methods to manage knowledge in various contexts. We might think of knowledge management as a means of dealing with knowledge for the purpose of improvement, whether personal self-improvement or from a business performance perspective. In reality, of course, knowledge management has been the rationale and foundation for developments in the sciences and in medicine for much longer.

According to Palmer and Platt (2005), knowledge management is: 'harnessing and applying all the knowledge in an organisation', a definition that encompasses the processes of creating, storing, organising, searching, sharing, and using knowledge. They also point out the importance of distinguishing between information and knowledge. Information can be recorded in reports and databases and is thus relatively easy to organise, share, and apply to specific problems. Knowledge, on the other hand, is a blend of information and reflective experience that is not readily storable but is more naturally retained in people's minds. In our view, the definition also suggests wisdom in that the knowledge is applied for a particular purpose. At the other extreme, we also consider *data* as an antecedent to information.

A useful starting point is to consider a knowledge transformation sequence.

11.3 Knowledge Transformation

It is important to distinguish between data, information, and knowledge, and then to understand that knowledge, through knowledge management, can be transformed into wisdom. Figure 11.1 shows a progressive and ascending transformation sequence, in a risk context, from data to information, information to knowledge, and then ultimately

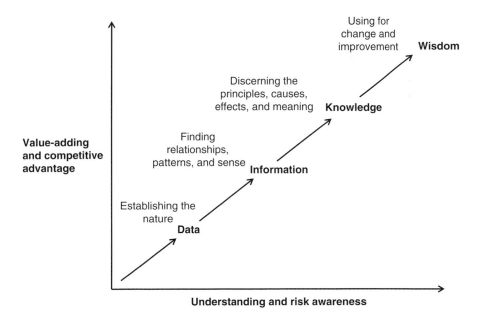

Figure 11.1 A knowledge transformation sequence.

to wisdom. The organisational outcomes of this transformational journey include better risk understanding and awareness on the one hand, and greater value and competitive advantage on the other.

A way to conceptualise this sequence is first to consider raw data. This is derived from observations and measurements that may be in the form of text, numbers, diagrams, pictures, video, or oral records. For the most part, each piece of data relates to single concepts.

Information can be thought of as the patterns identifiable within the data (e.g. the relationships, trends, and statistics). Here, multiple concepts are brought together in some way to make sense.

Knowledge is explanatory (and exploratory) in terms of the 'why?' perspective: understanding why the patterns, trends, and relationships exist, or, in the absence of knowing a reason, embarking on a journey of discovery (research) to find acceptable explanations for causes and effects. Meaning is added to the information. Such explanations may become paradigms (i.e. generally accepted approaches and methods for understanding and dealing with things), until new knowledge emerges to overturn the prevailing paradigm.

Knowledge comes not only out of formal transformations of data and information but also from personal experience and cognitive understanding. We will return to this distinction later in the chapter.

Wisdom can be considered as an accumulation of knowledge, or access to knowledge, and then knowing how, when, and where to apply it effectively. We may also understand it as deep insight or a justifiable belief (cf. Plato), capable of informing judgement (i.e. decision making). In the context of the organisation, the process involves gathering data, transforming it into information and then into identifiable knowledge, retaining or storing it, and then sharing it or making it accessible to apply in future circumstances. All this begins to make the organisation wise. It is knowledge management at its best. In gaining wisdom, however, understanding needs to be at its sharpest, as otherwise we may be considered simply as 'smart'. Being considered smart is not necessarily undesirable, but being wise is more highly esteemed in society.

11.4 Types and Forms of Knowledge

Knowledge management deals with more than just structured and static information. It encompasses 'living' information and experiences that circulate through an organisation and its people in all its daily activities. From the outset, therefore, we also need to conceptualise knowledge as being 'alive'. It grows through application, with new data and new information, with new experience, and then being combined with (or refuting) other knowledge.

Knowledge is thus considered to have two key forms: explicit and tacit. Explicit knowledge is more easily communicated and managed. It can be identified, recorded, stored, and shared through formal processes. Tacit knowledge is associated with experience. It is often subtler, less formal, usually found within individuals, and used intuitively. The knowledge that leads to wisdom may be tacit or explicit, or a combination of both. Experience, however, does not guarantee wisdom, but that may emerge from deliberate reflection upon experiential knowledge.

Tacit knowledge usually resides within the 'knowledge workers' in an organisation. Sometimes, they may not even be aware that they hold this knowledge, which makes management of their knowledge more challenging. These knowledge workers are at the forefront in the creation of new knowledge, and are the driving force behind growth, innovation, and continuous improvement. The loss of knowledge if they leave represents a significant risk to an organisation. Effective knowledge management mitigates that risk.

At the core of knowledge management is an effort to convert tacit knowledge into explicit knowledge so that it too can be stored, organised, shared, and reused. We will see how this becomes the kernel of the knowledge creation cycle.

From another perspective, we can also consider knowledge as theoretical or practical. Theoretical knowledge is propositional, derived from hypotheses that have been tested and are found to support the proposition. It is explicit knowledge that endures only until the hypotheses are contradicted by new ones and the old ones are discarded.

Practical knowledge is a cognitive understanding that develops from practice in doing something. If we have become used to doing something in a particular way (i.e. following a particular routine for carrying out a task), we gain practical knowledge that may be adaptable for other tasks. Practical knowledge may be partially based on theory but is mostly derived from experience.

Knowledge management strategies have also been described as 'mechanistic' or 'organic' (Anumba *et al.* 2005), whereby the emphasis on explicit knowledge is considered mechanistic and tacit knowledge is seen as organic. If the emphasis is placed mainly on explicit knowledge, a strong technological increase may ensue. If tacit knowledge prevails, then storytelling and communities of practice will develop. Key to such development is organisational culture.

11.5 Organisational Culture and Knowledge Management

There are two key determining factors in the success of knowledge management within an organisation: content and context. 'Content' refers to the knowledge itself that is to be managed, while 'context' largely refers to the organisational culture (see Chapter 12): the way in which knowledge is regarded, managed, and used in the organisation. In this book, all this is related to project risk management.

Military institutions, from the earliest times, have understood the strategic importance of collecting and storing knowledge properly and quickly. At the end of each military action, a so-called 'debriefing' is carried out. This comprises a factual review of events and an individual and collective reaction to these events, thus providing an opportunity for officers to sit down with soldiers, reconsider what occurred, and draw lessons for the future. In the process of reviewing events, feelings can be expressed and problems can be identified and addressed. An important step is to review the chronology of events, giving participants the opportunity to clarify where there is confusion, facilitate a new healthy cognitive adaptation, and enable the integration of their collective experiences into strategies for future action. In addition, this process is a way to harness the knowledge created through the action (or project) and transform it from individual ownership into organisational knowledge. Knowledge is converted from tacit to explicit because the organisational culture encourages this to occur.

Successful knowledge sharing depends upon the opportunities that people have to communicate and interact, and on the willingness of people to share their knowledge with one another. Knowledge-intensive organisations have to create and maintain an environment (i.e. culture) that encourages, and is open to, sharing knowledge. An intrinsic contributor to this environment is the knowledge creation cycle.

11.6 The Knowledge Creation Cycle

The transformation of tacit knowledge to explicit knowledge is one of the most important activities in knowledge management. It enables such knowledge to be stored and made available to others in the organisation. When others use that knowledge, they are effectively testing it by putting it into practice. This verifies the transformation and validates the knowledge. Validated knowledge delivers even more value to the organisation.

It is important to understand organisations as collective repositories of their knowledge, technology, history, and culture and not simply as asset maximisation machines. Organisations can use their knowledge resources to improve their long-term strategies and their overall effectiveness. Product and technical innovation are often considered the primary consequence of good organisational learning and knowledge management practices.

In terms of project risk management, after retrieving risk data, the aim should be to organise the data to produce information that can be collated and organised, following certain patterns that can be reused as knowledge by the organisation on each new project.

Integral to the knowledge transformation sequence (Figure 11.1) is the knowledge creation cycle. This cycle is a continuous or semi-continuous process of dealing with tacit and explicit knowledge (Nonaka and Takeuchi 1995). In addition, such an ongoing process of capturing dynamic interactions between tacit and explicit knowledge, testing the product against theory and using it in practice, may actually deliver more than the sum of the parts by creating new knowledge. It supports innovation and contributes to competitive advantage for an organisation. Figure 11.2 illustrates the cycle, which has four stages each comprising two activities.

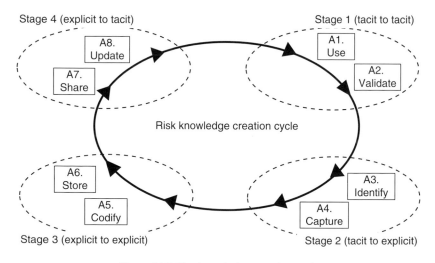

Figure 11.2 The knowledge creation cycle.

We can think of this cycle as an engine flywheel turning to drive the knowledge transformation process (Figure 11.1) up the hill towards wisdom. For our purposes, this is 'risk wisdom' in the service of effective project risk management.

11.6.1   Stage 1 (Tacit to Tacit): Use and Validate

In the first stage (tacit to tacit) of the knowledge creation cycle, tacit knowledge is used and validated. Project participants use their existing risk knowledge in a tacit way (Activity 1) and seek to integrate that knowledge with the similar knowledge of other team members involved in the project in order to manage project risks. In this way, they validate the tacit knowledge they are using and augmenting in an experiential way (Activity 2). The validation process is assisted by observations, data, and information as well as by perceptions and beliefs.

11.6.2   Stage 2 (Tacit to Explicit): Identify and Capture

In the second stage, occasions will be sought, usually through 'lessons learned' sessions, to identify the additional and experiential tacit knowledge of project participants (Activity 3) and capture it in some formally acceptable manner (Activity 4). This moves the knowledge from a tacit to an explicit state.

11.6.3   Stage 3 (Explicit to Explicit): Codify and Store

For Stage 3, the identified and captured explicit knowledge is codified (Activity 5): it is organised and arranged to suit the organisation's knowledge management strategy, and then stored (Activity 6) in its KMS.

11.6.4   Stage 4 (Explicit to Tacit): Share and Update

In Stage 4 of the knowledge creation cycle, through the KMS, new explicit knowledge is shared across the organisation (Activity 7) in various ways including via seminars and training. The knowledge then becomes tacit for those exposed to it. Through their feedback, the KMS is updated with additional knowledge (Activity 8) that may also be acquired from other internal and external sources, and the augmented and updated database then becomes available and accessible for tacit use on new projects, thus commencing a fresh knowledge creation cycle.

In Case Study A (a public–private partnership [PPP] correctional facility), the main contractor has an online KMS platform (an intranet), and access and consultation are mandatory when the contractor is considering whether to respond to a call for expressions of interest (EOI) for a new project. The stored knowledge is reused to support a go/no-go decision, and the tacit knowledge generated in this decision making process is harvested as part of the organisation's knowledge creation cycle.

For Case Study F (an aquatic theme park), the civil engineering contractor uses a similarly prescriptive approach (Table F.2) towards all new projects, with a template derived from past experience. In this case, however, because the organisation lacks a formal KMS and robust knowledge capture procedures, its knowledge creation cycle remains almost entirely tacit.

The activities associated with the knowledge creation cycle shown in Figure 11.2, together with the evidence found in the two case studies, suggest that a strategic perspective is important for managing risk knowledge in an organisation. While Chapter 16 (Strategic Risk Management) explores this in greater depth, it is useful to consider it here in relation to the stage activities of organisational knowledge creation.

11.6.5   Using and Validating Knowledge

Existing explicit knowledge starts a transformation into new tacit knowledge as soon as the knowledge is used. When explicit knowledge is linked to new experiences for the user, it becomes transformed into new knowledge in a tacit way. The example of cooking with a new recipe can be used to illustrate this. Initially, all that may be known is the recipe (explicit knowledge) in a cookery book. Each time the recipe is used, however, the experience is likely to reveal to the cook things that should not have been done and things that could be done differently. The recipe is being validated. For many cooks, the experience also provides an opportunity to try new ingredients as a variation to the recipe that will introduce a new taste sensation. All this is tacit knowledge learned by experience. Nowadays, computer simulation often achieves the same purpose by providing users with the experience of virtual reality, but unfortunately without the desirable aroma and taste sensations of actual cooking! Note also that failure may deliver as much value as success in this regard!

In the same way, project participants should be incentivised to use the KMS and provide further feedback. This is what fuels the knowledge creation cycle. After using the project risk knowledge that is available from the KMS, the experience of doing so may be different for each situation or project. A knowledge validation process occurs which essentially frames user feedback in an appropriate context. Incentives and encouragement to use the KMS are a matter for strategic management, as also is the process of contextualisation. Prescriptive policies should not be the only strategy.

11.6.6   Identifying and Capturing Knowledge

Knowing what tacit knowledge should be captured underpins the activities that an organisation employs to harvest and retain risk-related information and knowledge created or acquired whilst carrying out activities associated with achieving its objectives. As noted in this chapter, it is in this process that the transformation of tacit knowledge to explicit knowledge also begins.

As part of its strategic risk thinking, an organisation needs to understand what knowledge is important to its continuing existence and to the people carrying out activities on its behalf. Identifying relevant knowledge areas is, therefore, an important first step.

The identification of the key knowledge requirements should be linked to the objectives of the organisation. Generally, for project risk management an organisation will want to gather information and knowledge relating to each aspect of the risk management cycle depicted in Figure 4.2 (Chapter 4). The risk knowledge areas will therefore largely relate to:

- Project, organisational, and external contexts
- Identified risks and the processes of risk identification

- Risk analysis and evaluation processes and their outcomes
- Risk response and treatment options and associated decision making
- Risk monitoring and control procedures and their outcomes.

Acquiring project risk knowledge through the information held in project risk registers (PRRs) is essential, but it is not enough. Organisations need infusions of fresh knowledge in order to grow, and this means not just relying on knowledge acquired from current and past projects, but also a willingness to obtain new knowledge from external sources. These sources can be diverse and include other stakeholders, universities, industry seminars and conferences, professional and industrial associations, public debate, journal articles, and news media.

Some practical steps can be used to guide the strategic capture of project risk knowledge by organisations that do not yet have fully developed knowledge management and risk management systems in place. The steps comprise a series of questions that follow our preferred interrogative approach to this type of organisational management:

- How is risk knowledge currently managed throughout the organisation?
- How is knowledge management currently practised in the organisation?
- What are the key knowledge areas applicable to the activities of this organisation?
- How do these areas contribute to the objectives of the organisation?
- How do these areas affect the project risks that the organisation faces?
- Who are the experts in risk knowledge management in the organisation?
- Are they sufficiently expert; and, if not, who are appropriate external experts who can be consulted or recruited?
- What roles and groups in the organisation could best benefit from better risk knowledge management?

Since tacit knowledge resides in people, encouragement and opportunities are needed for it to be harvested. An organisation should implement processes (particularly through induction training or learning sessions) for individuals to assimilate, in a tacit way, the knowledge that they need for the tasks they are expected to perform. Induction is an important first step in the learning process. In-service training courses and 'lessons learned' workshops service ongoing knowledge sharing and post-project knowledge acquisition.

Storytelling is one appropriate way to encourage the surfacing of tacit risk knowledge. Debriefing at the end of a project in a workshop environment is useful where a written report can be produced to include the lessons learned and then be stored. Hearing participants' stories is one of the most powerful ways of achieving this. In these 'lessons learned' workshops, people share their experiences and what they think about them. They learn the best way to explain their thoughts in ways that create new knowledge that can be reused. Emotional content in project stories may be a strong indicator of veracity.

A workshop approach also allows opportunities for immediate feedback, as an essential part of effective communication (Chapter 19) to achieve shared meaning.

While the identification and capture process helps to transform the knowledge status from tacit to explicit, the intrinsic data may still be quite raw, particularly in terms of language and content. It is usually necessary to codify it in some way to render it capable of being stored in a KMS.

11.6.7 Codifying and Storing Knowledge

Project risk knowledge should be codified and stored in a way that makes it more understandable, more organised, and thereafter more accessible. Tacit knowledge has to be put into an explicit format. The stories cannot remain indefinitely as stories; otherwise, there is a danger that they will become folklore that is eventually degraded in terms of value and veracity.

Tacit knowledge is combined and shared in concepts that can be transformed into explicit knowledge in readable and understandable formats, thus allowing wider and more effective communication. Knowledge that is already explicit may need to be made even more explicit, hence the need for 'explicit to explicit' knowledge creation activities. Outcomes can be stored in different ways depending upon the structure of the organisation's KMS, the capacity of the information technologies used to service it, and the capabilities of the system users. All of this requires strategic thinking in the organisation.

A desirable first step would be the conversion of passive PRRs into 'smart' registers so that information is recorded once in the primary register and then automatically updated on all other registers, such as the organisational risk register (ORR), in order to avoid duplication of knowledge management activities. This is a matter for the technical design and architecture of the KMS.

The process of codifying project risk knowledge is essential. Knowledge may remain tacit in nature simply because no one wants to take time or energy to codify it into rationally explicit knowledge. If this situation arises, the chance to add value through knowledge transformation is missed. To some extent, this is demonstrated in Case Study F (an aquatic theme park).

Codifying risk knowledge is a challenging task. Different organisations may use different labels for what may essentially be 'fragments' of explicit knowledge, and there may be no consensus about this – even within organisations. Ideally, intra-organisational agreement should be reached about internal criteria and procedures before actually starting the processes of knowledge codification. For example, in Chapter 2, we discussed risk classification and alluded to the merits of alternative approaches such as generic or customised categorisation. While it should be possible to code project participants' tacit risk knowledge into either or both classification systems, there needs to be organisational agreement about which may be accorded priority, or which is essential and which is optional. The same issue applies to knowledge relating to qualitative measures for risk analysis and evaluation (Chapter 8), for response and treatment options, and for risk monitoring and control procedures (Chapter 10).

Ideally, project risk context knowledge should also be codified. Codifying adds value in the knowledge creation cycle through the capacity it provides for discovering and analysing patterns and relationships in information.

Storing explicit and codified risk knowledge is a matter of arranging, designing, and structuring the organisational KMS in a way that facilitates system access, uploading information, further transformation, updating, and knowledge sharing. Much of this requires strategic thinking and decision making about the KMS.

11.6.8 Sharing and Updating Knowledge

After transforming, codifying, and storing project risk knowledge, organisations need to establish appropriate ways to distribute, share, and update the stored explicit knowledge. A key step is to identify the most relevant activities of the organisation, and the means that people usually employ to look for knowledge about these activities when they need it. Knowledge should always be available in the right place and accessible at the right time.

Information technology has an important role in facilitating ways of sharing and updating mechanisms for risk knowledge management. The use of media such as intranet and internet platforms for project management communication permits access that can be controlled with a relatively high level of security.

The availability of a comprehensive risk knowledge database is critical to the successful maturing development of any PRMS. Risk knowledge should be integrated in a way such that it can be quickly disseminated and used throughout the organisation. A system that is not easily accessible will eventually become ignored – a casualty of the system decay mentioned later in Chapter 17. All the effort to identify, capture, codify, and store project risk knowledge will have been expended in vain.

Risk knowledge must be presented in a way that makes it user-friendly. The form of knowledge communication is vitally important. The success of YouTube as a contemporary knowledge sharing platform demonstrates this. Written documents may be less effective than diagrams. Diagrams may be less easily understood than videos, and visual animations may actually communicate better than video. In achieving effective access to, and communication of, risk knowledge, it is therefore vital to know who the audience(s) and users will be.

For sharing and updating risk knowledge, the following strategic considerations are necessary:

- Choosing appropriate techniques to support knowledge management activities that contribute to organisational objectives
- Ensuring the availability of resources such as people, time, equipment, and information technologies
- Committing the organisation to regular knowledge system review and performance improvement
- Continuously enhancing organisational risk knowledge and identifying the specific benefits and sharing of project risk management success and failure stories
- Monitoring knowledge management activities to identify and ameliorate barriers to their effectiveness.

Knowledge needs to be in the right place, but it cannot be allowed to remain static. Organisations should adopt a strategically proactive policy towards the dissemination of new knowledge as quickly as possible after it is created. While this can be done electronically through desktop alerts and email communication, physical gatherings such as seminars and 'communities of practice' meetings are valuable contributors to the learning process. They help people to use and validate the newly created knowledge and provide almost certain impetus to drive knowledge creation into another cycle.

While we have discussed the organisational knowledge creation cycle in considerable detail, and have advocated its strategic use in project risk management, there are aspects associated with it that need to be considered. These include additional issues relating to organisational culture, KMS alignment and information redundancy, the tools and techniques for eliciting and creating risk knowledge, and the development of organisational risk wisdom.

11.7 Additional Issues of Organisational Culture

Organisations sometimes do not know what their employees know, or how they access knowledge when they need it. This issue largely relates to organisational culture (see Chapter 12). Traditional power and governance cultures tend to focus on control, and where this is stultifying, staff may be reluctant to look for guidance in using the KMS. Such cultures tend to view knowledge as something to be guarded rather than shared. Covert cultures and subcultures tend to develop in such organisational environments.

Cultural change may have to be initiated and negotiated in the organisation in order to ensure that knowledge use and validation take place, as well as other activities in the knowledge creation cycle, so that the KMS is able to add real value to the organisation.

Dealing with organisational cultures with respect to information technologies and knowledge management is a matter of strategic management and responsibility.

11.8 KMS Alignment and Information Redundancy

If it is to be effective, the whole process of knowledge management should be aligned with the organisation's objectives. The large databases needed to service KMS can be expensive to create and maintain. They must be well managed and clearly demonstrate value adding outcomes, especially where they underpin a deliberate knowledge creation strategy.

In assessing what knowledge is important, it is essential to reduce any unwanted 'noise' or 'interference' (see Chapter 19) that may be acting as barriers to effective communication in the KMS.

The value adding connection between the KMS and organisational objectives should also be verified and measured, with the verification process carried out in a way that requires minimal extra effort by the organisation. For organisational PRMS, this is considered more fully in Chapter 17 (Planning, Building, and Maturing Project Risk Management Systems), but it is also applicable to wider organisational systems of knowledge management.

Knowledge does not remain relevant indefinitely. Redundant information should be eliminated from the system in a process of 'waste removal' so as to maintain system efficiency and effectiveness. How this culling is done, and how knowledge that has not been updated for long periods should be dealt with, will vary between organisations as a matter of strategic importance.

Information redundancy is often strongly associated with technological change, and such change provides a reliable clue when deciding what information is no longer relevant to the organisation. The technology may no longer continue to be relevant, or it might even become proscribed.

In the early stages of KMS development in an organisation, knowledge editing may have to be carried out manually. Ideally, however, the process should become at least semi-automatic. One way of achieving this might be through system flagging of knowledge areas that have not been updated or accessed over a predetermined period. The KMS could alert a responsible KMS manager (probably via email) for a delete/retain decision. Since many people have a natural 'hoarding' trait, suitable training in KMS editing might be required.

11.9 Tools and Techniques for Eliciting Risk Knowledge

Many organisations, and especially those that are small or medium in size, still focus upon quite complicated means of collecting and storing information to improve communication because it seems faster and cheaper to them. However, using methods that will facilitate the capture and reuse of knowledge as soon as it is created can be an effective and timely way to capture and transform tacit knowledge. Tools that will assist in at least some of the knowledge creation activities are shown in Table 11.1.

The entries in Table 11.1 do not indicate that the tools can *only* be applied for each of those activities, nor that they can *always* be used for that purpose, but rather that they are usually effective at those points. For many of the tools shown in Table 11.1, noticeable gaps in the knowledge creation process activities occur mainly in the system to system management of explicit knowledge, thus reinforcing our arguments for implementing effective systems for managing project risk knowledge in organisations.

11.9.1 Brainstorming Sessions

The effectiveness of brainstorming has been promoted in several of the previous chapters in this book. In project risk management, it is most often used to identify risks and to explore alternative risk treatments, but it is also a useful people-to-people tool in knowledge creation up to the point of storing the explicit knowledge. Unless the outcomes of brainstorming are deliberately captured, codified, and stored, even their tacit knowledge quality is likely to become degraded and rapidly lost.

11.9.2 Storytelling

Storytelling can be used to create more engagement and encourage more exciting sharing of knowledge among project participants. It is an effective way to share (and capture) tacit knowledge. People are invited to share their experience by recalling and recounting the 'stories' of their project. Since this is likely to reflect emotional intelligence, voice tone, language choice, and body language, audio and video-recording (with the knowledge and permission of the participants) are useful methods of capturing this tacit knowledge. Stories can be inspirations for change.

Storytelling may be conducted in a semi-structured manner through the use of template formats, but it is often best left to the reflective memories of participants. The extent of engagement of the audience with the presenter should be carefully monitored.

However, if the full knowledge creation cycle is not diligently pursued, storytelling just remains what it is – peoples' stories.

Table 11.1 Correlation matrix between tools and activities for risk knowledge creation.

| | Knowledge creation activities | | | | | | | |
| | Tacit – Tacit (people to people) | | Tacit – Explicit (people to system) | | Explicit – Explicit (system to system) | | Explicit – Tacit (system to people) | |
Tools	Use	Validate	Identify	Capture	Codify	Store	Share	Update
Typical in KMS								
Brainstorming sessions	✓	✓	✓				✓	✓
Storytelling		✓					✓	✓
Communities of practice	✓	✓	✓				✓	✓
Networking	✓	✓					✓	✓
Project reviews, debriefings, and 'lessons learned'	✓	✓	✓					✓
Mentoring and apprenticeships	✓	✓	✓				✓	✓
Induction and training courses	✓	✓	✓				✓	✓
Workplace design	✓	✓					✓	✓
People finders	✓	✓	✓				✓	✓
Intranets and IT platforms			✓	✓	✓	✓	✓	
Search engines and alerts	✓	✓				✓	✓	
Organisational culture	✓	✓					✓	✓

	Knowledge creation activities							
	Tacit – Tacit (people to people)		Tacit – Explicit (people to system)		Explicit – Explicit (system to system)		Explicit – Tacit (system to people)	
Tools	Use	Validate	Identify	Capture	Codify	Store	Share	Update
Typical in PRMS								
Risk identification tools (see Chapter 7)	√	√	√				√	√
Risk assessment tools (see Chapter 8)	√	√	√				√	√
Risk response and treatment (see Chapter 9)	√	√	√				√	√
Risk monitoring and control (see Chapter 10)	√	√	√				√	√
Project risk item records (see Table 7.14)	√	√	√	√	√		√	√
Project risk registers (see Tables 4.1 and 4.2)	√	√			√	√	√	√
Organisational risk register	√	√			√	√	√	√

11.9.3 Communities of Practice

Communities of practice comprise networks of people with common interests where participants can interact on a collective basis. Some construction companies encourage this type of network because it is a valid mechanism for the management of specialised knowledge throughout the organisation. If electronic communication facilities are used, the geographical restrictions on physical meetings imposed by remote locations may be avoided. Email interest groups and dedicated social networking internet platforms overcome the tyrannies of distance.

The discipline-based forums adopted among the project engineers and managers in the 'packaged' contracts in Case Study B (a rail improvement project) are a good example of communities of practice.

In a similar but simpler approach, weekly or fortnightly on-site 'toolbox' meetings for construction workers serve a similar purpose.

11.9.4 Networking

'Networking' is an important knowledge sharing activity typically carried out through attendance at seminars and conferences, visits to other companies, and membership of professional institutions. It is an effective means to access knowledge through other people and organisations. As a people-to-people tool, it works only in the using, validating, sharing, and updating activities of the knowledge creation cycle.

11.9.5 Project Reviews, Project Debriefings, and 'Lessons Learned'

In a project review, people and their knowledge about a project are brought together. The technique is very similar to a 'lessons learned' session, but the agenda is always interrogative, along the following lines:

- What was done well? Why? What were the beneficial outcomes?
- What was done less well? Why? What were the adverse consequences?
- What was done badly? Why? What were the adverse consequences?
- What could or should have been done to avoid or ameliorate what was less well done or done badly?
- How would that have improved outcomes, and by how much?

The conditions of where knowledge sharing happens are important in encouraging the right identification, storage, sharing, and use of knowledge for the benefit of the organisation.

The review questions suggested above are general rather than project risk specific. To achieve a risk perspective, additional questions are needed:

- What were the major uncertainties encountered?
- Did they align with the uncertainties we expected?
- Were our response treatments and monitoring and control procedures effective?

There is no default location for such reviews; it depends upon how the organisation wants the workforce to develop and interact in order to achieve the company's objectives. Ideally, reviews should be appropriately designed and planned in terms of location, the distribution of people, the occasions and durations, and the desired ambience in order to maximise the exchange of ideas among people working in the same area. They should be more structured than other people-to-people tools.

Such exchanges can enhance a positive organisational culture, reduce stress, and increase productivity and morale.

In practice, the occasion and conduct of project reviews are rarely consistent in many organisations, and the identification, capture, codification, and storage of the explicit knowledge derived from them are even more sporadic. Case Studies C, D, and F are examples where such organisational inconsistency occurs. Unless a more robust approach is adopted, organisational learning and the development of wisdom may become endangered.

11.9.6 Mentoring and Apprenticeships

Mentoring and apprenticeships are the classic traditional way of ensuring that the knowledge and wisdom of older generations are passed on to younger people. These approaches are not restricted to trades knowledge, nor should they be seen only as a one-way transfer of skills. In contemporary society, knowledge about new IT technologies is most often passed on tacitly from young to old – as anyone who has had to ask a grandchild to fix their smartphone problems will attest! Knowledge created through mentoring and apprenticeships generally remains tacit and rarely becomes explicit.

11.9.7 Induction and Training Courses

Induction and training courses are the formal explicit knowledge delivery mechanisms adopted by most organisations. While they are almost entirely intended for sharing and updating explicit knowledge, they actually stimulate the commencement of the next knowledge creation cycle as participants start to use the new knowledge. Their intrinsic value lies in knowledge dissemination.

11.9.8 Workplace Design

While not itself a knowledge creation tool, the design of contemporary workplaces can contribute to successful knowledge creation in organisations. Individual privacy may be highly valued by office workers and professionals, but any enclosed 'cell' design should be augmented by convenient nearby access to sufficient 'quiet' space for small gatherings of three or four staff, as well as additional facilities for larger meetings.

11.9.9 People Finders

'People finders' are important in allowing everyone in the organisation to have access to colleagues and experts, internally and sometimes externally. Some organisations require each staff member to develop their own individual profile to be accessible by others. The importance of creating an appropriate individual profile is that staff capabilities become more widely known. Essentially, these are CVs and should be regularly updated. How they are stored in a KMS and accessed is a strategic decision for the organisation. Typical 'people finders' also include directories and internet resources such as LinkedIn.

11.9.10 Intranets and IT Platforms

Nowadays, intranets and other IT platforms are essential tools in knowledge management. They allow knowledge to be shared quickly and efficiently within the organisation or between authorised organisations.

Larger organisations, especially those with multiple locations or separate companies in a group configuration, can afford to invest in customised software and more complex IT platforms. For basic use, 'open access' IT tools (i.e. available in the public domain) can provide a sufficiently effective platform for sharing organisational knowledge, at least in a system 'start-up' situation. 'Cloud'-based internet platforms with larger storage capacity offer acceptable security in most circumstances, but organisations still anxious about the security of confidential information uploaded to the 'shared' servers on these platforms can elect to purchase private servers in order to exercise more complete control over their information and knowledge resources.

11.9.11 Internet Search Engines and Alerting Services

The power of internet search engines and the opportunity to subscribe to internet-based alerting services are attractive to organisations and professionals wishing to have quick access to global information and to be kept informed of the latest news about selected topics. Internet search engines are offered by a variety of global IT companies, whose business models are based upon the income from the commercial advertising that automatically accompanies each individual search.

However, the use of these powerful tools exposes users to the threat risk of being inundated with information that is largely irrelevant unless care is taken to include appropriate filters in the search enquiry. Even with filters in place, where the retrieved information is excessive, there is always a danger that careless selection from it may lead to poor project decision making. Alerting services expose users to similar but usually less serious risks.

11.9.12 Organisational Culture

Organisational culture is not a direct knowledge creation tool, but works indirectly by encouraging sharing and feedback, and by avoiding any overemphasis on seeking to assign blame for failure. Positive organisational cultures also influence knowledge creation through the development of trust. This is discussed more fully in Chapter 12.

For example, encouraging an 'around the water cooler' staff conversation culture allows valuable tacit to tacit knowledge exchanges to take place. Some of these inevitably 'spin off' into explicit knowledge transformation.

As with networking, this is largely a people-related tool that is rarely applicable to system-related knowledge creation activities unless the culture encourages employees to more fully use and exploit the knowledge-based systems of the organisation.

11.9.13 PRMS-Related Tools

In each of the stages of the project risk management cycle (Chapter 4, Figure 4.2), tools and techniques are used to support project risk investigation and decision making. As indicated in Table 11.1, each has been presented and discussed in earlier chapters, and detailed descriptions are not repeated here. The important point about PRMS-related tools is that, besides their use in proactive risk management, these tools also have a project risk knowledge elicitation purpose and thus service the needs of risk knowledge creation and management. Table 11.1 shows how extensively they can fulfil this purpose.

While the knowledge creation cycle drives the generation of new risk knowledge, it does not thereby inevitably lead to greater risk wisdom in an organisation.

11.10 Developing Organisational Risk Wisdom

An efficient and effective KMS creates new knowledge that allows an organisation to grow by interacting with its new and past experiences. For the purposes of project risk management, people in the organisation should be able to access past risk knowledge before starting a new project and subsequently transfer their experiential knowledge from the new project back into the PRMS.

If done well, this process should also contribute to a more robust development of the organisation's risk profile, influencing its strategic appetite for risk and its preferred methods of dealing with project risks. The strategic thinking involved here is foundational in developing organisational risk wisdom. Over time, and nurtured by planned and deliberate reflection, it should grow into a positive culture within the organisation that will substantially influence project risk decision making. Where people trust the emerging risk management culture, and the continuously developing organisational wisdom that underpins it, their decisions about project risks will be made with greater insight and confidence. The key to developing wisdom is the deliberate retrieval of, and reflection upon, existing knowledge, as well as an appreciation of what knowledge might be lacking.

Achieving knowledge benefits for project risk management is only possible if the project risk knowledge created in an organisation is successfully incorporated into its PRMS architecture.

11.11 Project and Organisational Risk Register Architecture

Each organisation develops its own methods to record and capture risk knowledge, and organisations are complex systems themselves. Many factors contribute to the success of their actions, and each organisation reacts in different ways to the risk situations they encounter. Their creation and use of risk knowledge are therefore also likely to differ.

Broadly speaking, an organisation will want to harvest knowledge of actual risk experiences, develop and use individual risk registers for each new project, and maintain a formal organisational repository of risk knowledge (the ORR) informed by its risk experiences and capable of servicing its PRRs. From a knowledge management perspective, the arrangement of such information and knowledge records is important.

Project risks can be addressed during each of the three project phases they embrace: procurement, operation, and disposal (see Chapter 3). In each phase, the sequence and nature of activities vary, according to the type and scope of the projects undertaken. These variations influence the uncertainties associated with decision making and have a direct bearing on the project risks.

11.11.1 Capturing Project Risk Experiences

At any time, but particularly as activities are completed or 'closed off', a 'lessons learned' intervention can be carried out. A semi-continuous approach to risk knowledge capture is often more effective than leaving it to a one-off review occasion typically held at the end of the project procurement phase. One advantage of the semi-continuous approach is that some of the explicit knowledge gathered in this way is then available for reuse later in the same project. Table 11.2 illustrates a typical content format for capturing project risk experience knowledge.

Table 11.2 A project risk debriefing record template.

Project risk debriefing and lessons learned record			
Organisation:			
Project:			
Capture date:		Recorded by:	
Project phase:	Procurement	Operational	Disposal
Project stage:			
Risk event	**Consequences**	**Treatment**	**Lesson learned**
Description Code ref:	*Description*	*Description*	*Description*
Description Code ref:	*Description*	*Description*	*Description*
Description Code ref:	*Description*	*Description*	*Description*
Description Code ref:	*Description*	*Description*	*Description*
Knowledge transfer to PRR/ORR	Date:	By:	Verified:

The template suits a semi-continuous risk knowledge capture strategy since it can be used at any suitable time during project execution. Occasions might include project progress meetings or sign-offs for the interim completion of project stages or installation of components. Ideally, the captured information should be transferred to the PRR, and thereafter to the ORR, as soon as possible.

11.11.2 Project Risk Registers

The PRR is probably the key management support tool for project risk management, but it is also important for risk knowledge creation and management. Therefore, it warrants careful design.

A typical spreadsheet-based template was presented in Chapter 4 (Tables 4.1 and 4.2), and other computer-based applications are discussed in Chapter 18 (Computer Applications). Principally, however, the PRR configuration must be user-friendly, be sufficiently comprehensive to incorporate the necessary stages of the project risk management cycle (Chapter 4, Figure 4.2), have calculation capacity, and display explicit project risk knowledge and decision making.

Given these requirements, readers must make up their own minds about the value of a pre-PRR individual risk item record when deciding upon the configuration architecture for an organisational PRMS. The risk item record was presented in Chapter 7 (Project Risk Identification Tools) as a template for recording risk identification workshop outcomes. It comprises a formal record of each identified risk (Table 7.14) that incorporates a complete and precise risk statement, together with other risk-related information.

We admit that, given our professional training and experience, we have an ingrained preference for retaining some aspects of the traditional 'paper trail' generated by such information records; but with the prolific use of handheld electronic devices in contemporary society, there is little reason not to transfer the risk item record into the PRMS by keying in information directly, nor to bypass it altogether and use the PRR as the primary risk data entry point. This may require changing the organisational culture as well as suitably configuring the PRMS. In our defence, we remind readers that such individual risk records may have an evidentiary role in formal enquiries or legal proceedings.

An important potential side benefit of PRRs lies in the opportunities they present to develop project risk case studies, scenarios, and disaster recovery plans. Because the risk information is so cogent to the organisation itself, it is capable of providing realistic material that can be developed for training and practice purposes. Planning and rehearsing disaster recovery procedures, using information gathered from real-life project risk experiences, can contribute significant risk management value and capacity to an organisation. In this regard, the benefits of learning from experiences of failure will often exceed those derived from stories of success.

Beyond the project level is the ORR.

11.11.3   Organisational Risk Registers

As we noted in this chapter, the ORR is the larger repository of risk knowledge. The ORR may constitute the entire PRMS for an organisation, or it may be part of a wider integrated KMS. It is fed by information from each PRR and the knowledge captured from the risk experiences of each project, as well as from external sources. The new and existing project risk knowledge is blended within the knowledge system and cascades down to new projects proposed by the organisation. The feeding, transforming, and cascading cycle characterises a knowledge system that should thus be dynamic. An ORR should be aligned with the risk management objectives and policies of the host organisation, and with its wider organisational knowledge management aims.

How risk knowledge is managed and retrieved will determine the success of the ORR. The challenge is to highlight the valuable knowledge when it is needed, and to retrieve accurately relevant project risk information and knowledge as quickly as possible from what is likely to become a huge database of past projects. Since this knowledge accretion will occur over a relatively long time period, issues such as semantics, languages, cultural meanings, pronunciations, and even spelling differences can represent serious obstacles to system efficiency. The information redundancy issue discussed earlier in this chapter may also become critical.

Ideally, an interactive project risk management process should begin to access the ORR as soon as an organisation starts a new project. This is shown conceptually in Figure 11.3. It is represented as a sequentially numbered cycle of interactions between a project team (using a PRR) and the organisational ORR. Note that the ORR may be a dedicated risk-based system (or PRMS) or be integrated into a wider organisational KMS. The model also assumes a high level of automated artificial intelligence (AI)-based system activities and responses.

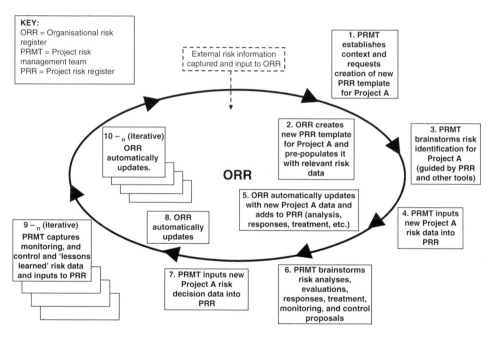

Figure 11.3 An interactive project risk management knowledge process.

The process follows the project risk management cycle outlined in Figure 4.2 (Chapter 4, Project Risk Management Systems) and starts when a project risk management team (PRMT) is appointed to manage the risks of a proposed new project and requests a PRR template (Step 1). The ORR automatically creates the new PRR template (Step 2) pre-populated with the relevant risk data available from the system and based on the information that the team has entered about relevant characteristics of the project.

With the new PRR template available, a brainstorming risk identification workshop is held (Step 3), using the template and validating the pre-populated information it contains. The team uses other tools for additional context establishment and risk identification (see Chapters 5–7), and double-checks for more recent external risk-related knowledge. The additional information is entered into the PRR (Step 4).

Once the PRR is updated, the ORR automatically updates itself (Step 5), adds further data (created by AI and machine learning technologies), and makes available additional knowledge about project risk assessment, risk evaluation, response and treatment options, and risk monitoring and control procedures. The team then proceeds with decision making about all these risk management tasks (Step 6), it updates the PRR accordingly (Step 7), and the ORR updates itself (Step 8).

From this point, the risk knowledge management process becomes more and more iterative as the project proceeds from planning and design to actual delivery and thereafter to operational readiness (Steps $9 -_n$ and $10 -_n$).

At each progressive interaction between the team and the PRR, full connectivity allows automatic updating to take place within the PRR and ORR. New knowledge from other sources is captured by the ORR and immediately shared with each relevant PRR. Knowledge in the ORR should thereby remain current in relevance and immediately available to benefit the risk management of new and existing projects undertaken by the organisation. In this sense, project risk knowledge in the organisation (and for its projects) is always 'live'.

Planning, designing, implementing, and improving such risk KMSs present challenges for any organisation.

11.12 Challenges for Implementing Risk Knowledge Management Systems

The challenges of implementing a risk knowledge management system (RKMS) in an organisation generally relate to issues associated with knowledge itself; the ways in which it is stored, accessed, and used; and the costs of system development and implementation.

11.12.1 Issues Relating to Knowledge Itself

The implementation of a KMS should embrace the kind of general knowledge and specific risk knowledge that the organisation wishes to manage. This will involve issues such as:

- Understanding what the appropriate knowledge is
- Understanding how that knowledge is acquired
- Understanding how organisations know what they know
- Understanding how organisations know what they need to know
- Understanding who needs to know what in the organisation
- Appreciating the usefulness of the knowledge stored in the system
- Having trust in the knowledge available.

Knowledge management generates immediate reactions about what knowledge is appropriate for the organisation and how it should be acquired and managed. Intrinsic to these are epistemological questions relating to knowledge itself: how we know what we know, and whether or not the knowledge created is trustworthy.

Each organisation needs to structure its knowledge management according to its needs. This will help to identify its most important internal knowledge areas and, to a large extent, the nature of that knowledge.

Addressing this challenge will also help to identify where risk management should be strategically located in the organisation. Risk knowledge management is important since it directly affects the capacity of an organisation to deal with uncertainty in achieving its objectives, particularly those relating to projects.

Furthermore, each department in the organisation needs to ascertain what its risk knowledge needs are in relation to the decision making it is required and empowered to carry out. Organisational scans are often needed to determine this.

11.12.2 Storing, Accessing, and Using Knowledge

Organisational KMS and PRMS can create challenges in the way they are structured, accessed, and used. Care is needed so that important knowledge is not lost.

As an example, in Case Study A (a PPP correctional facility), the contractor stakeholder in the special purpose vehicle (SPV) organises its knowledge by its main activities. These include: Commercial/Strategic and Tendering/Contract, Design (comprising buildability, constructability, durability, operations, and maintenance), Planning and Scheduling, Financial Management, Safety Management, Quality Assurance, and Environmental Management. 'Umbrella' labelling such as this can lead to specific knowledge becoming lost in the system or ignored because of the priorities accorded to some organisational activities.

Knowledge storage and access strategies will differ from organisation to organisation. Figure 3.11 (in Chapter 3, Projects and Project Stakeholders) illustrates another construction industry example, whereby activities are organised in different departments of a construction company, each under the responsibility of a director.

11.12.3 Knowledge System Development and Implementation Costs

When an organisation decides to implement a KMS, it will start to identify the core areas of its business and the level of investment necessary for developing the system. Large organisations are generally able to allocate sufficient financial and human resources for this exploration process, but that is not always the case for small and medium-sized entities. While there are many good reasons for the latter to adopt systems of knowledge management to capture risk initiatives, the system needs to fit the dimensions, capacity, and affordability limits of each organisation.

The balance between benefits and cost is actually a risk in itself, as the precise magnitude of any benefit is never certain. Organisational strengths and weaknesses form an important consideration here, and Table 11.3 indicates those for smaller organisations when compared to larger ones.

Table 11.3 Strengths and weaknesses of small and medium-sized enterprises (compared to large organisations) for knowledge management.

Strengths	Weaknesses
Internal structures tend to be more horizontal and flexible.	Fewer resources are available for testing new systems.
Management processes are more direct and less bureaucratic.	Long-term decisions are more difficult due to less strategic thinking.
Decisions can be made more quickly.	Organisational culture is often more reactive than proactive.
Faster reactions to market changes are possible.	Existing human resources are usually overstressed, with less time for self-assessment.

The table provides a good 'snapshot' check for organisations contemplating change, especially if this involves introducing or expanding a KMS.

Clearly, 'small' and 'medium' organisations may be strong in terms of flexibility but less so in regard to resources, including access to finance for knowledge system implementation.

Any organisation should have multiple strategies for implementing risk knowledge management, and these are influenced by the emphasis it places upon concerns such as return on investment, management efficiency, and the suitability of methods and tools.

11.12.3.1 Concern with Financial Issues and Return on Investment

Many organisations tend to regard issues related to project risk knowledge management as superfluous and thus not urgent. They can be postponed to the future. Costs and their effect upon the 'bottom line' usually take precedence.

However, this attitude fails to understand the importance of transforming the value of intangible knowledge and 'know-how' into more tangible assets. The transformed and tangible knowledge actually contributes as much as other assets to the goodwill value of an organisation. It does so by improving the chances of project success and by reducing the chances of incurring additional project costs through having to correct avoidable errors.

11.12.3.2 Concern with Time Management and 'Unproductive Tasks'

All organisations are concerned about the time that employees spend upon non-traditional and seemingly unproductive tasks. Time for employees to acquire new knowledge may even be regarded as an employee responsibility and therefore something to do outside working hours without cost to the organisation.

However, this attitude flies in the face of contemporary human resource management where career development is increasingly given greater prominence. An organisation that is overly reluctant to provide opportunities for staff learning might then have to deal with high staff turnover rates due to career dissatisfaction. This loses not only the tacit knowledge of employees who leave the organisation but also all the investment that was made originally in recruiting, training, and developing them.

On the other hand, some organisations might opt to seek new staff already equipped with the knowledge that is needed, arguing that it is cheaper and faster to do this. However, this option is likely to add further risks, as the integration of new staff into the culture and practices of the organisation is often difficult and time-consuming, and not always successful. Furthermore, the knowledge acquired in this way might not be exactly what was required.

Nevertheless, in both situations, an effective RKMS will help to capture new knowledge from the experiences of both existing and newly recruited staff.

Another challenge to implementing a KMS is that it will introduce additional management (i.e. oversight) activities to the organisation. Questions then arise as to who will be responsible for arranging and carrying out these activities. Some organisations respond to this challenge by creating separate systems for risk knowledge management and for other knowledge areas in the organisation. This is discussed more fully in Chapter 17.

Research into all the challenges has shown that 10 recommendations can be made for organisations that have decided to implement knowledge management (including risk knowledge management) systems (Vaz Serra 2011):

1) Establish a good culture of risk communication. Values of open communication and knowledge sharing need to be included in the vision and mission of the organisation.
2) Avoid information overload. The KMS must promote self-consultation. Automatic alerts and warnings need to be creatively designed and intrude only when needed in day to day activities, such as the introduction of risk alerts when starting new tasks.
3) Promote external access channels to link internal knowledge with external knowledge; the KMS needs to be connected to, and refreshed with, ideas from outside. The lessons learned from each project should be readily available for each new project, and links made available to external information about projects undertaken by others.
4) Appreciate knowledge as an intangible and dynamic asset, to be shared rather than hoarded. The organisation, through its KMS, should promote and value those who share knowledge.
5) Begin by implementing KMS in the area or areas of main business focus, and then grow the system according to the organisation's needs and acceptable pace of development. Attempting to implement KMS in all areas at the same time can increase the risk of failure. People, techniques, and planning are good starting points for a KMS focusing on project risk management.
6) Highly exclusive and separated KMSs fed only by dedicated experts have been identified as time and cost-consuming. Such implementation often transfers KMS gains away from those who create the tacit knowledge but who need to use the explicit knowledge. It may negatively affect KMS effectiveness in an organisation and create an unwanted subculture.
7) Unwarranted criticism of the system can inhibit participation. People should be able to easily see the advantages of KMS for themselves and not just for the organisation – a win–win process is needed.
8) Sharing knowledge in communities of practice helps to foster innovation. This could be included as an annual key performance indicator (KPI) for staff and organisational performance reviews.
9) Knowledge needs to be captured and reused in its 'live' phase as far as possible. The KMS processes should be incorporated into the normal activities of the organisation, being fed and consumed directly and immediately by those who create or share knowledge.
10) The KMS structure and processes should be as transparent as possible. 'Black-box' approaches tend to inhibit full understanding of content and outcomes. If an integrated PRMS is implemented, it should be capable of clearly distinguishing project risks and risk management through the inclusion of project and organisational risk registers.

To implement these recommendations correctly and efficiently, it is essential that communication in risk knowledge management is appropriate and sufficient.

11.13 Communication and Risk Knowledge Management

An important consideration in the implementation of a KMS relating to project risks is understanding how communication works (or should work) within the organisation and within each project. The relationships between project communication, organisational communication, and knowledge management must be effective for a risk management system to be successful. It is not possible to develop an efficient KMS or risk management system in a project or organisation where communication is not seen as a major contributory factor. Mature organisations place communication high on their strategic management agendas.

We discuss maturity levels for organisational risk management in Chapter 17 (Planning, Building, and Maturing Project Risk Management Systems), and risk communication is discussed more fully from a theoretical and systemic perspective in Chapter 19 (Communicating Risk), but it is essential to measure the quality and effectiveness of internal and external organisational communication before implementing a KMS.

In Chapter 3 (Projects and Project Stakeholders), Table 3.4 lists key decision elements and Figure 3.7 illustrates a typical process for project decision making. Chapter 3 also identified barriers to good decision making. Most of these relate to problems in communication: frame blindness, misreading historical evidence, and inadequate recording of decisions and outcomes. These areas of miscommunication usually arise from a lack of understanding about the importance of communication methodologies and the clarity and effectiveness (between senders and receivers) of the messages that they want to share.

At some time, we have all probably attempted to make or assemble things using instructions that were confusing or inadequate. Occasionally, we have been forced to seek the advice of others or consult internet sources for help. Organisational policies and procedural manuals can present the same difficulty, and need to guide and support employees in their decision making effectively. KMSs must adopt similar communication principles and processes.

Project performance is also improved when people communicate with each other and share good practices, lessons learned, relevant experiences, sensible advice, or opinions. An organisation seeking higher levels of performance depends on its capacity and capability to innovate, to conserve and augment its knowledge resources, and to disseminate them throughout the workforce. Employees are therefore communication receivers and senders in this regard.

It is also important for an organisation to learn from its mistakes, but the people involved will only feel comfortable about communicating and sharing such knowledge if a genuine culture of non-blaming exists in the organisation. Employees should be encouraged to share the full extent of their tacit knowledge, and organisations should openly acknowledge the value of such contributions.

The ultimate goal of knowledge management is to add value through its associated activities and processes. A strong emphasis on knowledge management in the organisation's strategic thinking and planning, and the integration of knowledge management activities with all of its management systems, are crucial contributors to the value adding chain. Effectively leveraging knowledge management can allow an organisation to become more adaptable, innovative, intelligent, and wise. All of this depends upon effective communication.

Therefore, a first step for KMS implementation is to develop a strong communication culture across the organisation, with sharing as a core value. As we will suggest in Chapter 12 (Cultural Shaping of Risk), entrenched cultures can be difficult to shift and change, and an organisational scan may be needed before embarking on such a venture. Since communication pervades every aspect of an organisation's activities, attempting change here can be a daunting prospect. However, when an organisation has confidence in the effectiveness of its communication procedures and tools, this will inspire similar confidence towards developing and implementing appropriate KMSs and processes.

Good organisational communication means not only that employees have all the information that they need at hand, but also that they can access it easily and quickly, when and where it is needed. Effective channels (media) for communication are therefore vital.

Internet search engines make an enormous amount of information available, but that does not automatically make everyone enormously knowledgeable. Nor does a large library, if it is seldom used. It is the availability, access, and *use* of such resources that make us more knowledgeable. If information and knowledge are hard to access or will require long periods of time to understand, our decision making is rendered less effective. A good communication system also needs to reduce communication barriers such as noise and interference (see Chapter 19) and minimise the intrusion of less relevant communication channels and messages.

In project management and project risk management, the effects of globalisation and different industry practices and terminology result in different ways of expressing the same thing. We alluded to this briefly in Chapter 4 in discussing standards for risk management. While terminology may be noise-free and interference-free when used in its particular context, communication barriers may emerge if such contexts have to be considered in combination. Graphical project communication media (plans, drawings, pictures) may be susceptible not only in this regard, but also in individual contexts since they often require greater and more highly trained cognitive interpretation – despite the adage of 'a picture saying more than a thousand words'.

Artificial Intelligence (AI) technologies help to reduce communication 'noise' and 'interference' barriers, and a method such as Natural Language Processing (NLP – using computers to understand languages, text, or speech) can help in the translation of terminologies into a common language that is easily shared. Machine translation, question answering, speech and voice recognition, and information retrieval technologies are also useful. An approach such as case-based reasoning (CBR), by implementing a 'four R's' process (retrieve, reuse, revise, and retain) for improving understanding, may assist with establishing a consistent rationalising approach.

In the architecture, engineering, and construction (AEC) industry sector, a contemporary and determined shift towards harmonising and developing collaborative digital tools in the form of building information management (BIM) systems is primarily intended to allow many construction problems to be minimised at the building design stage by improving overall information and knowledge sharing. This allows for project delivery to be made more efficiently with smarter systems, with better quality, safety, and environmentally sustainable outcomes. Such collaborative techniques rely heavily on effective communication.

Effective communication is therefore at the heart of effective project risk knowledge management.

11.14   Summary

The implementation of KMSs in organisations may require a profound change of culture, not only at the individual level, in learning the advantages of knowledge sharing, but also at the organisational level.

Organisations should create environments (tangible and intangible) that encourage the creation and sharing of knowledge as a means of adding value to their activities and outcomes. Good communication is fundamental to this aim.

Knowledge in an organisation needs to be managed systematically and made available for use in the right place at the right time, and in the right amount. Successful knowledge management depends upon achieving a culture in the organisation whereby people are willing to share their knowledge and experiences and are receptive to learning through the experience of others. Trust is an essential ingredient.

The inevitable pressures of time mean that people have to prioritise their knowledge needs. To some extent, this means that they place greater importance on their own tacit knowledge and may become resistant to augmenting that with explicit systemic knowledge. This dependency can only be addressed through the availability and accessibility of good KMSs capable of delivering knowledge that is highly relevant to the users' needs.

Knowledge relevance is dependent upon effective knowledge capture and transformation. The tacit knowledge that so often provides the essential lubricant for carrying out work must be harvested and transformed so that explicit knowledge is made more widely available. 'Lessons learned', performance reviews, storytelling, mentoring, and communities of practice are key ingredients in this process.

Good project risk knowledge management, through effective and efficient systems, enhances an organisation's capacity and capability. It encourages innovation and improves project performance, thereby adding value to the organisation and increasing its competitive advantage. Above all, it is an essential precursor to the development of risk wisdom.

References

Anumba, C.J., Kamara, J.M., and Carrillo, P.M. (2005). Knowledge management strategy development: a clever approach. In: *Knowledge Management in Construction* (ed. C.J. Anumba, C.O. Egbu and P.M. Carillo), 151–169. Oxford: Blackwell Publishing Ltd.

Davenport, T.H. and Prusak, L. (2000). *Working Knowledge: How Organizations Manage What They Know*. Boston, MA: Harvard Business School Press.

Drucker, P.F. (1993). *Post-capitalist Society*. New York, NY: Harper Business.

Fruchter, R. and Demian, P. (2005). Corporate memory. In: *Knowledge Management in Construction* (ed. C.J. Anumba, C.O. Egbu and P.M. Carillo), 170–194. Oxford: Blackwell Publishing Ltd.

International Organisation for Standardisation (ISO) (2018). *Risk Management – Principles and Guidelines* (ISO 31000. Geneva: International Organisation for Standardisation.

Nonaka, I. and Takeuchi, H. (1995). *The Knowledge-Creating Company: How Japanese Companies Create the Dynamics of Innovation*. New York: Oxford University Press.

Palmer, J. and Platt, S. (2005). *Business Case for Knowledge Management in Construction* (Publication C642). London: CIRIA (Construction Industry Research and Information Association).

Vaz Serra, P. (2011) *Knowledge Management at the Construction Company*. Unpublished PhD thesis, University of Lisbon, Portugal.

12

Cultural Shaping of Risk

12.1   Introduction

A common perception of 'cultural' risks in projects is to see them as the concerns of organisations operating, or planning to operate, in foreign countries. Thus, we hear about cultural issues facing a British company building a bridge in South East Asia, an American telecommunications company developing broadband communications facilities across a chain of Pacific Islands, an international conglomerate searching for oil reserves in foreign territorial waters, or a company wishing to target particular international countries with its latest hospital management system. There are many more examples.

In such instances, the concerns will generally relate to matters such as communicating effectively with foreign politicians, officials, and local workers; acclimatising and managing expatriate staff; dealing with differences in national customs; business practices and ethics; and acknowledging religious observances. The emphasis is on local sensitivities and their impact (generally anticipated to be potentially negative) upon the project.

While such concerns are entirely valid, we would argue that they are contextually skewed and constrained with regard to risk and culture. They look at project risk from the wrong end of the risk management microscope.

In Chapter 2, we proposed that risk is perceived and experienced by people. Our perceptions are influenced intellectually and emotionally by our world views, and by our beliefs and values. The risk construct is thus psychosocial. Since those views, beliefs, and values derive from cultural perspectives and practices, we can argue that our risk perceptions are culturally shaped.

In fact, all risks are culturally shaped to some extent. It is therefore misleading to create a separate category for 'cultural risks' in any risk classification system. For example, consider the threat risk of industrial action occurring during the project delivery stage, with consequences of delay and extra cost. If the project is being undertaken by an Australian contractor in another country, known for its radically socialist government policies and militant trade unions, the risk cannot be classified as a 'cultural' risk. It is still a 'political' risk, but influenced (i.e. shaped) by the political culture of the foreign country; and we might expect that the likelihood of the risk event occurring there is greater than in the contractor's less volatile home territory. Any response to the threat risk should recognise the cultural effect and the special situation. Note, however, that

Managing Project Risks, First Edition. Peter J Edwards, Paulo Vaz Serra and Michael Edwards.
© 2020 John Wiley & Sons Ltd. Published 2020 by John Wiley & Sons Ltd.

for a similar project undertaken in Australia, that contractor would still have to consider the cultural shaping of the risk as determined by the industrial relations policies and legislation of whatever flavour of Australian government was in power at the time (or likely to gain power during the delivery of the project).

It would be fair to claim, therefore, that project risks (threat or opportunity) in all human risk categories, according to the generic classification system presented in Chapter 2 (Table 2.2), are susceptible to cultural shaping. Table 2.2 should be corrected to omit its separate 'Cultural' risk category label. Ideally, the remaining category labels should each have a subtag '+Cultural Influence?' added.

Indeed, it can be argued that the cultural shaping perspective applies not only to human risks, but also to at least one of the natural risk categories proposed in Table 2.3. In some rural areas of Africa, old rubber tyres are placed on the corrugated sheet metal roofs of informally built houses according to a (culturally based) tribal belief that this will provide protection from lightning strikes (mitigation of impact) as the rubber will force the lightning flash to bounce back into the atmosphere. Thus, the natural weather risk of lightning strikes is culturally influenced. Note that the shaping of this risk response concerns the consequences of lightning strikes, not the likelihood of their occurrence.

In this chapter, we first consider culture and its influences in society (external to the project), and then examine it in the form of organisational cultures (internal to the project). The ways in which cultures shape risk in the project environment are considered, and suggestions are offered for dealing with this in project risk management.

12.2   Culture in Society

For many people, 'culture' in society is perceived as relating primarily to matters involving the arts (particularly the performing arts), music, and literature. This derives from dictionary definitions of culture as relating to the intellectual, artistic, and social development of a group. The more 'classical', 'serious', and prolific the level that these activities are considered to exhibit in a society, the higher their level of culture is deemed to be (at least by those living within it and assessing it) and the more greatly they are valued. Conversely, manifestly similar but less classical activities in the same disciplines, while possibly more often displayed, are regarded as less serious and of lower cultural value. The latter is often denigrated as 'pop culture'. However, societal values change over time, albeit slowly and unequally, across societies and their members. For the most part, 'pop culture' is far less often seen as a negative phenomenon in the twenty-first century than it was in the twentieth.

Culture in society is found in almost all areas and at all levels. Without limiting cultural diversity, we might, in addition to arts, music, and literature, find it in:

- Religion and religious practice
- Sport and recreation
- Public services
- Health practice and management
- Law and justice
- Education

- Research
- Industry and commerce
- Professions, trades, and other occupations
- Emergency services
- Security and defence forces.

Furthermore, cultural diversity in a society may be nationally, regionally, or locally differentiated, and is often distinguished by factors such as ethnicity, age, gender, and sexual preference. It is rarely perceived from a neutral perspective, but tends to attract positive or negative views that are extreme, thus making it important in risk management. It affects attitudes, behaviours, and decision making.

Given all this, a working definition of 'culture' might be:

> The influences that the ideas, assumptions, values, beliefs, experience, and knowledge of a society (or an identifiable part of it) bear upon its activities.

Culture in society is a far bigger field of knowledge and research than we are able to deal with in this book, especially with regard to understanding it and being wise about adopting and applying it in our everyday decision making. Our focus must now shift to organisational cultures and their relevance in project risk management.

12.3 Organisational Cultures

Since organisations, whether public or private, are essentially mini- or even micro-societies, it follows that they are likely to embrace mini- and micro-cultures. We can simply call them 'organisational cultures'. With some confidence, we can also expect to find *inter*-organisational similarities and differences in organisational cultures. Less frequently, we may encounter *intra*-organisational differences (i.e. particular cultures that may operate in one part of an organisation but not in another). For example, Friday afternoons are often treated differently by road workers in the public highways department of one state in Australia. Workers on a rural road project will officially finish work at 3.30 p.m. on Friday afternoons and then gather at the nearest local inn for after-work drinks and social interaction. Their counterparts on a city road project will also finish work at 3.30 p.m., but will then either go straight home or go individually to other entertainment activities in the city. The 'pub' culture of the rural environment has been replaced by a different social culture in the city (but we do not claim that one is better than the other).

Mimicking culture in society, organisational cultures derive from the assumptions, beliefs, and values that an organisation holds. An organisation may *assume* that it is operating within an appropriate governance structure or that its technologies are 'cutting edge'. It may *believe* that its workforce is satisfied, that its market share is unassailable, or that its competitors will never play fair. It may claim that its primary *values* include people over profit, that it desires to be 'sustainable' in its processes and products, and that none of its activities should render harm to the environment.

Organisational assumptions may be based (wholly or in part) upon objective, verifiable information; upon subjective opinion; or upon uncertain or faulty information. Organisational beliefs may be logically founded, historically grounded, or illogically and socially biased. Organisational values may be *espoused* or *enacted*. Espoused values are those that the organisation desires to hold, or those it professes to hold, but which are not self-evident. Enacted values are those that are manifestly held and practised in the organisation.

In the context of project stakeholder risk management, we can expand our earlier definition of 'culture' as:

> The influence that the ideas, assumptions, beliefs, values, experience, and knowledge existing in an organisation, together with those of the external society in which it operates or plans to operate, exerts upon the decisions it makes and the risks it faces and desires to manage.

Before exploring how organisational culture exerts its influence on project risk management, four further aspects must be considered.

Organisational cultures may be *overt* or *covert*. Overt cultures are transparent, open to the world. They are readily detectable on company websites, in mission and policy statements, and from other evidentiary or verifiable sources. One caution here is that the values underlying an overt culture should also be verifiable as enacted values. Espoused values may show potential, but they need to be demonstrated in practice. If an event project promoter claims to value spectator safety over event spectacle, then this must be demonstrably true. An example would be where a concert event with an internationally famous star, to be held in a large sports stadium, is cancelled after the discovery of evidence pointing to the likelihood of a serious terrorist attack. The concert promoter's spectator safety value is clearly enacted. If higher levels of security were implemented in lieu of cancellation, this would support an espoused value to some extent but would fall short of incontrovertible evidence for an enacted value, as priority is still being given to event performance.

Covert cultures, in contrast, are opaque and hidden from the world or even within the organisation itself. A covert culture is difficult to detect, especially if it exists at a high level within an organisation. The board of directors of a bank may explicitly claim to give priority to client protection in endorsing a new investment project, but actually adopt an implicit culture of prioritising profit performance over investment project risk. This is not to say that covert cultures are always bad – a company might, without declaring this openly, decide to give a small percentage of the profit on each of its projects anonymously to a nominated charity. That said, however, it should be noted that corruption is more often than not linked to a covert culture somewhere in the offending organisation. After the 2010 FIFA Soccer World Cup in South Africa, the Competition Commission of South Africa probed alleged bid rigging and anticompetitive conduct associated with the contracts for the construction of major football stadiums, as well as road, rail, and infrastructure projects connected with the event. Major publicly listed construction firms were implicated, and fines totalling ZAR1.5 billion have been imposed. The offences discovered by the Commission can only have been instigated by covert inter-organisational cultures operating at high levels within each of the colluding companies.

Organisational cultures may also exist as *subcultures* within an organisation. Assumptions, beliefs, and values in a subculture are distinguishable from those in its larger host culture and may even contradict them.

Governance structures and decision making power are areas where subcultures are typically found in an organisation, often appearing after companies have been transformed through the mechanisms of organisational change. Remnants of the old order may decide that they are 'the old hands who know what really goes on' and surreptitiously subvert new decision making processes to maintain their former power.

A third consideration is that all organisational cultures are, to a greater or lesser extent, influenced by the wider cultural context of the societies in which the organisations exist and operate. For example, if a nation, as part of a forgiveness and rehabilitation culture, offers direct support and encouragement for reintegrating and finding work for ex-offenders released from prison, an organisation might adopt this as part of its own culture and make specific employment opportunities available on the projects it undertakes.

Finally, we need to recognise that organisational cultures are dynamic, just as they are in wider society. Cultures are susceptible to change, much like the characteristics of risk events – likelihood and consequence – are dynamic and may change over time, as we noted in Chapter 2. However, cultures can take far longer to change, even when the intent is deliberate! Despite this seeming inertia, a prudent project stakeholder organisation will periodically review the wider society, and scan its governance structures and operational practices, to determine what cultural shifts may be happening. This can lead to 'cultural appropriation', whereby cultures from another context or organisation are adopted. For the most part, this may be a positive, overt, and enacted process to improve organisational effectiveness and efficiency. Sometimes, the appropriation is left as an espoused value that does not become enacted. For example, an organisation that appropriates a popular contemporary business value such as 'our people are our best resource' for its internet masthead, but then proceeds to retrench a significant proportion of its workforce, can hardly be said to be enacting that culture.

Figure 12.1 depicts the concepts we have discussed so far in a multidimensional model of the elements of organisational culture. It provides a model against which organisational cultures can be examined in more detail through the use of organisational scans.

12.3.1 Organisational Scans

Exploring organisational culture for the purpose of project risk management is best done by scanning the mechanisms through which the organisation conducts its business. Table 12.1 shows a list of typical mechanisms, but this list is not exhaustive.

The scan is intended to reveal areas of the organisation where the existence of an organisational culture will influence the ways in which the organisation undertakes and delivers projects. Once these influences are known, consideration can be given as to how they affect the project risks (threat and opportunity) that the stakeholder organisation faces. Strategic decisions can also be made about changing or removing those cultures.

It is usually not necessary to conduct comprehensive scans for every project. Where projects tend to be similar in type, scope, and nature, the frequency of organisational scans should be determined by the typical duration of such projects and the detectable rate and nature of any organisational change that is taking place. Thus, for a small organisation that undertakes perhaps fewer than 50 similar information technology (IT) projects each year, each valued at less than about $250000, where the organisation

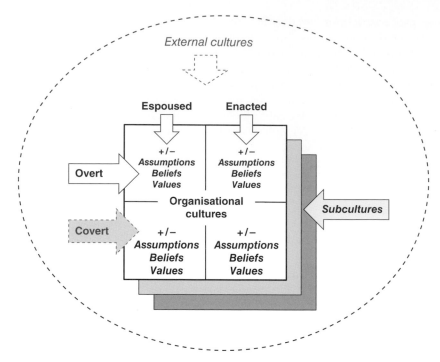

Figure 12.1 Elements of organisational culture.

enjoys a stable structure and workforce, and its project market is also relatively stable, then an organisational culture scan conducted every four or five years might be sufficient. However, should significant change be taking place in any of the underlying areas, such as increasing levels of market competition in a declining economy or facing a more volatile workforce in a booming economy, then the scan frequency should increase.

For stakeholder organisations undertaking much larger projects of much longer duration, for example multi-million-dollar infrastructure development projects, the interval between organisational culture scans could be extended to perhaps six to eight years. However, *mega-projects* (i.e. >$1 billion in value) might warrant an organisation scan for each one. Scan intervals beyond eight years are not advisable, given the dynamic nature of many project risks and the contemporary rates of technology innovation. A guiding precautionary principle might be: if the existing organisational culture, or some part of it, is likely to exert a highly significant influence on a particular project, then conduct the organisational scan as part of the risk management process for that project.

Scans of organisational culture should not only search for evidence of cultural influence, but also distinguish if it is negative or positive. While the preferred emphasis may be on discovering negative cultures with the intention of mitigating them in terms of their impact upon project risks, positive organisational cultures can be considered with a view towards reinforcing them, ensuring that they are spread consistently throughout the stakeholder organisation, and seeking to exploit them in the management of project risks. Examples of negative and positive organisational cultures are shown in Table 12.2.

Table 12.1 Typical areas for culturally influenced organisational practice.

Organisational structure and practices

Governance structures

Power distribution

Leadership style

Location(s)

Language preference and style

Appearance
- Uniforms or dress codes
- Logos, images, and branding
- Furnishings, etc.

Gender distribution

Recruitment and human resource management

Competitive practices

Environmental practices

Occupational health and safety

Intra-organisational communication

Public relations

Relationship management

Marketing

Corporate social responsibility

Table 12.2 Negative and positive organisational cultures.

Negative	Positive
Denial of liability	Acceptance of responsibility
Inflexibility	Flexibility and resilience
Distrust	Trust
Executive distance	Operational closeness
Oversimplification	Complexity acknowledged
Bare legal compliance	Ethical and legal compliance exceeded
Blaming and shaming	Learning through failure as growth
Knowledge as power and advantage	Strategic cooperation and sharing
Secrecy	Openness

Again, the list is not exhaustive, but we can offer an illustrative example that brings together factors from Tables 12.1 and 12.2. As an example of an organisational culture, consider the form of language typically used in most English speaking police forces when reporting incidents in written or verbal communications:

A male person of interest has allegedly discharged a firearm which has left a female person laying bleeding on the footpath. The female person has been taken to hospital but has been declared dead on arrival.

The 'formal' language culture adopts a verb conjugation that is quite convoluted. Instead of a simple past perfect style ('the man fired a gun'; 'the woman he shot died on the footpath'), the verbatim report actually implies a continuance of actions that is impractical and rather weird. Consider a similar verb construction in a phrase such as 'The actor has appeared in several productions'. This simply implies that she or he has appeared in at least three ('several') productions in the past, and *continues* to appear in productions now (but not necessarily the same plays). Of course, the connotation is different if the perfect tense is used: 'she appeared in several plays' would be appropriate if referring to an actor who has died. We are sure that the intention of the police report in our example was not to imply that the suspect fired a gun on several occasions in the past and is likely to do so in the future, nor that the victim is likely to die on the footpath again at some time, and certainly not that the hospital is likely to repeat the declaration of death for that victim again! The use of such a 'formalese' language culture, rather than encouraging the use of plain language, actually runs the threat risk of obfuscation. In communication theory (see Chapter 19), we might regard it as message interference or 'noise'.

12.3.2   The Organisational Scanning Process

Scanning an organisation in terms of organisational culture is an interrogative process of forensic enquiry. It is a relatively high-level strategic management task that should be led by, or at least include, representatives of senior management. However, such scans are likely to be neither complete nor conclusive unless other levels of operational management are involved at appropriate points. Senior management may encounter difficulty in detecting cultures operating covertly at the 'shop floor' project level, and, in the interests of organisational openness, lower level management should be able to explore the practices and decision making processes of senior management. It may even be necessary to involve line workers in scans. Consultants can be employed to facilitate scans and minimise selective bias, but the scan is essentially an internal activity of honest self-examination.

The scanning process is framed around seeking answers to questions that are addressed initially to the organisation as a whole and subsequently to areas within it, such as those indicated in Table 12.1. Wherever possible, a project focus should be maintained, since the eventual objective is to assess the influence of organisational culture(s) on project risk.

Typical questions might include:

- What are the assumptions, beliefs, and values that influence our organisational decision making (on projects)?
- What overt organisational cultural factors are involved?
- What covert organisational cultural factors can be detected?
- What organisational subcultures exist, and how do they exert influence?
- What cultural certainties exist?
- What cultural uncertainties can be detected?
- How much of each cultural influence is negative, and how much is positive?
- What organisational cultures should/could be changed/improved?

Since the process will include the collection of factual information together with perceptions and opinions, the use of internal surveys, with well-designed and carefully administered questionnaires that include 'open' questions, is probably a good starting point to explore at least the first four questions. Care is needed to ensure confidentiality and to provide 'whistle-blower' protection if needed.

After suitable analysis, the survey findings can be considered in intensive workshops conducted in each major department, together with the remaining four questions. Workshops should not become protracted, nor laden with other agendas.

There is a danger of intra-departmental bias occurring, particularly with respect to positive and negative evaluation of organisational cultures. Workshop participation should therefore always include at least one 'neutral' participant free of bias.

The outcomes of departmental culture self-examination workshops should be summarised and reported to senior management, accompanied by appropriate recommendations. This reporting, and the decisions flowing from it, can be formalised (and archived) and transformed into guidelines for relevant project risk management activities.

Before considering the application of cultural influences in project risk management, it is important to explore the nature and influence of external cultures (i.e. beyond the stakeholder organisation).

12.4 External Cultures as Project Risk Shapers

The exploration of external cultures and their influence on project risks is similar, but not identical, to the scanning process undertaken within the organisation. Media scans are an appropriate mechanism.

12.4.1 Media Scans

Carrying out media scans for investigating external culture requires nothing more than being consistently and selectively alert to the messages and information distributed in popular media such as newspapers, magazines, journals, television, and film. All of them reflect, in various ways, the nature of culture and the cultural activities in the local, regional, and national environments where you live and work. This includes business and industry cultures.

With minimal training, and maximum encouragement, management staff in an organisation will easily and quickly learn to identify and capture knowledge about cultures that will inform the organisation's context setting approaches to risk management.

While these external cultures are usually deep-rooted and resistant to change other than in the long term, around their fringes they are often more dynamic and volatile in the short term. 'Pop culture' is a good example – even if its relevance to risk management may seem remote!

Cultural alertness scanning of this nature allows organisations to answer the question: What is happening in the world in terms of culture that could affect our risks on this project?

Follow-up questions are needed to properly frame the cultural references observed in the media:

- What local, regional, or national cultures are portrayed?
- What business or industry cultures are revealed?
- To what extent are these cultures positive or negative?

External cultures are generally overt, in the sense that they should be easy to find through the popular media. It is far more difficult to detect covert external cultures, at least with any reliability, but any organisation should be alert to this possibility. Questionnaire surveys would be an impracticable and expensive method. Critical reflection upon the media observations, aimed at 'reading between the lines', is probably the best approach. This can be done individually or in workshops similar to those advocated in Section 12.3.2 for assessing organisational culture scans.

As an example, 'cronyism' (Neilsen 2017, p. 121) is a form of corruption that favours friends at the cost of ignoring or manipulating the processes of fair competition. Cronyism rarely operates in an overt manner, but is more often practised covertly. Despite this, rumours are likely to abound, and a prudent project stakeholder should consider them carefully in its project risk management.

For dealing with the influence of external cultures on project risk management, the processes of reporting and escalating workshop outcomes are similar to those adopted for examining organisational culture.

In terms of frequency, media scanning of external cultures may be semi-continuous, but not rigidly pursued. Also, tapping into a variety of media sources is better than relying on only a few conventional channels.

Where an organisation proposes to embark on a project that is significantly beyond its arc of experience in terms of project nature, scope, or location (i.e. where uncertainty is likely to be high), then specific consideration of external culture should be included as an integral part of the risk management process for that project. Apart from the external cultural perspectives noted here, the organisational cultures of other project stakeholders should be considered.

12.5 Organisational Cultures of Other Project Stakeholders

Care is needed in project risk management to ensure that the organisational cultures of other project stakeholders are not overlooked. They cannot properly be regarded as external cultures since, if they were not stakeholders in the project, your organisation would not be particularly interested in them as they could not affect your involvement and the risks to which you are exposed. Nor, however, can they be considered under the cultural umbrella of your organisation, since you can do little or nothing about changing them.

Identifying the organisational cultures of other project stakeholders is no simple task. You can hardly scan their organisations, and there is likely to be little information available through broad-stream media. Previous experience gained from working with a project stakeholder on other projects is clearly an advantage, as that may provide valuable pointers to its organisational culture as practised in a project setting. So too, if you have not had direct experience, is information obtained from other stakeholders who have worked with the organisation in question. However, acquiring such information might be difficult, and its reliability and accuracy might be tainted. Some information might be retrievable from the corporate websites of other project stakeholders, but questions might arise about espoused and enacted cultural values. Certainly, the covert cultures and subcultures of another project stakeholder will not be broadcast to the world. The information is therefore likely to be incomplete and possibly uncertain.

Despite all these difficulties, the organisational cultures of other project stakeholders do warrant consideration in project risk management.

Next we turn to the process of applying cultural influences, in terms of the ways in which they shape project risks.

12.6   Applying Cultural Shaping in Project Risk Management

Once the cultural influences for project risks are identified, consideration of them must be incorporated into the project risk management cycle. Failure to do so can exacerbate risks. For example, an architectural firm was invited to submit designs for an arch feature to span a major road at the entry to a major Australian city. The arch was intended to make a visual but temporary contribution to the celebration of a significant event in the city. The architects were renowned for their radical approach to design and believed that they should demonstrate this on all their projects (an overt culture at least partly enacted). While the arch design carefully accommodated all conceivable traffic situations in terms of height and width, the innovative use of materials included steel and PVC tubes of varying diameters and lengths arranged in a 'porcupine' array. Unfortunately, the array left the structure vulnerable to accidental damage that could compromise its integrity. The failure to consider all the implications of the radical organisational culture raised the level of accident risk in this project.

In another example, the governing council for a large private high school maintained a culture that advocated the supremacy of uninterrupted continuity in all teaching activities. Adherence to this culture resulted in a major building alteration project going substantially over budget due to the effects of the culture on scheduling construction activities.

In both examples, threat risks were adversely shaped by organisational cultures. For each project, a more considered approach to culture at the outset, as part of the risk management process, could have enabled the project stakeholders to reach suitable compromises.

With reference to the risk management cycle described in Chapter 4 (Project Risk Management Systems) and illustrated in Figure 4.2, the initial consideration of cultural influences would be undertaken in Stage A (establish internal and external contexts), and their application would occur in Stages C_1 (analyse risks) and C_2 (evaluate risks), Stage D (respond to risks), and eventually Stages E (monitor and control risks) and F (capture risk knowledge). These stages were also described in more detail in Chapters 5 and 7–10.

The approach to dealing with cultural risk shaping should be deliberately undertaken as proceeding progressively from the general to the specific.

Figure 12.2 draws on Figure 12.1 and Tables 12.1 and 12.2 to show graphically how consideration of the influences of organisational and external cultures flows through to project risk management. This figure is quite complex, so we will provide a 'walk-through' explanation.

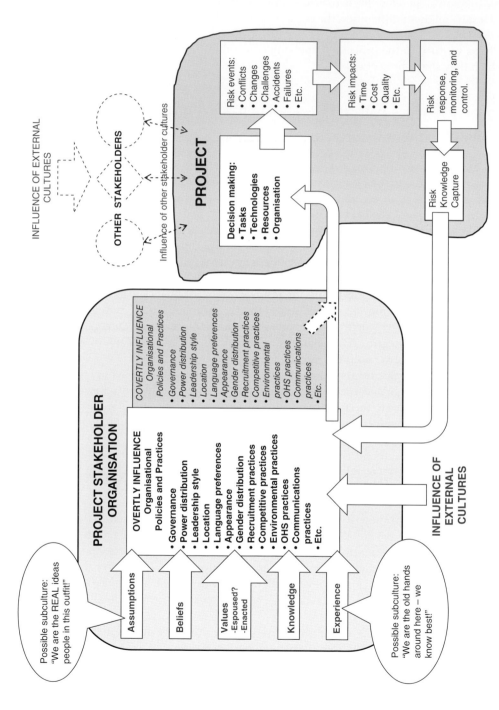

Figure 12.2 Stakeholder-to-project cultural risk shaping and management.

From Figure 12.2, we see that the knowledge and experience of an organisation, together with the assumptions, beliefs, and values that form its organisational culture, influence its organisational policies and practices. These are also influenced by subcultures operating within the organisation, and by the effects of cultures emanating from diverse external local, regional, national, industry, and professional sources. However, besides the overt influence (that is evident to the organisation itself and to the world), there is also likely to be a covert influence (known at least in part to the organisation but hidden from the world). The cultural influences are brought to bear upon the projects the organisation undertakes as a project stakeholder, through the decision making associated with project tasks, technologies, resources, and project organisation, and thereafter to the processes of project risk management. The known organisational cultures of other project stakeholders are taken into account. The risk management loop is completed when specific project risk knowledge is captured and absorbed back into the stakeholder organisation.

If we use Figure 4.2 (Chapter 4) to examine the role of culture in the project risk management process more closely, we might find that in Stage A (establishing internal and external contexts) cultural awareness is explored by addressing questions such as:

- What external cultural factors (local, regional, national, business, and/or industry) are relevant for this project?
- Which external cultures are negative and which are positive in effect?
- What organisational cultures are relevant for this project?
- What organisational subcultures are relevant for this project?
- Which organisational cultures and subcultures are overt, and which are covert?
- Which organisational cultures and subcultures are negative, and which are positive?

If other project stakeholders are known at this stage, then consideration can be given to their organisational cultures and the influence these might have upon your project risks. By virtue of what we have argued in this chapter, this consideration will be limited by the information available to you. If the other stakeholders are not yet known, then this task must await that discovery, which could occur as late as Stage E (monitor and control risks) in the project risk management process.

During Stage C_1 (analyse risks), and following the identification of project risks in Stage B through techniques that explore project task planning, technology use, resource needs, and proposed organisation arrangements, the line of questioning now moves from the general to the specific for each identified risk:

- How do external cultural factors a, b, c, and so on influence uncertainty in terms of the likelihood and consequences of risk events that threaten the successful outcomes intended for this task, technology, resource, or project organisation?
- How do organisational cultural factors x, y, z, and so on influence uncertainty in terms of the likelihood and consequences of risk events that threaten the successful outcomes intended for this task, technology, resource, or project organisation?

A similar approach is suitable for assessing *opportunity* risks:

- How could external cultural factors d, e, f, and so on help to exploit or enhance the benefits, in terms of this task, technology, resource, or project organisation, that may arise from this opportunity?

- How could organisational cultural factors p, q, r, and so on help to exploit or enhance the benefits, in terms of this task, technology, resource, or project organisation, that may arise from this opportunity?

This stage of project risk management is usually the appropriate point to consider the organisational cultures of other project stakeholders, if they are known. Questions include:

- What other project stakeholders can be identified?
- What can be discovered about their organisational cultures?
- How would these cultures influence the characteristics of any of the risks we have identified for this project?

The task of considering the influence of organisational and external cultures on the project risks you have identified is often made easier by dealing first with groups of risks in terms of the risk *classification* system adopted by your organisation (Chapter 2). For a generic system (e.g. Chapter 2, Table 2.1), you could consider cultural influences on project *financial* risks as a group, since all the identified risks included in this group for a project are likely to be influenced in the same way.

We can also examine cultural risk shaping at a more detailed level through another example. Consider the following threat risk statement for a project:

> There is a chance that foreign or migrant workers will not be sufficiently proficient in English and not fully understand, or fail to follow, required safety procedures during work operations on the project, leading to accidents involving injury or death, damage to the works or environment, program delay, additional cost, legal action, punitive fines and compensation payments, adverse media attention, and reputation damage.

External cultures that could shape this risk might include the prevailing attitudes to work safety in the home countries of the employees or the English language requirement monitoring actually carried out (as distinct from that legally stipulated) by immigration authorities for this particular entry visa category.

Relevant organisational cultures might include those relating to English language training, safety induction courses, and worker supervision.

In Stage C_2 (evaluate risks), the outcomes of the analysis of the influence of external and organisational cultures (yours and those of other project stakeholders) upon specific risks are considered for their ability to shift risks from one assessment category to another (e.g. Chapter 8, Table 8.1, from 'Low' to 'Medium'). This is vital when the assessment categories are used to guide strategic risk management policies.

Once the effects of the external and organisational cultural factors have been assessed, and any risk assessment category shifts made, the effects can be considered in terms of their influence upon the decision making for the risk response option (see Chapter 9):

- What changes to our project decision making should we consider in order to mitigate the cultural shaping effects for these risks?

Now, the full range of risk response options is brought into play. Figure 12.2 does not display these, but it is possible that partial or complete risk transfer responses may have to be considered with respect to the ways in which the organisational cultures of other project stakeholders shape the project risks your organisation faces. For example, if another project stakeholder is renowned for neglecting environmental issues in its project activities, you may be able to incorporate protective liability and remediation conditions into any formal project agreement with that stakeholder. Similarly, if a client stakeholder is known to persistently delay payment for invoiced work beyond the period set out in a contract agreement (e.g. from 30 days to 90 days or more), then your response to the threat risk of delayed payment may be to insist upon significant interest accrual on amounts outstanding beyond the agreed period for payment.

During the execution of the project, in monitoring and controlling project risks (Chapter 4, Figure 4.2, Stage E), consideration of the shaping effects of external and organisational cultures should only be required if any of these cultures are found to have changed significantly during the intervening period. However, the same processes of cultural risk-shaping may have to be applied to any new project risks that emerge.

Objective knowledge about organisational and external cultures, gained directly from experience on a project, can make a valuable contribution to the risk wisdom of an organisation. It not only informs decision making for future projects, but also points to aspects of the organisation that could and should be changed. The process of capturing, recording, and adapting that knowledge should be implemented as part of the organisation's systematic approach to project risk management (Chapter 4, Figure 4.2, Stage F). The architecture of the organisational risk register was discussed in Chapter 11 (Project Risk Knowledge Management), but the value of knowledge about the risk shaping effects of the influences of external and organisational cultures is maximised if the reliability and validity of the underlying information are assured during the capture process.

12.7 Summary

In this chapter, we have argued that our perceptions and assessment of risk are culturally shaped. For project risks, this shaping is influenced by personal views and by the combined effects of the organisational culture of the project stakeholder, the cultures of the external environments in which the stakeholder operates, and the organisational cultures of other stakeholders involved in the project.

Identifying cultural influences is important so that they can be properly considered in the systematic processes of project risk management.

We close this chapter with some brief observations about organisational culture and risk.

Overt positive organisational cultures are easily subverted by covert negative cultures.

Overt negative organisational cultures can be extremely difficult to change.

Covert negative organisational cultures may be almost impossible to detect and to change.

Covert positive organisational cultures may be hard to distinguish but, when surfaced, can be used to shift negative cultures.

Assessing, responding to, monitoring, and recording the effects of cultural influences on project risks may be difficult and at times unrewarding, but it is necessary if we desire to mitigate those effects on the project threat risks we face, or to exploit them for opportunity risks.

Reference

Neilsen, R.P. (2017). Viable and non-viable methods for corruption reform. In: *The Handbook of Business and Corruption: Cross-sectoral Experiences* (ed. M. Asslaender and S. Hudson). Bingley: Emerald Publishing Ltd. ISBN: 9781787148970.

13

Project Complexity and Risk

13.1 Introduction

We frequently hear of, read about, or even work on projects that are regarded as 'complex'. What do we actually mean by this? What makes some projects more complex than others, and what are the implications for project risk and risk management?

Beyond our relatively narrow project perspective lies the much larger field of systems complexity, which continues to attract the interest of systems theorists worldwide and has led to the establishment of complexity science departments and research centres in many universities. That wider field is beyond the intended scope of this book, although we recommend it for further study to readers who are interested in this topic, since systems theory is an important contributor to project management.

As we have already seen in Chapter 3, each of the three phases (procurement, operational, or disposal) of a project will comprise elements relating to tasks, technologies, and resources, brought together through organisation and linked by decision making. In turn, each of these elements usually involves a larger number of sub-elements. There will be task sub-elements, technology sub-elements, resource sub-elements, and organisation sub-elements. Each sub-element will be subject to decision making, and each associated to some degree with uncertainty and risk.

We also know that projects generally involve multiple stakeholders, each one thus adding to the potential web of project complexity.

In this chapter, we discuss project complexity, mainly through the concept of two complexity states, and consider how this might affect project risk management.

13.2 The Concept of Complexity

'Complexity' derives from 'complex', which most English dictionaries define as: 'consisting of several closely connected parts, complicated; intricate'. In the context of projects, we would probably replace 'several' (= more than two) with 'many', since most modern projects surely comprise far more than three parts!

The term also has a mathematical connotation ('complex numbers') which need not concern us here.

Managing Project Risks, First Edition. Peter J Edwards, Paulo Vaz Serra and Michael Edwards.
© 2020 John Wiley & Sons Ltd. Published 2020 by John Wiley & Sons Ltd.

Complexity is thus a state of being complex, implying an intricate and complicated combination of parts that may prove difficult to deal with. Modern theorists attempt to disassociate complexity and complication, but the distinction is not sufficiently robust to be sustainable.

The early contributions of Baccarini (1996) and Williams (1999) have helped to clarify and extend our understanding of complexity as it relates to projects. They propose that, conceptually, the complexity of a project is influenced by the level of uncertainty associated with two states relating to its constituent elements and sub-elements (parts). These are the conditions of 'differentiation' and 'interdependency'. The latter was explored even earlier by Thompson (1967).

This chapter is based upon their arguments, but in fact few theories of complexity are contradictory: they are simply presented from different perspectives.

For example, Remington and Pollack (2007) typify complexity as structural (variability in the 'what'), technological (the 'how'), temporal ('when' effects), and directional (change going from 'where to where').

Kahane (2004) pursues a social networking view, suggesting that complexity arises from differences in perspectives and interests, where cause and effect gaps are wide, where problem solutions are not necessarily found in advance, and where people must participate in dealing with them.

Stacey (1996) explored complexity from two dimensions, certainty and the measure of agreement between constituents, and showed graphically how the relationship between them could indicate situations ranging from 'simple' to 'anarchy'.

The latter actually stems from the antecedents of complexity in chaos theory, which proposes how meaningful order is found among chaotic systems, and conversely how extensive networks and behaviours can arise from simple ideas and connections. In social networks, for example, a small group of 10 participants can generate 45 separate channels of communication $[(10 \times (10-1))/2 = 45]$.

Snowden and Boone (2007) created a Cynefin framework which essentially reorganises Stacey's model into scenarios categorised as: 'Simple', 'Complicated', 'Complex', and 'Chaotic'. In Table 13.1, we bring Stacey and Cynefin together to suggest a model or prism for assessing complexity in the very early stages of a project.

The resemblance to the five-point risk severity assessment scale in Table 8.7 is deliberate, and similar caveats apply. Setting and defining the assessment scale intervals are strategic management tasks. Practice in using the model is essential. More often than not, applying it is personal, intuitive, and subjective.

'Perceived uncertainty' in Table 13.1 may relate to each or all of the task, technology, resource, and organisational requirements of a project. 'Resolution space' relates to the perceived extent of the stakeholder's alternatives in terms of project decision making. It is thus possible to explore complexity in each of the project elements and to assess overall project complexity against the combined results of any individual analyses. As with multiple risk impact measures (Chapter 8, Table 8.5), the highest level for any indicator would form the accepted measure.

Projects exhibiting high to very high uncertainty and high to very high resolution space in the early stages may be considered 'chaotic'. They are likely to be high-risk, volatile, and extremely difficult (but not necessarily impossible) to manage. At the other extreme, projects with very low to low uncertainty and very low to low resolution space might be considered low-risk and straightforward.

Table 13.1 An uncertainty/resolution space complexity matrix for projects.

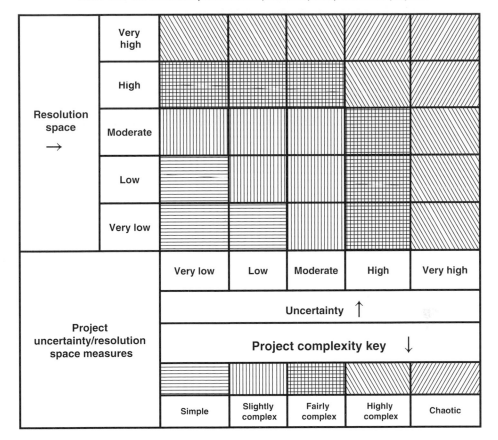

Interpolation between these extremes is likely to depend entirely upon judgement based on experience. Ideally, it should be influenced by the reflective wisdom derived from organisational risk knowledge management (Chapter 11).

Four examples serve to demonstrate this conceptual complexity model. The first is a project presented in Chapter 3 (Projects and Project Stakeholders). Figure 3.8 displayed the organisational structure for an informal settlement upgrading project in southern Africa. High differentiation is present in the number of labour-only subcontractors involved.

The second example is Case Study B (a rail improvement project), where high differentiation is also indicated in the number of existing railway level crossings to be removed (52 sites) and the 11 packaged procurement 'bundles' involved. We might therefore describe both projects as 'complex'. However, an initial assessment against the parameters of the matrix in Table 13.1 may shed more light on this.

In our view, the rail improvement project warrants a 'High' rating for both the level of uncertainty it presents and the extent of the resolution space involved. The uncertainty rating derives from the potential reactions of the multitude of stakeholders involved and the capacity to manage their expectations, plus the need to provide different and changing traffic diversions and alternative public transport arrangements over different periods for each site. Apart from the basic design alternatives (i.e. road over rail, road under rail, rail over road, rail under road, hybrid combinations, or road realignment), the resolution space is expanded by the large variety of design features and technical requirements involved. These ratings would place Case Study B in the 'Highly Complex' project band, according to our interpretation in Table 13.1.

For the settlement upgrade project, we would rate the associated uncertainty and the resolution space each as 'Low–Moderate', thus rendering it as a 'Slightly Complex' project. The uncertainty of employing many largely unknown labour-only subcontractors, in terms of the failure risk this presents, is mitigated by learning curve effects and by the diffusion of the risk itself – the chance of failure occurring for all subcontractors is very low, and the individual consequences of failure for one or two of them will not be high. Stakeholder management should not present undue difficulty, and the desired outcomes – better sanitation and public health – are not controversial. The resolution space is not large, since the basic toilet upgrade design is simple, and the materials needed to construct them are unvaried. Indeed, for this project, if the experience and capability of the programme manager and consulting engineer are known and trusted, then the uncertainty and resolution space ratings might be revised to 'Low–Very Low' and the whole project assessed as 'Simple' in terms of complexity.

The third example differs from the first two. In Case Study C (an aid-funded civic project), the project itself is fairly straightforward in terms of the 'client–project' 'manager–main contractor' relationships and the procurement system used. It is made more complicated for the funding government's senior advisor, firstly in terms of the multiple relationships he has to manage between the project stakeholders and secondly by the relatively high level of uncertainty in several aspects of the project. For the senior advisor, we would rate the project as 'Fairly Complex', with a 'High' level of uncertainty and a 'Moderate' solution space rating.

We go back to Chapter 3 again for the fourth example. Figure 3.7 depicts the organogram for a hospital refurbishment and extension project in South Africa. If the project proposed to build a brand-new hospital, we might place it at 'Moderate' to 'High' for the uncertainty and resolution space scales; and then as a matter of prudence perhaps regard it as a 'Highly Complex' project. However, because of the additional uncertainty that always accompanies alterations as opposed to new works and the high number of stakeholders with their significant power relationships and extensive communication needs, we could be forgiven for imagining that this project could verge on the 'Chaotic'! It is not beyond the bounds of possibility, and this rating should heavily influence our project risk awareness and management.

The presence of uncertainty, as demonstrated in these examples, within the concept of complexity places that state directly within the purview of risk.

For our purposes, however, the differentiation/interdependency model proposed by Williams (2002) provides a more considered direction for proceeding further with our discussion of complexity.

13.2.1 Differentiation

The condition of differentiation complexity in a project represents the extent to which the project's constituent elements and sub-elements *should* be treated as individually identifiable, or, more simply, the *number* of distinctive parts of the project that we should recognise as such.

This is not necessarily the same as the number of parts that we *could* count. If we can make an objective, reliable count, then there will be little or no uncertainty involved. What we *should* count separately infers some degree of uncertainty, as our opinions about this might differ.

If we take an 'old technology' example, in a riveted steel bridge project it might not be necessary to identify and count each fastener. For most purposes, a ratio of x kilograms of rivets to y tonnes of steelwork might suffice. On the other hand, it may be important to distinguish between different types and configurations of rivets or between similar rivets fixed in different situations. The 'should recognise' connotation of differentiation is thus an important consideration here and must be guided by the project context and the individual project stakeholder's project and project risk management perspectives.

In the steel bridge project, the number and type of rivets needed are unlikely to be a concern for the client stakeholder, whose major concern is for the satisfactory delivery of a properly functioning bridge. Nor is this information likely to interest the construction contractor, who may be satisfied with a quoted price per tonne of steel supplied from the fabrication subcontractor responsible for delivering completed bridge elements to the construction site. The steel fabricator, however, will need to determine the numbers and types of rivets required in order to obtain them from the manufacturer/supplier and to organise the requisite labour and equipment for making the bridge components.

This simple bridge example shows us that, to some extent, complexity is 'in the eye of the beholder'. It also illustrates only one aspect of projects – resources in terms of the materials required.

For any project, we will have to differentiate not only between resource sub-elements, but also between the various task, technological, and organisational requirements for the project (Chapter 3).

We may wish (or need) to distinguish between the activities of particular tradespeople; between construction process activities; between technologies; between different materials handling requirements (vertical and horizontal transportation) and storage; between locally supplied and imported resources; between task scheduling requirements; or between the ratios of management, supervisory, and line staff in the stakeholder organisation. These are just a few of the many distinctive separations that inevitably become necessary to make the project 'work' in the sense of a particular stakeholder being able to bring it to effective completion.

For an information technology (IT) software application design and development project, differentiation complexity might be assessed by the number of separate and distinct algorithms needed to accommodate all the possible combinations of application routines envisaged by the system design. Organisationally, different specialists might be needed to develop different pieces of system design and computer code, so that each piece can be assembled with others to create the complete application. Some of these may be deliberately sourced from different global time zones in order to maximise productivity or exploit particular sources of expertise.

High levels of differentiation tend to affect project planning and execution, through the sheer number of different things which must be dealt with, and through the extent and level of decision making associated with that.

We can complicate matters more (thus increasing the complexity!) by considering different circumstances for differentiation.

Spatial differentiation occurs in projects where elements or sub-elements have to be prepared in (or sourced from) several places, and then brought to one location for incorporation into the project. Gaudi's famous Sagrada Familia church in Barcelona, Spain is an example of this. In the ongoing completion project, parts of the stone towers are constructed at off-site locations (with stone sourced in the United Kingdom), transported to the city, and hoisted and fixed in place with cranes. The Airbus A380 jetliner is probably an even more extreme example of differentiation complexity. This aircraft is said to comprise over 4 million separate components, many of which are manufactured and part-assembled in factories all over the world.

Or it may be that a sub-element is developed or fabricated in one place, but has to be installed in many. Banking and other IT projects are good examples of this, where a centrally produced software application package (e.g. client account processing) may have to be deployed throughout hundreds of branch locations. Indeed, as we noted, spatial differentiation occurs frequently in contemporary IT applications development projects, where parts of the project development work are outsourced to specialists operating independently in different places around the world, each of whom may be known only by, and contracted to, the project manager in the commissioning organisation.

For a banking IT project, it may be possible to carry out the final deployment for all branches remotely from a central location via a dedicated and secure intranet connection. However, the project objectives will almost certainly require that each package be tested in situ to ensure that it also functions properly at each physical bank workstation for staff use when responding to customer enquiries. Besides testing from the 'bank side' of the application, exhaustive testing from the 'customer side' will also be needed, to minimise 'glitches' and protect the system from unauthorised intrusions.

Spatial differentiation, in the form of *dispersion*, may thus contribute significantly to project complexity.

Temporal differentiation occurs when project tasks, technologies, resources, or organisation has to be separated in time. Such situations, however, usually arise as a consequence of interdependency between the various parts. We will deal with this later in the chapter.

Differentiation complexity affects our capacity to manage projects efficiently and effectively. Since project management is a human task, there are limits to how much a project manager can deal with at any one time, despite contemporary claims about multitasking! Paradoxically, however, we tend to address differentiation complexity by *reductionism*: breaking projects down into smaller units, to make them more easily understood and manageable, but thereby also increasing the overall level of differentiation! The paradox suggests that complexity itself may have complicated facets.

Breaking down a project into smaller parts allows each unit to be managed separately in terms of tasks and other requirements, but under a coordinating *supra*-management umbrella. Case Study B (a rail improvement project) is a good example of this in terms of the rail crossing removal authority's role.

A commercial online sales product-branding project might be dealt with separately for each product for the design and development stage of this IT project, and then brought together only at the final roll-out online. On the other hand, a very large tunnel project might begin with a unified planning and design stage, but then be separated into separate contracts to allow tunnel boring and lining to begin more or less simultaneously at each end, before eventually being integrated again, after the breakthrough conjunction, for the fitting-out work and commissioning.

From our perspective in this book, it is uncertainty in the differentiation surrounding project tasks, technologies, resources, and organisation that will influence a stakeholder's risk management for a project. Such uncertainty usually arises through conditions of interdependency.

13.2.2   Interdependency

Interdependency is the nature and extent of any dependent relationships that may exist between the differentiated parts of a project. If one part cannot be completed without another, or one part cannot function without another, or one technology cannot be applied without another, or one resource is unobtainable or unusable without another, then interdependency exists.

Such dependencies can arise not only *between* the task, technology, resource, and organisation elements and sub-elements of projects, but also *within* them. Higher levels of interdependency complexity are found most often in the task elements, and the impacts of interdependency complexity are most profoundly felt in the time planning (e.g. scheduling and programming) aspects of project management.

Relationships giving rise to interdependency may take one of three forms: pooled, sequential, or reciprocal.

For *pooled* interdependency, if the differentiated parts required in a project element or sub-element can be dealt with one at a time, but not necessarily in any strict order, and each without interference from others, until all elements of the project are complete or a distinct point of integration is reached, then a minimum interdependency relationship exists. In such a case, for example, completion times for each sub-element are simply pooled. In essence, the total time required for the whole element would be determined by the time needed to complete the slowest part. Such relationships are rare in contemporary projects, where activities tend to be sequentially planned.

In *sequential* interdependency, the conditional sequencing relationship between constituent project parts is the influencing factor. One sub-element must follow or precede another (or be undertaken at the same time) as part of an essential and deliberately engineered sequence. This aspect of complexity is reflected most clearly in the critical path network (CPN) approach to project scheduling, where it is important to identify discrete activities (tasks) that must be completed before a following activity can be commenced, as well as those that can (or must) be carried out in parallel. The time required to complete a project is represented by the length (duration) of the critical path, and not necessarily by the pool times needed for each element and sub-element within it.

Bar (Gantt) chart scheduling also serves to illustrate sequential interdependency, but it is generally less informative than a comprehensive critical path method of project planning.

In the room renovation project described in Chapter 7, Figure 7.1 illustrates where the outcomes of CPN modelling of sequentially interdependent activities are shown in a bar chart format.

In sequential interdependency, a change in one part (activity) does not necessarily require a change in its dependent partners, as long as the sequence remains unaffected.

However, some adjustment may be required for the start and finish times of each activity, and the overall critical path may change. Uncertainty is managed by the introduction of 'float' time to provide more scheduling flexibility.

Reciprocal interdependency occurs when a change to, or turbulence occurring in, one element or sub-element of a project has a flow-on effect and necessitates change to, or induces turbulence in, one or more of the other elements or sub-elements. The reciprocity extends to factors other than adjustment of activity durations.

On a construction project, for example, changing the internal ceiling design for an office building might mean changing the ceiling suspension system, the type of light fittings to be fitted in the ceilings, and even the wiring, switches, and circuit controllers required for them. It might also invoke changes to the ceiling ducts for heating, ventilation, air conditioning, and other engineering service systems (such as fire sprinkler valve configurations) for the building.

Another example might be a conversion project, whereby creating two smaller hospital wards from one large one might, in addition to the otherwise relatively straightforward building and services alteration work, entail changing signage throughout the whole floor (and perhaps other floors and foyers) and amending centrally controlled door keying, security systems, and identification registers.

For an IT software application project, the discovery of a fault ('bug') in a piece of programme code for an algorithm might require amendment to, or the replacement of, code in other algorithms and sections.

Complexity arising from interdependency is not as easily addressed as simple differentiation. Increasing interdependency levels by further separation is not a paradoxical management solution here! It is impracticable. Design simplification might be one option if it is feasible, but enhancing programme/schedule flexibility is often found to be the only practical approach. This approach may be an anathema to project clients.

Given the nature of complexity, how then should it be compared among projects?

13.3   Relative Complexity

The difficulty of comparing complexity in projects has already been mentioned. In many instances, it may be neither practicable nor desirable to make such comparisons on any precise, absolute, quantitative basis, although this is an aim for much of complexity science.

If quantitative analysis of project complexity is difficult, then relative comparisons between projects become a matter of subjective assessment, and thus susceptible to all the errors and biases of human judgement. We noted this for the early stage complexity assessment approach shown in Table 13.1.

Now, for example, let's assume Projects X and Y are similar in type and vary only in scope, detail, and location. Each has the characteristics shown in Table 13.2.

The number of project stakeholder organisations is included because, if for no other reason, it complicates matters in terms of inter-stakeholder communication and the uncertainties associated with the effects that their decision making might have upon the project risks that you are trying to manage.

Table 13.2 Differentiation complexity in projects.

Project element	Project X	Project Y
Tasks	15	45
Technologies	3	6
Resources	40	65
Stakeholder organisations	2	4
Totals	60	120

Purely on the basis of comparing the number of constituent elements (tasks, technologies, resources, and stakeholder organisations) involved, Project Y is prima facie more complex than Project X in all respects. Should we assume therefore, on the basis of the differentiation count totals, that Project Y is twice as complex as Project X?

Comparisons are rarely as straightforward as this, since the *intra*-constituent characteristics of the two projects might not be sufficiently similar. Thus, although Project Y requires the use of six technologies, and Project X needs only three, if those three for X are less well developed, more prone to failure, or susceptible to higher variance in their outcomes, or require scarcer and more highly specialised operatives than the six well-understood, well-tested, well-practiced, and widely available technologies of Project Y; then, from a technological point of view, Project X might be considered more complex than Project Y and is likely to be more risky, especially if the project stakeholder is unfamiliar with the three technologies of Project X but highly experienced with the six for Project Y.

Case Study B (an aquatic theme park) demonstrates this to some extent in terms of supervision of the construction of the complicated floor slab for the wave pool. In the absence of a supervisor highly experienced in this form of construction, the civil engineering contractor appointed one they knew from previous experience to be adept at dealing with concrete floor construction in many different situations.

While this comparison deals only with project technologies, we should not necessarily expect comparisons between the other elements of each project to yield similar results.

More importantly perhaps, the limitations of our comparison expose a fallacy of differentiation complexity theory. It is not necessarily the *number* of separable parts that drives complexity, but the nature of the parts themselves, the nature of the separation between them, and the relationships (interdependencies) between them that must be considered. Differentiation alone may therefore be an unreliable measure of complexity, as interdependency invariably comes into play. It is also why our preliminary uncertainty/solution space complexity assessment model (Table 13.1) is somewhat crude.

This caution also suggests that comparing differentiation complexity, even in similar types of projects, is not easy. For dissimilar projects, it is likely to be impossible, but in any event is not usually necessary.

Interdependency complexity in projects may be easier to assess than differentiation, since it is visually discernible though inspection of CPN analyses where these are available. For pooled and sequential interdependency, Gantt charts can also reveal comparative levels of complexity. However, neither of these project scheduling techniques readily facilitates comparison of reciprocal interdependency unless substantial analysis and iterative reprogramming are undertaken.

Our interest in, and concern for, project complexity should always be in terms of trying to gain a better understanding of it, with a view to improving our management of the projects we undertake, and more particularly the management of project risks. We are therefore less concerned about theories of complexity, and more interested in its manifestation on the projects we manage and its effect on project risks. To this end, all that we covered in Chapter 11 (Project Risk Knowledge Management) comes into play. Two simple post-project debriefing questions stimulate the project complexity knowledge quest: 'How difficult was that project to manage?' and 'What complications arose?'

Better understanding of project complexity arises largely through experience and judgement. Tacit knowledge about this can and should be captured. The transformation of tacit knowledge into explicit knowledge and then into project complexity wisdom, however, is a process of careful reflection, preferably undertaken organisationally rather than individually (see Chapter 11).

Next we must address the question: How does complexity relate to project risks and risk management? The answer lies in the uncertainty associated with project decision making.

13.4   Uncertainty and Project Complexity

In Chapter 2 we noted that, by definition, uncertainty is some state that is short of certainty: something is not fully known; information is incomplete. We have also accepted that risk is the effect of uncertainty upon objectives.

In projects, many decisions are made on the basis of forecasts of outcomes or events occurring at some time in the future. The future cannot be known with certainty. Therefore, each forecast is vulnerable to some degree of uncertainty in terms of the input factors to the model used to produce it, or uncertainty in the performance of the model itself. Depending upon the nature of the forecast, this uncertainty may relate to the likelihood of the event occurring, its timing, or the magnitude of any consequences. This aspect of project decision making was discussed in Chapter 3.

While most projects are subject to at least some degree of uncertainty, project uncertainty states are rarely static. The level of uncertainty in a project changes over time, in both nature and degree, as time passes and project elements and sub-elements are completed, or changed to suit new circumstances. The dynamic nature of project uncertainty was shown in Figures 4.3–4.6 for the pharmaceutical manufacturing facility and IT project examples described in Chapter 4. The relationship between project uncertainty and project complexity is therefore also dynamic. As project uncertainty diminishes over time, so too should project complexity. Debriefing a complex project, however, in order to capture project knowledge can itself be a difficult process.

At the outset of a project, particularly during the conceptual or early development stages, project complexity is perceived more or less intuitively. It is guided largely by experience, as the lack of detailed information about the project elements means that any dependency relationships between and within them are largely unknown at this point. Potential indicators of complexity, such as bar charts or critical path analyses, may not yet exist, although some help may be available in the form of project methods statements. We have to rely on interpreting complexity through subjective two-dimensional models such as that shown in Table 13.1.

More realistically, complexity in projects is highly multidimensional and dynamic. Figure 13.1 illustrates this, if at the cost of some deliberate oversimplification.

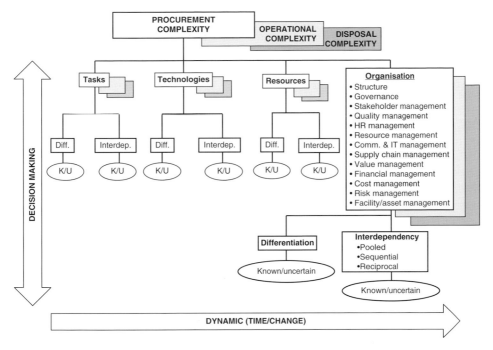

Figure 13.1 Project elements, environments, and complexity factors.

On the right-hand side of the figure, the 'organisation' element in the project procurement environment is expanded to list potential sub-elements, in this case the different areas of management likely to be found within the project stakeholder organisation. None of the other elements in Figure 13.1 displays the unique subdivisions that would certainly exist in reality, and we have chosen the organisation element for expansion in this way simply because this is likely to be universally similar even for different types of projects. The vertical broad arrow signifies that decision making occurs across all the project elements, while the horizontal arrow reflects the inevitability of change occurring over time. Most readers will recognise similarities with their own circumstances. Notwithstanding the limitations of its graphical simplifications, the diagram is useful for illustrating and understanding the concepts of complexity and uncertainty, how they are associated with decision making, and to some extent how they are dynamic in terms of changes occurring over time.

Taking the example of the organisational element in the procurement environment for a project stakeholder, differentiation complexity is immediately apparent in the number of sub-elements identified, and interdependent relationships will almost certainly occur between them. The extent of these relationships may be uncertain, and they may change, or be affected by the addition of new sub-elements, over time, hence their dynamic properties.

Individual stages within each element may present their own complexity situations. For example, in a construction project where the procurement stage includes planning design, tendering, and construction processes, complexity can arise in any or all of these.

Figure 13.1 does not tell us anything about complexity and its influence upon specific project risks. Attempting to show this would make the diagram far too difficult to understand visually (an unwanted communication complexity!). For the purpose of indicating project risk complexity, 'risk complexity mapping' may be a possible approach.

13.5 Identifying and Mapping Complexity

The use of the work breakdown structure (WBS) technique for the purpose of risk identification was introduced in Chapter 7 (Project Risk Identification Tools) and displayed in Table 7.2 and again in Table 7.13. Table 7.2 showed a matrix of project activities (tasks) against their respective technology, resource, and organisational requirements.

Should a large number of entries be found in any (or all) of these project WBS fields, it could indicate that project differentiation complexity is occurring in one or more of those dimensions, especially if the number is greater than anticipated. For example, many entries in the 'technology' field might suggest a project that is technologically complex. Any identified project risk displaying substantial association with a technology, if it were 'tagged' in some way, would thus attract special risk management attention (in terms of analysis, treatment, monitoring, and control).

In Table 7.13, the matrix was extended to include a few of the generic categories of risk. In this table, cell mapping entries record identified risks with coded labels. Using this table, any risks directly associated with a field displaying differentiation complexity could be recorded with a 'd' suffix to indicate this. Such risks are thus 'flagged' for special attention.

We noted in Chapter 7 that mapping risks in this way makes it possible for us to see more clearly how risks (and particular types of risks) are distributed throughout the project. Patterns or clusters emerge. A valuable picture of the 'riskiness' of the project is obtained, which can be subsequently compared with the risk maps for alternative projects or archived historic projects, using the organisational risk register (Chapter 11, Project Risk Knowledge Management) as an information source. Adding differentiation complexity to this risk knowledge may increase its value substantially.

Identifying and recording complexity arising from interdependency is also possible. Again using the example of Table 7.13, further suffixes could be added to the coded labels for identified risks. Dependency relationships might be recorded with 'p' (pooled), 's' (sequential), or 'r' (reciprocal) suffixes according to the nature of the dependency.

The problem with attempting to record interdependency in this way is that it is not a straightforward matter of comparing counts of items (as with an initial approach to differentiation complexity). Each instance of interdependency has to be considered on its own merits, and this could render the whole mapping approach impractical and unwieldy. A CPN analysis diagram might establish interdependencies and their effects, at least for activities and schedule time, but this tool is unlikely to be available when needed in the project risk management cycle, and it will not directly show technology, resource, and organisational interdependencies.

A better approach is probably to retain the WBS table to establish a picture of project differentiation complexity, but then to adapt the risk register item record (Table 7.14) so as to record at least the presence of differentiation and interdependency. 'Tick-boxes' should suffice for this. If a WBS is not likely to be available, then the context establishment stage should attempt at least a rudimentary assessment of project complexity (beyond that found in Table 13.1) which can be flagged on each risk register item record.

13.6 Influence of Complexity on Risk Management

Following context establishment and risk identification, where it is possibly 'flagged', complexity usually influences project risk management in two ways: in the risk analysis process, and in the consideration of response and treatment options.

Of the components of complexity, reciprocal interdependency is likely to exert the largest influence, since it can affect both risk probability and impact analyses, and the nature and extent of treatment alternatives.

For the likelihood of occurrence, consideration has to be given to which end of the complexity reciprocity dependency will be the source of the initial risk trigger event, and then to whether or not additional probability will be associated with the dependency 'flow-on'.

Such levels of analysis are generally beyond the capacity of computerised applications. Analysis, inevitably qualitative, has to be conducted individually with robust logic for each risk circumstance and interpreted carefully with respect to its implications.

The same caution applies to risk impact assessment, perhaps more so since multiple consequences may escalate rapidly in nature and extent.

In exploring and deciding upon appropriate risk responses and treatments, the dependency logic has to be considered again, especially for treatment options, as the effects of treatment sequencing are likely to be important.

All this is tantamount to saying that complexity begets complexity. If the project is complex, so too will be the risk management required for it. This is important to remember in planning and resourcing risk management activities for particular projects, especially with regard to time. Case Study B (a rail improvement project) indicates how this was addressed effectively on a complex project.

We conclude this chapter by offering some thoughts about complexity in relation to 'mega-projects', the increasing phenomenon of contemporary development.

13.7 Complexity and Mega-projects

There is an 'effort versus reward' balance to be considered in assessing project complexity. It is important for all projects, of course, but particularly so for 'mega-projects'. The balance we refer to is the trade-off between the time and information needed to conduct project complexity assessment, and the benefits arising from it.

On the one hand, we can develop a much 'richer' picture of the project, one that yields a deeper and more sophisticated understanding of the project intricacies in terms of task, technology, resources, and organisational requirements and arrangements, together with the relationships between them and the decision making that they entail.

On the other hand, there is really no 'quick and dirty' approach to complexity assessment (beyond the initial approach suggested in Table 13.1), and few other shortcuts are available.

The complexity wisdom we referred to earlier in this chapter takes time to develop and accumulate. Reflection requires sustained intellectual effort and insight, and applying such reflection to future projects takes creative energy for it to be effective.

Through their sheer scale, and often their scope, mega-projects are complex undertakings. They are usually defined in terms of cost to procure (e.g. projects exceeding $1 billion), but this is a crude indicator. It ignores any operational or disposal costs and does not seem to ever be indexed to inflation. Despite these limitations, 'mega-project' denotes a project that is very large, substantially beyond the scope of normal expectations, and capable of exerting influence and impacts beyond its objectives and the activities required to achieve them.

Case Study A (a public–private partnership [PPP] correctional facility) is probably a mega-project if the discounted present value (PV) of operational costs is added to the AU$650 million construction price tag. It displays complexity in several respects. Firstly, temporal complexity arises from the project concession period of 25 years. Procurement complexity is found in the governance, administration, and stakeholder agreements and relationships needed to embrace this period. Scope complexity emerges through the construction and operational requirements involved. An example of reciprocal interdependency is found in the capacity scope changes involved in the negotiation period: not only because of the increased extent of physical accommodation contemplated by them, but also through the concomitant changes needed for future operational staffing ratios, medical treatment capacity, and required expansions to education and training facilities and programmes. Interestingly, post-delivery risk reflection carried out for this project noted how the original staffing levels estimated for planning, design and construction proved to be inadequate. Early experience in the operational phase indicates that understaffing will also occur in this phase of the project.

At an estimated $8 billion cost, Case Study B (a rail improvement project) qualifies directly as a mega-project (if programme = project). Differentiation complexity is found in the overall programme, requiring the removal of many railway level crossings across a wide metropolitan area, some with associated works to new or existing stations; extensive track modification; new elevated tracks; and upgraded communication systems. Dispersion complexity arises from the widespread locations of sites. Procurement complexity arises through the decision to 'package' the work, each package comprising a 'bundle' of individual crossings and other associated work. Management and resource complexity occurs where extensive temporary road traffic diversions and bus replacements for rail services have to be put in place in different locations for varying periods. Procurement complexity is also found in the nature of each of the individual packages. Packages where it was thought that uncertainties would be relatively low were tendered on a design and construct (D&C) basis, thus transferring most risks to the contractor. Where more stakeholders were involved (more differentiation complexity) and high uncertainty (= complexity) was expected, 'alliance' arrangements were set in place whereby professional consultants, contractors, and subcontractors worked together in a unified way to share risks. Additional procurement complexity arose in the package tendering procedures as they were not standardised across the programme. Early packages were tendered competitively; some subsequent packages were partially price competitive or negotiated.

Case Study B uses the classic reductionist approach to differentiation complexity by splitting the larger project up into smaller, more easily manageable projects. However, this has to be done on a reasoned basis to justify how the split should be made and how overall management control can be maintained. Since the complexity of the latter would tax the administrative resources of the public client if it were carried out 'in-house', the strategy of creating a temporary rail crossing removal authority (with clearly defined reporting channels and requirements) was adopted to 'distance' and contain project administration from other day-to-day government activities.

The question might be asked: Would the same level of complexity arise if a similar programme was implemented to remove traffic light controls at 50 major road intersections across a similar metropolitan area in order to improve traffic flows? There would be no associated rail complexity, far fewer technology issues, and fewer stakeholders since designated major roads are a state roads authority responsibility (and not that of local councils). No replacement bus services would be needed, and temporary traffic diversions should be more easily managed. The question is pertinent, as the removal of multiple rail crossings flowed from the pre-election promises of a state government and was delivered after it came to power. The traffic light removal programme has since emerged in the manifesto of the main opposition party for a forthcoming election in the same state. Politics always plays a part in such projects, as we shall see in Chapter 14.

Three important effects come into play with respect to the management of complexity in mega-projects and, to some extent, in all large projects. These are the 'learning curve', 'tolerance', and 'familiarity transfer' effects.

Human beings generally show great adaptability. Faced with complicated, unfamiliar situations, our initial responses may be those of psychological anxiety and stress. However, as long as they threaten no physical danger, we quite quickly become accustomed to them, and the harmful psychological effects (e.g. anxiety arising from project complexity) subside. Our growing familiarity with the intricacies of the project serves to lessen our negative perceptions of its overall complexity. This is tantamount to a 'learning curve' effect. It may evolve from rationalising the complexity through simplifying its constituents (i.e. reductionism), from developing greater confidence in our ability to deal with it, or from both of these.

For project risk management, the complexity learning curve effect serves to heighten the psychological perceptions of the magnitudes of risk exposures occurring early in project stages and to diminish those perceptions for risks eventuating later in project stages. Care is thus needed to avoid bias or error in those risk assessments (Chapter 8, Project Risk Assessment and Evaluation).

After we have been through the complexity learning curve with a few projects, usually over several years, we can develop a 'tolerance' to the demands of complexity. Psychologically, we may come to believe that we are fully equipped to deal with complexity on any project. 'Conditioning' has developed. While this could be an admirable trait in terms of personal development and staff recruitment, with it comes the threat risk of overconfidence leading to neglect of proper and timely consideration of the specific complexities of each new project. This could then lead to failure to identify risks or to inadequate assessment of them.

'Familiarity transfer' occurs when staff new to a project are exposed to the complexity knowledge of those who have already been through the learning curve for that project. Such experience, if properly transferred and inculcated, drastically shortens the project complexity learning curve for the newcomer and improves his or her project risk management assimilation process. If the transfer is incomplete or falsely premised, project success will be threatened.

13.8 Summary

Complexity has been a growing area of academic interest over the past three decades, and has attracted additional interest from project management theorists and practitioners since it is inevitably associated with many projects.

Among other theories about complexity, that of differentiation and interdependency is attractive, particularly when related to the WBS and CPN analysis approaches to project planning and scheduling. Before these come into play, however, a simpler qualitative method could be used as a gauge for project complexity by assessing levels of uncertainty and the likely size of the solution space available in terms of the tasks, technologies, resources, and organisation that will be needed.

Neither differentiation nor interdependency is straightforward to assess, however, since each is contingent upon the nature and extent of its association with a project. The relative complexity of different projects is therefore difficult to assess with any great confidence.

Complexity is dynamic and driven by the uncertainty inherent in decision making. Administratively, it is usually addressed by reducing a project to more manageable parts; but in risk management, project complexity presents a problem for risk identification, for risk analysis, and for choosing risk response and treatment options.

Mega-projects always present an array of complexity issues. Although not intractable, they call upon substantial reserves of experiential knowledge and wisdom that can severely tax the capacity of project risk managers to deal with them.

Our main point in this chapter is that complexity risk arises in projects in the uncertainty pertaining to differentiation and interdependency in their association with project decision making.

In Chapter 14, we explore politics in projects. They also present challenges for risk management.

References

Baccarini, D. (1996). The concept of project complexity – a review. *International Journal of Project Management* 14 (4): 210–214.

Kahane, A. (2004). *Solving Tough Problems*. Oakland, CA: Berrett-Koehler Publishers.

Remington, K. and Pollack, J. (2007). *Tools for Complex Projects*. Aldershot: Gower Publishing.

Snowden, D.F. and Boone, M.E. (2007). A leader's framework for decision making. *Harvard Business Review* 85 (11): 69–76.

Stacey, R.D. (1996). *Complexity and Creativity in Organizations*. Oakland, CA: Berrett-Koehler Publishers.

Thompson, J. (1967). *Organizations in Action*. New York: McGraw-Hill.

Williams, T.M. (1999). The need for new paradigms for complex projects. *International Journal of Project Management* 17 (5): 269–273.

Williams, T.M. (2002). *Modelling Complex Projects*. London: John Wiley & Sons.

14

Political Risk

14.1   Introduction

Are any projects entirely free from politics? In our view, there are none in the public domain and few in the private sector. Politics exert a major influence on projects of all types, hence the need for this chapter.

Just as risk has an affinity to projects, in that both are clouded by uncertainty, so too do we find that political influence is no stranger to projects. Politics are rarely clear-cut and transparent, but more often than not are manifested in shifts of opinions and perceptions and by partisan perspectives. They may be glaringly obvious or artfully camouflaged. On the other hand, politics can give rise to clearer distinctions between threat risks and opportunity risks for many projects.

It is also important to note that Paul Slovic's psychological *affect* heuristic (Chapter 2) looms large in political risk. There are no dispassionate politics – emotion is an intrinsic factor that must be recognised and, where possible, addressed.

Our starting place here is a definitional exploration of what in English may be a singular or plural noun.

The traditional definition of 'politics' is 'the science or art of government', and the Greek linguistic roots relate to 'the affairs of cities'. Thus, there is a traditional connotation of politics as relating to public government, going as far back as the obsessions of ancient Greek and Roman philosophers, and no doubt even further back in eastern philosophies. However, both definition and connotation have shifted over the past century. Such linguistic development, in this and for many other terms, is not unusual. Language is never static.

A contemporary understanding extends the meaning of the word beyond the traditional to assert that politics may arise in any situation where power is exercised to influence decision making or decision outcomes. It is also used, as in the phrase 'politically correct' (PC), to question or evaluate actions, influences, and even written or spoken language itself, in terms of perceived contemporary social acceptability or behavioural norms. In this context, it is more often stated in the negative ('politically incorrect' or 'non-PC').

The shift in meaning can be largely ascribed to developments in public and corporate governance and power generally occurring after each of the two world wars. The bloody Russian revolution of 1917 saw enormous and rapidly dramatic change in national governance from a hereditary monarchy to social communism. The First World War itself

Managing Project Risks, First Edition. Peter J Edwards, Paulo Vaz Serra and Michael Edwards.
© 2020 John Wiley & Sons Ltd. Published 2020 by John Wiley & Sons Ltd.

brought about changes in European hegemonies and societal standards. The effects of both events rippled across Europe and beyond. The decade before the Second World War brought further changes through the rise of fascist dictatorships in Germany, Italy, and Spain, and was followed by the post-war emergence of Eastern and Western European power blocs. Even more recently, global power shifts on the Indian subcontinent and throughout Asia have wrought further changes. China's implementation of its international Belt and Road Initiative in government policy is a classic example, drawing on the physical trade routes of the past to forge or strengthen international relationships that could bring new opportunity and threat risks to the receiving or neighbouring countries.

Private sector corporate politics have also come into far greater prominence, mainly as a result of the increasing globalisation of commerce and the development and rise in importance of modern communications technology. What were originally the political machinations of a few eighteenth-century trading companies, such as the British and the Dutch East India Companies, and later the Hudson Bay Trading Company in North America, with their dependence upon simple means of communication, have since exploded into the political activities and interventions of national and international mega-corporations and conglomerates, each exerting global reach and power through virtually instantaneous communication channels. The ramifications of these huge modern organisations can influence (and alarm) governments worldwide. That many of them are closely related to information technology and telecommunications is not without significance.

A contemporary connotation of politics is that of power exerted as a means to achieve ends. While not exclusively so, the connotation is more often negative than positive – that of 'having to play politics to get things done'. Sadly, this probably derives from the Westminster model of parliamentary governance that, while it claims to present a noble 'contest of ideas', is inevitably conducted in an environment of divisive speech, controversy, and adversarial behaviour. Politics in any truly cooperative sense is still a rarity in contemporary Western society.

This chapter is not the place to expound further upon this background and the philosophical frameworks that underlie it. It suffices to say that the connotation of politics has 'morphed' into two dialectic opposites: a preparedness to compromise ('the art of the possible'), either willingly or unwillingly; or the adoption of extreme positions (a 'game' to be won at all costs). Threat and opportunity risks may be found in either stance.

The influence and presence of politics in projects have also spread from a purely public governance viewpoint to one that embraces both public and private perspectives in ways that are not necessarily mutually exclusive. Nor is the situation static. As we will see in Section 14.2, the spheres of political influence are expanding as the connotation broadens.

We therefore offer a working definition of politics in the context of project risk management as 'the exertion of power to influence (or control) decision making and actions intended to achieve project objectives'.

In our experience, political risks tend to lie at the more extreme ends of the threat–opportunity continuum of project risk. In terms of threats, they can be pernicious, entrenched, and difficult to manage. Political opportunity risks, on the other hand, may be harder to discern with confidence and to exploit successfully.

It is also important to note that our assumption here is that the influences upon decision making and the actions relating to political risks are all legal. Any evidence of illegality would transform them into criminal acts of corruption, bribery, and so on. This would then allow them to be categorised as social or legal risks.

However, politics are never far away from the threat risk of corruption (Chapter 10), and any combination of the two will have a corrosively damaging effect on the achievement of project objectives. More often than not in such juxtapositions, political pressure is exerted to engage in corruption or to condone it.

We also contend that the responsibility of 'speaking truth to power' as a means of holding politicians to account has in the twenty-first century been substantially usurped by 'speaking alternative truths to retain power'. In this era of so-called 'fake' news, the unrelenting demands of the 24-hour news cycle, and the 5-second news 'grab', it has become easier to replace reliable factual information with any pronouncement that will help to retain or gain power. In this sense, it is probably fair to say that the gap between politics and corruption is becoming ever narrower.

Given these cautions, in the remainder of this chapter we consider the spheres of activity influenced by politics, the complexity associated with political risks, and their management in project settings. The chapter concludes with some thoughts about the more severe extremities of political risks.

14.2 Political Spheres

Following our claims in Section 14.1 about the growing ubiquity and portability of 'politics' as a word, we can nominate many spheres of decision making and action with which they are associated. Table 14.1 lists some of the areas where politics are commonly found and practiced.

Table 14.1 Spheres commonly associated with politics.

'Politics' is often associated with:	
Agriculture and food security	Infrastructure
Arts and entertainment	International affairs
Climate, environment, and sustainability	Justice
Commerce	Local and regional government
Communications	Literature
Defence force and defence industry	Manufacturing industries
Development and land use planning	Minerals extraction
Economics and finance	National government
Education	Oil and gas resource exploration and exploitation
Emergency services	Press and media
Employment	Professions
Energy generation and utilisation	Public transport
Fisheries	Religion
Gender, ethnicity, age, and disability	Science and research
Health	Sport
Housing	Universities
Indigenous affairs and culture	Water supply
Industrial relations	Welfare

The spheres of interest in this list were easily identified in less than 10 minutes, and we think that readers would not be unduly troubled to expand the list further. Indeed, we could recommend it as an interesting and distracting exercise when the serious burdens of project risk management begin to fall heavily on your shoulders – you will be surprised by the places or spheres where you discover politics in play!

Risk examples drawn from some of these areas are presented in Section 14.3.

14.3   Dimensions of Political Risk Factors

In terms of their presence in projects, political risks are multifaceted. Table 14.2 indicates some dimensional factors and suggests a combinatorial extent for them. We have defined short, medium, and long-term impacts for this table, but obviously these can be modified to suit organisational and project circumstances.

Table 14.2 Factors associated with political risks.

	Political risk variables		
	Intra-organisational	Inter-organisational	External
Decision or action	By you?/By others?	By you?/By others?	By whom?
Deliberate impacts	On you?/On others?	On you?/On others?	On you?/On others?
Unintentional impacts	On you?/On others?	On you?/On others?	On you?/On others?
Positive impacts	On you?/On others?	On you?/On others?	On you?/On others?
Negative impacts	On you?/On others?	On you?/On others?	On you?/On others?
Immediate impact	On you?/On others?	On you?/On others?	On you?/On others?
Short-term impact (<3 years)	On you?/On others?	On you?/On others?	On you?/On others?
Medium-term impact (3 - 5 years)	On you?/On others?	On you?/On others?	On you?/On others?
Long-term impact (>5 years)	On you?/On others?	On you?/On others?	On you?/On others?

The combinations of possibilities suggest that well over a thousand situations could require exploration for a political risk in a project setting, with inter-organisational politics occurring among project stakeholder organisations and external politics arising from parties beyond the project system boundary (Chapter 5, Figure 5.1).

The particular dimensions of political risks associated with particular projects should be established at an early stage in project risk management, preferably as part of the contextualising process.

An organisational scan should reveal those parts of your own organisation that are most prone to engaging in internal politics and those that are most susceptible to the impacts. Beyond this, the context establishment process of project risk management should allocate some time to discovering the propensity of other project stakeholders and external agencies to engage in a political manner to influence project decision making and actions. At this early stage in a project, knowing (or suspecting) the locus and type of political influence might be sufficient in terms of project risk management, as this at least raises political risk awareness in your organisation. More detailed knowledge is likely to emerge about the uncertainties associated with particular political risks as you proceed further around the project risk management cycle. From Table 14.2, the 'what', 'who', and 'when' questions framed in the list should then provide useful triggers for identifying and assessing these risks.

It is also important to remember that you are aiming to explore the *uncertainties* associated with political risks for your project and their effect on project objectives for your organisation. Avoid becoming mixed up in someone else's politics if they do not affect you!

Before examining the processes of managing political risks for projects, we offer some illustrative examples drawn from observation and experience.

14.4   Examples of Political Risks

The examples provided in this section are not exhaustive. They should help you to increase your project risk awareness by indicating the nature of political decision making and actions that may influence risks arising in different categories on projects.

Organisations that undertake projects in foreign countries quickly become aware of what are often termed 'sovereign risks' (although this term is not always helpful). These are mainly threat risks and denote risk situations that are peculiar to that context where special conditions or treatments apply to foreign companies working in a host country.

Sovereign risks typically comprise situations where the expatriate project stakeholder is likely to encounter difficulties in the host country, such as:

- Permit approval delays
- Profit transfer and repatriation restrictions
- Involuntary transfer of intellectual property and technology
- Discriminatory tax regimes for foreign companies and expatriate staff
- Local participation or part-ownership prescription
- High tariff barriers for imported materials and components
- Constraints on the employment of expatriate staff and workers
- Community opposition
- Expropriation or nationalisation
- Barriers to the repatriation of expensive equipment
- Licencing restrictions
- Excessive demands for royalty payments
- Hostile press attention.

While many of these situations may be known (with substantial certainty) beforehand, and appropriate management treatments arranged and implemented, there will always be residual uncertainty about whether or not political action will occur that would influence the severity of these risks. There is even a possibility that political decision making and action in a third or fourth country will affect the project risks of your stakeholder organisation. We also hesitate (being mindful of the need to avoid boring our readers) to point out that corruption might play a part in any of these risks!

Where projects in foreign countries are financed through aid funding provided by the contractor's home country, by a third-party country, or by an international organisation such as the World Bank, political perceptions and attitudes in those countries or organisations can circumscribe a contractor's ability to carry out the project successfully and efficiently. The political drivers of key stakeholders can therefore shape almost every risk the contractor and project sponsors face.

Industrial relations form another political risk, especially for large projects. Labour unions may negotiate with project sponsors or contractors, bartering limits to industrial action for additional benefits to members or the right to refuse entry to project sites for non-union workers. Depending on the political ideology of the national or regional government in power, policies may be implemented to bar such negotiations and concessions on public sector projects, on the grounds of encouraging greater competition among project bidders. Such is the political threat risk 'minefield' of industrial relations.

At the core of this type of activity is the desire for employers and employees to reduce uncertainty surrounding work that involves unions. Since it occurs on individual projects, there is no guarantee that agreements negotiated on one project will flow through to another, so the reduction in uncertainty (if it is achieved) is often quite limited and only temporary.

Political risk can be found further up the 'food chain', of course. A national government can (and usually does) exert influence and even final approval over projects intended to benefit communities in regional government areas, particularly in terms of project funding. Where the political flavour of the national government differs from that of a semi-autonomous regional government, project approvals or funding contributions may be unnecessarily delayed, diverted, or even refused. In some instances, this type of political activity may even diminish the efficacy of economic developments that are already in place and well-matured. For example, a national government may refuse to extend the landing rights for international airlines operating flights into airports in major regional cities (or it may actually reduce those already in place) where the political persuasion of the regional government is different from, and unsupportive towards, the central government. Or the national government may offer additional landing rights to airports where the regional government is highly supportive of its policies – in essence, where its electoral power bases lie. This type of political activity effectively contracts or expands regional economic growth in a 'chokehold–release' fashion that leaves some regions exposed to greater uncertainty in terms of their developmental objectives and proposed projects. Such situations mean that even where projects are funded and go ahead, they are still faced with ongoing uncertainties, especially if the funding is supplied on a 'drip feed' basis.

Another example, this time arising at a local level and then escalating to a national government level, concerns the intention of an international mining company to go ahead with a project to install new machinery at its ore extraction concession in a Third World country. The company's aim is to improve productivity, yield a better quality of ore for processing, and increase profit in the medium to long term. Despite the company's assurance that the new machinery will lead to more jobs and upskilling of mineworkers, the local community does not trust this assurance. It regards the upgrading project as a further threat to employment in an already impoverished area and fears the flow-on impacts on the regional economy. The mineworkers' union seeks greater benefits for its members in return for what the mining company perceives as somewhat lukewarm cooperation. However, the local member of parliament (renowned as a political firebrand) has threatened to take the issue to parliament and force the national government to cancel the mining concession unless the company offers cast-iron guarantees for no redundancies among general mineworkers, provides appropriate skills training, hires a minimum of 50 extra skilled workers each year for the next 10 years, does not relocate its administrative offices and staff from the local town, and pegs the royalty fee per tonne of mined ore to the national Consumer Price Index. Already we can see the intricate web of politics the company must navigate between the workers, their unions, the local community, the politician, and the national government. As an international company, it may even consider the possibility of closing the mine. The national government will be aware of this and will wish to avoid the possibility.

Case studies in this book also show evidence of the influence of politics in projects.

In Case Study A (a PPP [public–private partnership] correctional facility), the politics of justice and privatisation emerge. Accommodation pressures on the existing prison system caused the state government to twice increase the accommodation scope for the proposed new prison. The government insisted on pursuing a privatised PPP procurement approach, despite considerable doubt being raised about this type of arrangement in overseas instances of failure and despite the procurement difficulties associated with a previous PPP prison project in the same state. To some extent, the opposition party in the state parliament has been politically 'wedged' by not wishing to be seen as 'soft' on law and order issues and by desiring to sustain a 'small government' ideological stance.

Case Study B (a rail improvement project) was initiated through political action at the state government level during an election process. In the execution of this project, state government political comment and advertising have emphasised the public benefits that will flow, while opposition party politics have focused upon the disruption to public services, nuisance caused to residents by construction activities, prolonged inconvenience to public transport users, and aesthetic dis-benefits of 'sky-rail solutions'. Within the project packages, stakeholders have tried to exert political influence over project decision making, in order to advantage local communities and lobby groups. A more overt political shaping is found in the programme schedule, whereby half of the crossing removal projects were due to have been completed by late 2018, at about the same time as the due date for the next state election.

For Case Study C Part 2 (an aid-funded Pacific Rim civic facility), a clear objective for the funding donor government was to enhance relationships with small island nations in the region. On the other hand, it would not be politically expedient to be too closely involved in the project and risk being seen as trying to 'buy' influence.

In Case Study D (a train mock-up project), the state government required 35% of the value of the work to be undertaken within the state, as a contribution to employment and the economy. This would have influenced how process uncertainties were treated. From a completely different perspective, intra-organisational politics have almost certainly played a part in the 'silo' knowledge culture found in this case.

These examples provide a key to the first principle of dealing with political risks: know the stakeholders involved.

14.5   Political Stakeholders

If political risks are to be managed on a project, it is essential to discern the parties involved and to clarify their purposes. This may be easier said than done. Where the political presence and its influence are obvious, identifying the parties should be a straightforward matter. However, if their identity is concealed or camouflaged (as we noted in the introduction to this chapter), tracking the source or sources of political influence can be difficult and time-consuming. Given the rapidly increasing use of social media networks by contemporary governments and administrations, while one link in the network may be obvious and identifiable, other contributors to it can deliberately choose to conceal their true identities yet still be capable of engaging in political comment and persuasion, especially if the contribution represents diatribe rather than reasoned debate. Evidence arising after the 2016 presidential election in the USA has shown how data illegally drawn from social media network sources could be analysed, manipulated, and used to yield slight but significant electoral advantage to particular contestants, without necessarily disclosing (unless by accident or whistle-blower action) the identity of the political manipulator or even the precise nature of the manipulation.

Time spent in investigating, identifying, and considering the instigators of political influence and action is rarely wasted in project risk management. The knowledge not only is valuable for the project at hand, but also will almost certainly result in longer term benefits for the project stakeholder organisation. 'Know your enemies well (and your friends even better)' is good advice for project risk managers!

14.6   Managing Political Risks

The management of political risks on a project can be complicated, difficult, and sometimes intractable. Throughout this section, management of these risks is considered against each of the stages of the risk management cycle (Chapter 4, Figure 4.2). Our focus is mainly upon inter-organisational and external political risk sources.

14.6.1   Contextualising

We have noted in this chapter the importance of the context establishment stage in increasing knowledge about the project and its setting, and in raising risk awareness in the organisation. We have reiterated this view in terms of political risks and have suggested that even just identifying the potential sources of political influence will contribute to that awareness. In our view, failure to contextualise each project that an organisation undertakes is poor risk management. For political risks, establishing the context allows an organisation, at the very least, to discern where its objectives for the project are likely to be compromised. In this regard, an 'NQR aroma' – literally the intuition that something is 'not quite right' – is often a good starting indicator.

For international projects, risk guides are available that rank and compare countries in terms of criteria that relate to issues such as their government and administration and their industrial economies, trade, and finance. While these meta-guides will help to sharpen the NQR sense, they are probably too coarse-grained to sufficiently contextualise political risks for one project in one foreign country. For this, the context establishment process should consider a range of factors including:

- Geopolitical issues relating to that country
- Turbulence in its national politics
- Respect for the rule of law
- Discernible national cultures
- The nature and effectiveness of its governance and administration
- The strength and consistency of its anti-corruption measures
- Any restrictions on foreign trade and investment, including tariff barriers
- The strength of its economy and fiscal regime
- Degree of harmony in its foreign relations (particularly with the project stakeholder's home country)
- The independence of local industries and markets
- The role and power of local press and media
- Response to local demonstrations and protests.

The assessment and weighting of each factor will be guided by the relative importance of each one to the project stakeholder. Previous project experience in the same country, or even on the same continent, is useful for assessment but should not be given undue prominence. History warns us about the volatile and often dramatic nature of political activity.

For domestic projects, the approach is likely to be substantially different. It will also be largely guided by the currency of local and national information, and focus far more on establishing the 'who' of political influence and the extent and effectiveness of their powers.

A useful interrogative tool is found in the first stanza of Rudyard Kipling's well-loved poem in 'The Elephant's Child' as part of his 'Just So Stories' series (Kipling 1902):

> I keep six honest serving-men:
> (They taught me all I knew)
> Their names are What and Where and When
> And How and Why and Who?

Asking the six questions at appropriate points will help to elucidate political risk factors.

14.6.2 Identifying Risks

As part of the process of identifying project risks (see Chapters 6 and 7), in addition to the other questions raised and in conjunction with any specific identification techniques, workshop participants at this stage of a project should also ask: What power or influence is being exerted (or could be exerted) over this decision that can introduce uncertainty into the achievement of desired objectives?

As with other types of risk, the risk statements for identified political risks should be as precise and complete as possible, as this will inform and guide the following processes of analysis, response, monitoring, and control. Given that the 'who?' and 'what?'

questions have been asked, the political risk statements should, as far as possible, also address the 'where?', 'when?', 'how?', and 'why?' questions.

Scenario testing (Chapter 7, Section 7.7) is often a valuable tool for identifying and describing political risks. Comprehensive analysis of politically framed scenarios can yield good clues for risk assessment and risk response.

14.6.3 Analysing and Assessing Risks

For the most part here, we will focus upon external political risk events (those arising outside the project system boundary) and those emanating from other project stakeholders.

In analysing and assessing political risks, caution must be exercised to avoid the bias of the 'availability' heuristic (Chapter 8, Section 8.2). For example, care should be taken with relying upon assessments that proceed along these lines: 'Three out of the last five projects we have undertaken for this type of client in this type of location have experienced political interventions by x stakeholders that have led to y uncertainty in z aspects of the project'. Using the information available for previous experience is acceptable, but only if investigation shows that no critical factors or environments have changed during the interim period.

For the great majority of political risks, qualitative analysis is the only feasible technique, since the nature of any 'hard' data, even if they can be collected, will rarely be appropriate for any reliable quantitative assessment. In any event, if political influence is already being exerted over the project, there is no uncertainty about its likelihood of occurrence – only about the potential impact. If nothing political has happened so far, then the probability of a political risk occurring is essentially a judgement call for the project stakeholder and will depend on perceptions of the likely source and nature of the event.

Nor is there much advantage gained by trying to quantify the impact precisely. Other than possible delays to the project, other impacts may be more intangible in relating to difficulties and barriers rather than to direct effects on project costs. Political risks, if they eventuate, more often than not lead to disruption and turbulence. Both are difficult, if not impossible, to assess reliably and confidently in terms of their magnitude.

For example, the promoter of a new 'fracking' project to extract natural gas from a rural underground location using high-pressure water injection might foresee the likelihood that an environmental or agricultural lobby group will carry out some form of protest action to threaten the project itself or the extraction rights approval processes of the responsible authority. The uncertainties associated with this form of political risk will relate to questions such as: What group(s), when, where, how often, what form, what intent, and what direct and indirect impacts? Protest actions that occurred in the past may provide only limited and unreliable guidance, since the strategy of most protest groups is to achieve their objectives in radical and news-grabbing ways that will garner public support or attention.

Furthermore, it is not safe to assume that political risk activity will always happen in the upfront stage of the project cycle. Many 'landmark' projects have been associated with politically based risk events at their opening ceremonies. Others have been similarly affected at key milestone points as opposition forces gather strength and momentum and as such points become newsworthy.

14.6.4   Responding to Risks

Response options for political risks on projects are generally limited to mitigation by reducing the likelihood of occurrence, the impact, or both. Simple retention (i.e. ignoring the risk) is often unwise, as it can convey an impression of indifference to the potential situation and become a reputation threat risk. Risk avoidance, transfer, or sharing alternatives are usually not available, or they are impracticable to undertake in the context of the project, although bipartisan approaches may be effective in mitigating the likelihood of occurrence of political risks that involve protest lobby groups.

As some of the examples in Section 14.3 demonstrate, it may be possible to take pre-emptive action to reduce the probability that a political risk event will arise on the project. Negotiation with the relevant parties is a key reduction strategy, albeit through concession or compromise, but this requires that the instigators are known to the project stakeholder and that negotiation is actually possible.

Proactive mitigation of the consequences of political risks should focus upon two areas: maintaining appropriate backup resources (e.g. alternative sources of supply) or workaround routines (e.g. arranging alternative project access routes that will bypass protest action locations).

Key to any response will be good public relations capacity and capability. This also helps to counter any reputational threat to the organisation. Candid public statements, appropriately timed and avoiding hyperbole, comprise the most effective countermeasure, especially if they are framed to avoid indifference or aggressive behaviours. Where possible, any direct involvement in political conflict should be avoided, and a neutral stance adopted as far as this is possible and appropriate. Any organisational or personal sympathy with political protesters should not be openly manifested.

There is an important corollary to mitigating political risks for a project in this way. Wherever possible, the reduction treatment should be rehearsed (i.e. through scenario testing) and even stress-tested. This strategy can then be followed by post-response reassessment and consideration of further reduction measures if deemed necessary.

14.6.5   Monitoring and Controlling Risks

Watchfulness through awareness is the essential key to monitoring and controlling political risks associated with projects, along with being alert to any new risks that might arise through changes in political actors and scenarios. Media sources and contact networks can play a valuable part in this.

14.6.6   Knowledge Capture

Ideally, knowledge capture for political risks on projects should take place as soon as each one is 'closed off' (i.e. the risk can no longer occur or any impact has ceased). The fresher such knowledge is, the more value the knowledge retains in terms of learned wisdom for the project organisation. Miniature 'lessons learned' debriefing sessions are a good means of capturing the knowledge. Much of this will be tacit knowledge arising from individual experiences. Bearing in mind that we have emphasised the unique nature of many political risks, any transformation to organisational risk wisdom is likely to be in the form of useful principles rather than specific remedies or actions.

14.7   In-house Political Risks

It would be remiss of us not to give some attention to political risks emanating from within the project stakeholder organisation itself. Intra-organisational influences of a political nature, and their impacts upon projects, are not uncommon. Indeed, it would be surprising if they were.

In-house politics are inevitably driven and shaped by the cultural forces operating within an organisation, and these were extensively explored in Chapter 13. Where the politics are overt, and reflect alignment between the espoused and enacted values of the organisation, their effects on the projects undertaken by the organisation are usually benign. At times, they can not only demonstrate best practice but also contribute significantly to project success, thus embodying corporate claims that 'our best resource is our people'. Typically, positive organisational politics might see one department trying to outdo another in their support for, and contribution to, a project. This would be the politics associated with good-natured rivalry and would be encouraged by senior management – unless the good nature were to disappear and be replaced by intra-organisational nastiness.

Matters become difficult when negative covert cultures operate within an organisation and its values are misaligned. Because of their hidden or disguised nature, covert cultures are difficult to discern, especially in terms of project risk management. Good risk awareness will help, particularly the NQR sensitivity we have referred to in this chapter, and the internal context establishment processes discussed in Chapter 5. Typical clues might lie in sudden re-prioritisation or diversion of organisational resources for a project, deterioration in inter-departmental cooperation, repeated absences from key project meetings, frequent delays in providing important information – the list is long. While any one of these might be innocuous and explainable, combinations of two or more should be regarded as suspicious and likely to be due to political action occurring within the organisation.

There is little point in proactively trying to assess the likelihood of occurrence of internal political risks such as these, although good internal contextualisation will help. If they arise, they will inevitably have consequences that will threaten project success in some way.

Where negative internal political risks are detected or suspected, straightforward and open confrontation is the only effective response to mitigating or avoiding them. Any attempt at subtle 'game playing' is likely to exacerbate the situation and is just tantamount to participating in the political fracas.

Knowledge capture of negative intra-organisational political activity on a project is a tricky issue. None of the actors will want their 'dirty linen' laundered in this way, and, unless the response confrontation has been successful and ongoing, there may even be attempts to manipulate the capture process or the information that is collected. The sole remedy lies in firm but fair leadership by senior management in the organisation and the elimination of covert political activities that subvert or threaten successful management of the project.

14.8   More Extreme Political Threat Risks

While their intent and impact are aimed more at societies and other groups of people, rather than projects per se, the more extreme forms of political threat risk cannot be ignored completely from a project perspective.

We refer here to risks that involve some sort of violent or forceful actions intended to cause harm to people or damage to property or systems, with a view to achieving political ends or demonstrating political views. The violence may be initiated by opponents in order to satisfy political demands or weaken existing power holders. However, it should be remembered that history shows us that it can also be instigated by authoritarian governments to enforce their hold on power and to deter protests against the regime.

Included among such risks would be terrorism (the use of violence to induce fear), lethal assault, sabotage (the destruction of property or systems), and hijacking and kidnapping (unlawfully detaining people or denying them liberty in order to extract political advantage or ransom).

Cyberattack (denial of service, theft of data, or deliberate disruption) upon public or private computer and information systems may also be intended to achieve political ends as well as to hold an organisation to ransom.

War is the ultimate extreme political action.

While extreme political risk activities and events (other than sabotage) seldom target projects directly, the indirect consequences for projects or project stakeholder organisations may be extensive and disastrous. In effect, such activities have a driver effect similar to that depicted in Figure 5.2 in Chapter 5, in which 'social' risk drivers could be amended to include a sociopolitical subset of influences.

Furthermore, extreme political risk events should not be regarded as being associated solely with projects carried out in politically unstable foreign countries. Events that have already occurred in the twenty-first century show that an organisation might face them in its home country.

Scenario testing and more specifically focused context establishment, incorporating the six 'Kipling' questions, are usually the most effective approaches for identifying extreme political risks.

Assessment of extreme political threat risks is invariably qualitative and limited. Remember that these are likely to be risks with very low likelihood of occurrence but catastrophic impact, and their severity is extreme (as indicated by the right-hand column in Figure 8.7 of Chapter 8). The biggest issue with assessment is the high level of uncertainty involved with the likelihood of occurrence and with risk consequences. Furthermore, an organisation cannot afford the reputational risk of having been shown to be wrong in its earlier judgement in any post-event enquiry following an extreme political risk event, should it have chosen to ignore (i.e. simply retain) the risk, especially if this was done on an uninformed basis.

Response by avoidance may mean having to abandon the project. Mitigation measures may be similar to those described in this chapter for non-extreme political risks, but the measures will inevitably have to be more stringent, and much attention will have to be given to escalating existing security measures.

Cyberattacks are particularly difficult to counter pre-emptively. Pre-emptive system protection measures and post-attack data and system recovery are complex and costly, but the costs should be set against the long-term losses that might be suffered.

Heightened risk awareness should be instilled in all project staff: being alert is key to making rapid responses. Monitoring and control of extreme political risks must be undertaken frequently and influenced by intelligence gathered from all available sources. Emergency staff evacuation (including medical evacuation) plans should be in place and rehearsed. It may be necessary to ensure that protective clothing, equipment, and communication devices are conveniently located, and that code or password-protected alert facilities have been installed and tested. Communication channels must be reliable and resilient. High levels of cooperation should be established with external authorities and emergency services. Skilled public relations expertise should be at hand. However, each of these measures comes with its own problem: that of maintaining vigilance and preparedness throughout the project life cycle.

Knowledge capture from any real-life experiences of extreme political risk events and activities is vital, in order to raise organisational awareness even higher and to inform any additional response measures that may be contemplated. The knowledge capture process should be implemented as soon as possible after any extreme political threat risk event has occurred, and the process should be paralleled by appropriate staff counselling. This applies as much to cyberattacks as it does to physical violence and injury.

14.9 Summary

Politics are encountered on almost every project. Because of the shifts in meaning over the past century, politics are no longer confined to the affairs of public government but have spilled over into every field of human activity. Despite this ubiquity, political threat risks can be difficult to detect and difficult to manage on projects. The difficulties of managing them are exacerbated by the fact that they can arise from external, inter-organisational, and even intra-organisational sources.

The strong association between politics and emotion means that political attitudes tend to be passionate and fixed. Elements of irrationality may be detectable in many political threat risks, and together with the awareness of 'something being not quite right' there may be a perception that the political action towards the project does not make complete sense.

The likelihood of occurrence and the consequences of political threat risks are rarely amenable to quantitative analysis. Even qualitative assessment may not be entirely practicable. Given this, some form of risk management action is usually required.

The impacts of such risks generally include disruption and turbulence to any or all of the processes of project delivery.

Response options to political risks are limited. Watchfulness is a key to monitoring and controlling political risks, particularly as new risks can arise with little or no warning.

Political risk knowledge capture on a project should not be unduly delayed, and the knowledge transformation should focus on principles of managing such risks rather than on risk details, as the latter will probably be different on the next project.

Political risks emanating from within the project stakeholder organisation are likely to originate from, and be shaped by, the prevailing organisational culture. Where the culture is covert and negative, the task of managing internal political risks escalates exponentially. Addressing such risks means addressing the culture itself, and this may be difficult, if not impossible, in the short-term environment of most projects. Direct and open confrontation may be the best response, but will require the support and leadership of senior management in the organisation.

Risks at the violent extremity of politics cannot be ignored, albeit that their intended consequences generally affect projects indirectly rather than directly. If any such risks are anticipated, a high level of readiness must be maintained. High uncertainty, limited response options, and the impracticability of sustaining awareness and watchfulness in the longer term all render these risks difficult to manage.

As a welcome shift from the foreboding topic of political threat risks, in Chapter 15 we explore the nature and treatment of opportunity risks in projects.

Reference

Kipling, R. (1902/2017). *Just So Stories*, republished with intro. by J. Stroud. Melbourne, VIC: Puffin Australia.

15

Opportunity Risk Management

15.1   Introduction

In Chapter 2 (Section 2.3), we noted that the main focus throughout this book would be upon risks that present uncertain threats to the achievement of project objectives. However, the possibility of more positive project outcomes cannot be ignored completely, and this chapter attempts to address that situation.

One chapter (out of 20) to deal with opportunity risks might appear parsimonious, but the proportion probably represents the amount of attention spent in real life dealing with opportunity risks compared with that devoted to threat risks, particularly with respect to projects. While we have noted (in Chapter 3) that all projects are a response to recognition of a need or opportunity, the management of those that are initially aimed at exploiting opportunities quickly develops into protecting the fulfilment of the objectives set for them rather than the exploration of further opportunities.

It is thus fair to say that opportunity risks are largely neglected in project risk management. Case Study B (a rail improvement project) supports this view. While this is understandable – after all, no one wants their project to fail – every organisation could usefully devote some attention to the potential benefits of identifying and exploiting opportunity risks.

This situation tends to flow naturally from the educational backgrounds of the professional staff involved in many projects. The curricula for most degrees in project-related professions, such as engineering, construction, and project management itself, place substantial emphasis in their problem solving approaches on issues such as careful and economic design, prudent control, safety, and adherence to other compliance requirements. Even architecture, perhaps the most creative of these professions, is constrained in large measure by the realities of projects. The educational curricula cannot be overly criticised in this regard, as they are intended for the ongoing protection of the public as well as preparing professionals for their careers. Not only are the cautionary approaches prominent in education curricula, but also they provide the essential framework for the practice standards promulgated by all professional associations.

In subtle ways, the solutions to problems have become increasingly 'vanilla' flavoured since they are likely to present fewer threat risks to project stakeholders. Correspondingly, however, less time is made available for learning about, and practising, creative exploration of project opportunities, particularly in a risk setting.

Managing Project Risks, First Edition. Peter J Edwards, Paulo Vaz Serra and Michael Edwards.
© 2020 John Wiley & Sons Ltd. Published 2020 by John Wiley & Sons Ltd.

Not every opportunity risk will bring added benefit to the project stakeholder on the project in which it is identified, although most should do so. For various reasons, such as impracticality, resource and scheduling limitations or priorities, technology constraints, and so on, full exploitation of opportunity risks might have to be deferred or declined for that project. However, if the opportunity risk information is captured by good risk knowledge management, it may well be developed, taken up, and exploited on future projects.

In this chapter, we will consider the definition and concept of opportunity risk, how it arises in projects, and how it may be managed.

15.2   Concept of Opportunity Risk

Most dictionaries define 'opportunity' in terms such as 'a favourable time for action' or 'a favourable chance'. When risk is associated with opportunity, then uncertainty must be present in any or all of the characteristics of time, availability, and nature of the opportunity, and in the nature and extent of the potentially favourable gain. Opportunity risk management action may flow from a decision that either has been made or can still be made.

In Chapter 2, we referred to risk as a two-sided coin: threat and opportunity. Technically, this is correct. If we had such a coin and tossed it 200 times, we could expect each side to land face-up on about 50% of those occasions. But we have already noted that we do not regard projects in this light because we do not toss the coin in this way – in fact, we do not toss the coin at all! Instead, we probably spend 95% of our project risk management time searching for and dealing with threat risks, and only 5% looking for and exploiting opportunity risks. The chapter allocation for this book shows that we have deliberately 'weighted our coin' in this way.

While the 'opportunity' definition clearly has a positive connotation, we must be careful in tying it too closely to similar-sounding terms such as 'opportunistic' and 'opportunism'. The former denotes conduct based on prevailing circumstances rather than on principles – the antithesis of good risk management – while the latter denotes the practice of seeking and taking profit from all actions and outcomes (a matter for ethics). In contemporary society, the social connotations of both tend to be more negative than positive.

In the introduction to this chapter, we alluded to factors that help to explain and justify the negative threat risk perspective that prevails in project risk management. Other factors help to reinforce this grip.

The concept of risks as events with outcomes that threaten the project objectives of a stakeholder organisation is well entrenched. Indeed, a negative view of risk is prevalent throughout society. It is reinforced in almost every book written about risk and risk management, and about projects and project management. Our book is no exception. Even where authors are careful to mention a connection between risk and opportunity, the treatment is usually brief, and the majority of the risk examples presented is still slanted towards those with adverse outcomes. Again, our book is no exception.

Contrast this situation with the addiction to gambling manifested in almost all groups in modern society. This suggests that we are interested in, and keen to exploit, opportunities to wager money in a desire to gain more. It corresponds to the psychosocial understanding of risk discussed in Chapter 2. Yet that predilection is not truly analogous to exploring opportunity risk on projects. A gambler seeks the thrill of putting a relatively small stake at risk for the chance and hopeful expectation of winning a commensurately far greater reward. Unless the gambler is a professional, knowledge that the actual chance of winning is probably far less than 0.01% and that the corresponding chance of losing the stake is thus greater than 99.99% is ignored in preference to the emotional thrill associated with the experience. The professional gambler would never accept such odds, of course.

This is not the case with opportunity risks in projects. The thrill is rarely addictive, and any opportunities that emerge tend to be regarded conservatively and soberly.

In our experience, many people (and most project stakeholder organisations) tend towards risk aversion in their project decision making, especially in relation to follow up decisions after an initial opportunity risk seeking choice. We alluded to this in Chapter 2 by suggesting that a person's risk attitude is not necessarily consistent over the whole range or cycle of decision making. For example, construction contractors are said to exhibit risk seeking (e.g. opportunistic) behaviour when bidding for new work, but quickly shift to risk aversion (protection of the anticipated profit) after winning a tender. Their risk aversion is thus framed by opportunism. To be fair, the research underlying these findings is not recent, nor is it extensive; and Case Study A (a public–private partnership [PPP] correctional facility) and Case Study F (an aquatic theme park) show that the main contractor in each case conducts diligent threat risk investigation, in compliance with company policy, before submitting any project bid.

In a different project context, someone deciding to sail solo around the world, or walk to the North Pole, has made an initial risk seeking decision to exploit an opportunity. However, almost every decision after that is made with the risk-averse intention of avoiding or minimising threats to the achievement of his or her objective.

Stakeholder organisations tend to follow either of these patterns in projects of all kinds, although to be fair not all stakeholders are involved in the initial opportunity seeking project decision.

In Chapter 3, we referred to the claim of Parkin (1996) that much of project decision making, at least for engineers, is guided by codes and specifications, and that there are two contrasting modes of organisational management: the ordered approach of the administrative mode, and the power and politics interests of the enterprise mode. Project organisations tend to adopt the administrative mode in the belief that it has greater clarity and is easier to implement in what may be seen as a business management environment. Since the aims of the administrative mode are self-evidently intended to protect fulfilment of the project objectives of the organisation from threat, it is hardly surprising that project risk management focuses almost exclusively on threat risks. However, that is not tantamount to saying that an opportunity perspective should always be ignored in project risk management.

15.3 Opportunity Risk in Projects

Writers who challenge the essentially negative threat view of project risk management argue that it emphasises the management of bad luck and fails to capitalise on good luck. In our view, this trivialises the aims and importance of management.

Another argument states that a focus on threat risks does not permit consideration of potential trade-offs between project objectives – that opportunities levered off one objective cannot be used to enhance achievement in another, simply because the management approach focuses exclusively upon potential threats to each objective. This really belittles the skills of those involved in risk management, as it assumes that they are incapable of 'cross-over' thinking.

It is also suggested that risk management is usually delayed until the project and its scope are sufficiently defined, so that the information then available will be better and permit more effective risk management. Such delay thus limits any early systematic search for opportunities to exploit in terms of improving project performance. However, this suggestion flies in the face of the previous argument, which assumes that risk management can be implemented as soon as project objectives are established (i.e. before the project scope is fully determined).

All these arguments are somewhat reductionist, since they assume that every opportunity must have positive outcomes and that every threat will be completely negative. They rarely acknowledge that exploiting opportunities will almost certainly introduce more risks and change the characteristics of others, and they fail to consider that treating the negative effects of threats may actually produce other positive results.

Delaying the use of risk management on a project until sufficient information is available also runs counter to the intrinsic purpose of risk management and fails to appreciate the benefits of a dynamic project risk management system (PRMS). As we argued in Chapter 4, risk management is not a one-off intervention at a single point during a project, but a systematic and ongoing approach to dealing with risks over the whole project life cycle. The project risk register (PRR) should be implemented at the commencement of the project (i.e. the conceptual stage), so that it is capable of reflecting and influencing the outcomes of decision making at the earliest possible stage.

Two observations flow from all of this. The greater the focus on threats in risk management, the greater will be the number of opportunities that may be missed. However, the greater the focus on opportunities, the greater will be the exposure to threats, since every opportunity may bring with it a 'downside' potential in some form.

Consideration of several project examples should help to consolidate our thinking about opportunity risks.

15.4 Examples of Opportunity Risks

The examples presented here are drawn from projects referred to in earlier chapters or from the case studies included in the Appendices.

15.4.1 IT Brand Product Personalisation Service

The IT brand product personalisation project was described in Chapters 3 and 4. Readers will recall that the original project concept was for one business unit in a commercial group organisation to develop and implement a personalisation service to offer to its online customers. For a small additional cost, buyers could order their products to be embroidered with personal names or initials.

During the project development stage, it became clear that the presentation of the personalisation option on the order page of the website would require expansion of the parameters of the brand company's customer database. This was a quite straightforward adjustment, but system designers realised that this also presented a further marketing opportunity. It was originally envisaged that the personalisation choice for customers would be limited to names or initials in a single style. With a reconfigured database, however, substantially more options might become available as long as product manufacturing processes would not present constraints. Customers could be offered more online style choices. It might even be possible for logos or graphics to be made available, or ultimately for customers to upload their own personalisation designs.

In the event, time pressure on the IT project itself meant that the opportunity could not be fully assessed and exploited. However, extra capacity was designed into the new customer database system, and the product embroidery machine checked and found capable of accommodating more variations in product personalisation design. The opportunity was thus partly exploited, but largely deferred.

15.4.2 Botanic Gardens Special Display Project

The schedule for a botanic gardens special display project was described in Chapter 7 as an example of an associated/representative risk identification tool. While the project schedule (Table 7.9) was primarily used as a tool to identify project threat risks, the question arose as to what other means might be used, outside the physical location of the gardens, to promote the display project. The audience capacity for the opening ceremony could not be expanded, so additional promotional activity could occur before or after the opening date. Perhaps without even realising it, the project stakeholder was engaging in opportunity risk exploration. During the project design stage, some thought was given to proposals for cooperating with several local authorities and offering to synchronise display promotion activities with local events that they might be planning. Eventually, however, those ideas were shelved as it was thought that they required too much additional time and resources to negotiate and exploit properly. In this example, the opportunity risk decision was deferral, partly because of the additional threat risks to the current project. Currently, the client organisation has appointed a small team to make contacts with local councils and other agencies, so the opportunity risk has not been completely rejected.

15.4.3 Case Study A (PPP Correctional Facility)

Case Study A describes a PPP correctional facility project. The organisational structure for the project is shown in Figure A.1. In this case, the design and construct (D & C) contractor in the original consortium recognised an opportunity for ongoing participation as a contractor to the facilities management (FM) stakeholder. Although not familiar with this type of work, the D & C contractor's parent company believed it was fully capable of undertaking the maintenance work, and the opportunity was exploited by submitting an appropriate bid that was subsequently accepted.

After a relatively short period of carrying out the construction-related FM work, the parent company foresaw further business potential in this role and established a separate specialist FM division.

In this example, the opportunity was more fully exploited than on the IT project and even more fully exploited as a deferred potential. Besides the profit motive, a significant benefit achieved from the exploitation of the opportunity has been through the upskilling and capacity building of existing staff.

15.4.4   Case Study C (Aid-Funded Pacific Rim Island Civic Project)

In Case Study C, uncertainties arose in the requirements for the information and computer technology (ICT) installation for the new Parliament building. Through the Foreign Affairs Department of the funding government, the senior advisor was able to access expertise from the ICT specialists in the FM section of the Parliament of the funding government. These experts were able to clarify the ICT brief and advise the ICT D & C subcontractor. This opportunity risk was subsequently further exploited in the early planning for a similar project proposed for another Pacific Rim island government.

These examples show how opportunity risks can be identified and managed to add value. However, all of them reveal a lack of any deliberate, systematic, and proactive search for opportunity risks. More often than not, opportunities were recognised only after they had arisen or had derived from threat risk management.

Typically, not all the opportunities were fully exploited, largely because of the additional threat risks they might bring to the projects (particularly in terms of capital costs or project completion times). Their potential benefit was recognised, however, and some opportunities were deferred for future adoption rather than abandoned completely.

15.5   Managing Opportunity Risks

The proactive management of opportunity risks has implications in terms of the personnel involved and the PRMS.

15.5.1   Implications for Personnel

Importantly, the people tasked with managing threat risks may not be best suited (at least initially) to explore and exploit opportunities. Not everyone involved with risk management will be equally adept in dealing with both extremes of the risk continuum. The psychology of risk, and the variability in risk profiles and attitudes, explains this.

Further evidence of this dichotomy is found in the increasing proliferation of specialist (subcontractor) project stakeholders in the contemporary world of projects. The inability of a single stakeholder to deliver all the elements of a project (tasks, technologies, resources, and organisation) under one organisational umbrella (i.e. in-house) has led inevitably to the introduction of others who, because of their specialised knowledge or skills, are able and willing to take advantage of the opportunity to deliver specific project sub-elements. In a sense, therefore, subcontracting is as much opportunity seeking behaviour on the part of subcontractors as it is a means of transferring risk by head contractors. The risk profiles and attitudes of each are essentially situation (or context) specific.

We do not advocate a separate opportunity risk management 'team' for each project. Not only would that be an unjustifiable luxury for most organisations, but also it could lead to a divisive risk awareness environment that could be considered unhealthy. Nor do we support recruitment of opportunity risk 'specialists' – even if such people were found to exist.

A better approach would be to 'grow' an integral opportunity risk culture within the organisation, gradually raising opportunity risk awareness by expanding the risk mindset of staff, and thereby increasing organisational risk management capability.

A starting point for this type of cultural change might be to task one or two members of senior management to undertake 'opportunity risk reviews' on selected current projects, and to ensure that this strategy is communicated to the whole organisation ahead of its introduction. Feedback on any positive outcomes should also be communicated in this way. From this point, depending upon reception of the strategy and its outcomes, reviews could be extended to all projects undertaken by the organisation.

If the strategy is received enthusiastically, willing members in the organisation's project team could then be invited to participate in opportunity risk management training sessions. The training would orient participants to the desirability of exploring project opportunity risks, show them how this would fit into the existing PRMS, and use historic projects as training simulators. The simulations would first look at opportunity risks ab initio (i.e. without reference to the original risk management process), possibly by using the objectives of the original project. Subsequently, trainees could explore opportunities that might be found in the documented response decisions for the original threat risks. Practice with the alternative use of familiar risk identification tools would be useful, and training should also include consideration of the threat risks that might arise from opportunity risk exploitation.

Where the strategy receives only a lukewarm or negative reception, then the senior management project review process should be patiently extended, and any successes more fully celebrated, until eventually a more positive shift in culture can be detected in the organisation. One indicator would be when innovative opportunities are perceived in the management of safety hazards, since that is probably the most dominant area of threat risk management yet the one where new solutions are hard to find.

A strategy such as this recognises the difficulties associated with cultural change. It attempts to expand mindsets in a participatory way. However, it is not a 'quick fix' solution to raising opportunity risk awareness in a project stakeholder organisation and has to be aligned with any entrepreneurial culture already in place.

15.5.2   Implications for the PRMS

Consideration of opportunities does not replace threat risk management but is a desirable element to be appropriately integrated into overall project risk management. While effort expended on exploring opportunities may be only one-twentieth of that spent on threat risks, the outcome value equation is not necessarily the same. A well-exploited opportunity might yield huge benefits to the project stakeholder, not only for the target project but also for many future projects.

With the inclusion of opportunity risks, the stages of the PRMS would remain the same: context establishment, identification, analysis and evaluation, response, monitoring and control, and knowledge capture. However, some differences will occur within each stage.

During the context establishment stage (Chapter 4, Figure 4.2, Stage A), the risk management workshop participants could adopt a three-column page or whiteboard format for noting considerations. The first column would list the internal and external contextual issues to be considered. The second column would record negative (threat-related) observations about each issue, while the third column would note any positive (opportunity-related) aspects of the same issues. This approach should reveal the threat and opportunity risk 'drivers' on that project for that stakeholder organisation. For this stage, both types of driver can be addressed at the same time.

In the risk identification stage, the techniques for recognising opportunities are essentially similar to those used for identifying threats to the project. For opportunity risk identification, the interrogative questions (Chapters 6 and 7) would be positively, rather than negatively, framed.

Bearing in mind the 1:19 ratio proposition made in Section 15.1, threat risks should be identified first, since project objectives are unlikely to be seriously affected if opportunities are ignored or overlooked. We do not advocate simultaneous exploration of threats and opportunities, partly on the grounds that some threat risks might thereby be missed, but also because a simultaneous approach is more fatiguing mentally and the demands upon the risk management workshop are already intense. Furthermore, opportunity risks might be more difficult to grasp conceptually, especially if this approach is unfamiliar to risk workshop participants. Opportunity risks rarely emerge as fully formed conceptually as do threat risks. Also, they may not fall into neat generic or customised risk categories. Opportunity risk classification is an under-researched area in project risk management.

Precise opportunity risk statements should flow from the identification stage, for example 'There is a chance "p" that benefit "b" will be gained if opportunity "a" is exploited during the period "t".'

Note that this statement differs from that suggested in Chapter 7 (Section 7.9) for threat risk events. More often than not, for opportunity risks the main probability attaches not to the occurrence of an event leading to a consequence, but to the attainment of a particular beneficial outcome if a decision is made to exploit the opportunity. However, some opportunity risks do initially arise as trigger events.

A precise opportunity risk statement will at some point, but preferably as soon as possible in the risk identification stage, include the words 'There is a chance that ...'. The single-sentence statement should contain references to:

- A situational opportunity
- The uncertainty (likelihood of occurrence) if the opportunity is identified as a trigger event
- The period of exposure to the likelihood of occurrence for the opportunity event (if that is identified)
- The consequences of exploiting the event
- The uncertainties and time constraints associated with the consequences
- A risk category and a risk identifier code.

For the risk analysis stage (Chapter 4, Figure 4.2, Stage C_1), some of the qualitative indicator scales proposed in Chapter 8 for threat risk assessment must be adapted for assessing opportunity risks.

The three-point likelihood and impact indicator scale shown in Table 8.1 (Chapter 8) remains unchanged, but, as noted in this chapter, it would now represent the chance that a particular benefit could be gained from the opportunity that is identified and exploited.

The five-point probability indicator portrayed in Table 8.2 could also be used for opportunity risk analysis, as long as it is remembered that this would more often apply to the probability of gain occurring if an opportunity were to be exploited, and not to a likelihood of occurrence for the opportunity. In that sense, we might argue that the opportunity gain is now the event and the opportunity a condition precedent to it.

For the customised probability indicator presented in Table 8.3, one small amendment would be necessary for the lowest ('rare') interval descriptor. The 'only occurs as a 100-year event' reference is appropriate for natural threat risk events such as floods or volcanic eruptions, but not for opportunity exploitation through human agency. A more suitable replacement might be 'Occurring only once in a lifetime'.

Table 8.4, with its negative impact factor scale descriptors, must be completely replaced for opportunity risk assessment. Initially, a scale addressing only financial benefits might be sufficient, and this is shown in Table 15.1. Note that for the highest level descriptor, 'exceptional' replaces the now unsuitable 'catastrophic' of Table 8.4; also note that arithmetical vector symbols are used to more carefully delineate financial amounts.

Table 15.1 Five-point interval scale for opportunity risk financial impact.

Beneficial impact or consequence	$Gain
5. Exceptional	>$xxxx
4. Major	≤$xxxx
3. Moderate	≤$xxx
2. Minor	≤$xx
1. Insignificant	≤$x

Table 8.5, with its multiple measures of negative threat risk impact, must be completely reframed. Table 15.2 presents an example of multiple impact measures for opportunity risks.

The measures indicated represent what we believe could be meaningful gains for a project stakeholder organisation intending to exploit opportunity risks. Each stakeholder, however, should establish (through strategic risk management) the measures best suited to its involvement in projects.

Importantly, these linguistic measures also reflect a reality about opportunity risks in that they are typically less specific than the contrasting threat risk impact measures. It is usually found that more uncertainty attaches to almost every aspect of opportunity risks. While this can be accommodated in the linguistic measures of qualitative risk assessment, it must be carefully considered if quantitative analysis of opportunity risks is contemplated. The wider ranges of probability data used as inputs for quantitative simulation techniques mean that modelling outcomes may lack sharpness and definition. This can affect decision making for opportunity risks in terms of confidence and reliability.

Table 15.2 Multiple measures of opportunity risk impact.

Rating	Financial gain	Safety	Environmental	Societal	Organisational	Reputational
5. Exceptional	>$xxxx	Safety performance improvement 51–100%	Large improvements in all aspects	Large positive impact on all sectors of society	Large improvement to culture across all departments	Large increase in national and international reputation and demand for organisation's products or services
4. Major	≤$xxxx	Safety performance improvement 26–50%	Large improvements in several aspects	Good impact on several sectors of society	Large improvement to culture in some departments	Noticeable increase in international reputation and demand for organisation's products or services
3. Moderate	≤$xxx	Safety performance improvement 11–25%	Small improvements in all aspects	Positive impact on society generally	Culture improved throughout organisation	Noticeable increase in national reputation and demand for organisation's products or services
2. Minor	≤$xx	Safety performance improvement 5–10%	Small improvements in several aspects	Positive impact on some sectors of society	Culture improved in some departments	Increased local interest shown in organisation's products or services
1. Insignificant	≤$x	Safety performance improvement less than 5%	Small improvement in one aspect	Little or no benefit to society	No observable improvement to culture	Small improvement in local reputation or demand for products or services

For evaluating opportunity risks (Chapter 4, Figure 4.2, Stage C_2), the matrix shown in Table 8.6 requires slight amendment. Table 15.3 shows that 'Potential' should replace the previous 'Severity' label. The shaded 'Low', 'Medium', and 'High' cell indicators remain, as long as it is remembered that the actual placement of them also requires strategic risk management attention.

Table 15.3 Three-point opportunity risk potential matrix.

	High	OR1, OR 2	OR3	OR4 *High*
Likelihood →	Medium	OR5, OR6, OR7	OR8, OR9 *Medium*	OR10, OR11
	Low	OR12 *Low*	OR13, OR14, OR15	OR16
Opportunity potential	/	Low	Medium	High
			Impact ↑	

Similar adjustment is required for the five-point risk severity matrix shown in Table 8.7.

Table 15.4 shows how this matrix can be configured for opportunity risk evaluation, subject to the strategic risk management determination of the project stakeholder organisation.

Given the application of the five-point rating scales suggested here and in Chapter 8, it should be possible to score both threats and opportunities during the risk analysis stage. Then, any opportunity risk achieving a score substantially higher than its threat counterpart could be accorded priority in terms of treatment. However, the validity of this suggestion, and the priorities derived from it, should be carefully tested and justified in practice.

The labels used in Figure 9.1 (Chapter 9), to describe alternative risk treatment options in the risk response stage of the PRMS (Chapter 4, Figure 4.2, Stage D), are inappropriate to indicate how opportunity risks might be dealt with. Table 15.5 shows the contrast between threat risk treatment options and those for the positive gain potential of opportunity risks. Clearly, exact correspondence between treatment options at either end of the risk treatment continuum is not possible. This supports our caution that assessment of threats and opportunities should not be attempted simultaneously in project risk management, especially since the risk response stage of a PRMS directly involves a critical decision making phase for the system.

Table 15.4 Five-point opportunity risk potential matrix.

		Insignificant	Minor	Moderate	Major	Exceptional
Likelihood →	Almost certain					
	Likely					
	Possible					
	Unlikely					
	Rare					

Impact ↑

Opportunity potential key ↓

Negligible	Low	Medium	High	Extreme

Table 15.5 Comparison between threat and opportunity risk treatment options.

Threat risk response options	Amplification	Opportunity risk response options	Amplification
Avoid	Take another course of action.	Exploit	Deliberately seek to obtain the maximum benefit from the opportunity.
Transfer	Pass the risk (or part of it) to another party.	Transfer/share/ defer/partly exploit and defer	Pass the opportunity to another party. Agree to share management and benefits of the opportunity. Defer the opportunity Partly exploit and then defer the residual opportunity.
Reduce	Mitigate one or more of the risk components, and retain the residual risk.	Enhance	Improve one or more of the risk components before exploiting or sharing it.
Retain	Retain the whole risk without treating it.	Ignore	Take no further action with respect to the opportunity.

Note how the 'Transfer' option available for threat risks should be substantially expanded for opportunity risks.

The ALARP (*as low as reasonably practical*) principle for strategic management of threat risks, as shown in Figure 9.1 (Chapter 9), could remain, requiring only the replacement of the text box responses with the options indicated for opportunity risks in Table 15.5 and some adjustment in text box placement along the continuum.

In our view, however, the value derived from strategic use of ALARP for threat risks is not matched when using it for opportunity risks. This is because we think that there are more uncertainties associated with opportunity risks. This is not the same as saying that these uncertainties are necessarily greater, although that may also be true since exploration of opportunities is likely to involve deeper steps into the unknown, but that more uncertainties must be considered. The actual opportunity may be uncertain in terms of existence, nature, timing, extent, and duration. Uncertainty will also be associated with the means of enhancement and exploitation, the potential benefit to be gained, and the practicability of any sharing options. For example, the greater certainty achieved by insurance transfer for many project threat risks is not appropriate for opportunity risks. The requirements associated with enhancement, exploitation, and sharing, in terms of time and resources, could also necessitate structural change within the stakeholder organisation itself. Attempting to impose a predetermined strategy such as '*as high as practically manageable*' (AHAPM) might therefore unduly influence the proper assessment and consideration of opportunity risks. Establishing such a strategy might in itself be a difficult task. Flexibility rather than rigidity is needed, and a case-by-case approach is usually more appropriate. That said, ignoring opportunity risks with lower levels of potential is not a strategy that should be pursued persistently by an organisation unless the potential opportunity gains are substantially outweighed by the costs of exploitation. Lower level opportunities can provide a valuable organisational training ground.

Just as responses to project threat risks may bring with them yet more threat risks, and possibly some opportunity risks, so too do responses to project opportunity risks. These paradoxes should be carefully considered in project risk management. Opportunity sharing is probably revealed in one of its highest forms in the increasing use of strategic alliances and joint ventures. These cooperative approaches to fulfiling project objectives are now used in many industries (e.g. construction, automotive manufacturing, aviation, and aerospace) to deliver major projects, and project benefits, that might otherwise be beyond the capacity of a single organisation. They are often promoted as 'risk pain/risk gain' procurement systems. However, although aimed at maximising potential opportunity benefits, alliances and joint ventures can bring with them additional risks that can threaten the harmony required for such relationships. Should things start to go wrong, a retreat to more adversarial relationships often ensues.

It is also important to remind readers that every risk treatment response – whether for threat or opportunity risks, and even for a 'retain' or 'ignore' response – comes with a cost attached to it.

The monitoring and control processes of a risk management system (RMS; Chapter 4, Figure 4.2, Stage E) could be used to encompass opportunity risk management, but the actual processes involved are likely to be substantially different from those needed for monitoring and controlling threat risks, especially in terms of monitoring frequency and performance indicator measures. Additional opportunity risks identified during this stage can be looped back through the project risk management cycle (Chapter 4, Figure 4.2), but their full exploitation potential may now be constrained by time for the project at hand.

Subsequent knowledge capture from risk opportunity management should be conducted separately from processes used to record threat risk outcomes and experiences, since the nature of the information and the manner in which it will be used are unlikely to be similar. The atmosphere of an opportunity risk debriefing session is also likely to be different from that for threat risk debriefing.

While a PRR (Chapter 4, Figure 4.2, Stages A–E) can be formatted to accommodate threat and opportunity risks together, in our view the organisational risk register (Figure 4.2, Stage F and following) should be suitably configured to yield greater knowledge benefit and easier post-project knowledge access within the stakeholder organisation. Keeping threat and opportunity risks together in the PRR, however, simplifies periodic reporting requirements and ensures that opportunity risks do not become neglected. These knowledge management issues were discussed more fully in Chapter 11.

15.6   Summary

Threat risk management and opportunity risk management are not treated with equal importance in terms of their influence upon the success of projects.

The success of projects is determined by the achievement of the objectives established for them. Realistically, most projects are undertaken not with the aim of maximising the exploitation of all possible opportunities but in order to fulfil predefined objectives and functions. A project which meets those objectives and delivers those functions effectively and efficiently can still be considered successful, even though a potentially better project, which could have delivered greater benefits, might have been within reach if identifiable opportunities had been exploited. The priority of a project stakeholder organisation is always to first ensure the attainment of its project objectives.

Thus, while threat and opportunity may be seen as opposite poles on a risk continuum, any balance of risk management in any project stakeholder organisation is always tipped towards focusing upon events that, through their consequences, could threaten the achievement of project objectives. If these objectives are not attained, the project cannot be entirely successful. On the other hand, if project objectives are achieved but could have been exceeded, the project cannot be said to be unsuccessful.

This does not infer in any way that opportunities are best ignored. It is simply a matter of setting appropriate management priorities in terms of available resources. Also, if the people engaged in risk management activity are not properly equipped to deal with opportunity risks, then the priority assigned to project threat risks should prevail until the situation is rectified.

While theories of behavioural psychology suggest that as people we tend to be either risk seeking or risk averse, the professional attitudes and skills of those involved in managing project risks should enable them to bridge effectively between threat and opportunity risks, given suitable acclimatisation and training. At the very least, awareness of project opportunity risks can be instilled and improved.

The principles and processes of systematic project risk management can accommodate both threat and opportunity risks, albeit with commensurate adjustment to assessment measures and treatment response options. However, opportunity risk management needs to become – like threat risk management – deliberately proactive and systematic. The analysis and assessment of opportunity risks may involve more uncertainties than for threat risks.

PRRs can be formatted to include threat and opportunity risk management, and this would help to encourage and sustain opportunity risk awareness through the periodic project reporting process. For a higher level organisational risk register, however, it is better to arrange this so that each type is dealt with separately. This approach is likely to yield greater long-term advantages for the organisation.

Finally, we prefer cooperative, rather than overly competitive, approaches to projects. Sharing opportunity risks can yield great benefits to the parties concerned. A project stakeholder organisation would do well to avoid being regarded as opportunistic and should shun opportunism. Such avoidance requires strategic risk management.

The implications of a strategic RMS in an organisation are considered in Chapter 16.

Reference

Parkin, J. (1996). *Management Decisions for Engineers*. London: Thomas Telford.

16

Strategic Risk Management

16.1   Introduction

A contemporary organisation uses strategic management as a defining approach to engage in continuous processes of planning, organising, monitoring, analysing, and evaluating all the activities needed for the organisation to fulfil its mission and achieve its objectives.

High-level strategies are formulated for all the areas of management (existing or contemplated) within the organisation and, just as importantly, for its relationships with other organisations. Strategic management then ensures that these are rolled out effectively across the organisation. Such management seeks to ensure that the 'grand plan' is instilled in everyone's consciousness across the organisation with a view to getting everyone 'on board'. This, of course, may entail having to deal with organisational cultures – overt and covert – and may lead to organisational change.

'Strategy' is yet another term encountered in this book where the meaning has shifted over recent decades. From its older military definition as a large-scale plan or method to win victory in battle or contest, the connotation has now widened to denote a technique employed to ensure that an objective is achieved. Nevertheless, given its popular analogous association with war, competitive sport continues to associate 'strategy' with winning (or at least not losing!).

Tactics (another word with military derivation) are the means of deploying a strategy. In rugby football, a player might wish to use a strategy to avoid a tackle by fooling opponents into running the wrong way. This may be achieved through a tactic of throwing a 'dummy' pass: pretending to pass the ball to another team member but pulling back at the last moment and swerving away to run in a new direction while still retaining the ball. 'Stratagem' is sometimes used as an alternative to 'tactic'.

As noted, strategic management is always in service of something – typically the achievement of objectives. Strategies are serviceable only as long as they are relevant and effective. Among the factors that can affect this serviceability are organisational change, organisational culture, and technology change.

Organisational change is inevitably accompanied by risks, and an organisation must consider this in the change planning and implementation processes. Change may also entail modification to the organisation's risk management system (RMS) itself.

In terms of culture, an organisation should:

- Know the overt cultures existing in the organisation.
- Attempt to discover any covert cultures.

Managing Project Risks, First Edition. Peter J Edwards, Paulo Vaz Serra and Michael Edwards.
© 2020 John Wiley & Sons Ltd. Published 2020 by John Wiley & Sons Ltd.

- Know which cultures are positive and which are negative.
- Know and verify the assumptions, beliefs, and values upon which the cultures are based.
- Understand the effect of the organisational cultures on its activities and practices.
- Identify organisational cultures that should be changed.
- Understand the external cultures in each of the external contexts within which the organisation undertakes projects.

Technology change is well-known as one of the more difficult areas of strategic management. It may introduce risks that are unfamiliar to the organisation. It is almost always disruptive to processes and for staff, and is often a fertile ground for negative, covert organisational subcultures to seed themselves and grow.

In strategic management, therefore, strategies are developed in the service of achieving objectives, while stratagems or tactics are the means of deployment for the strategies.

Strategic project risk management (PRM) primarily addresses the need for convergent risk thinking in an organisation that is project-driven. Its purpose is to ensure that an organisation is effectively and efficiently able to manage the project risks (threat and opportunity) that it faces. To this end, the organisation will deliberately implement strategies to ensure that it is able to:

- Identify as many project risks as possible.
- Properly assess and evaluate the identified risks.
- Explore appropriate risk response treatments and make suitable treatment decisions.
- Carry out appropriate monitoring and control of risks during its project engagement.

In addition, a wise organisation will want to know the contexts for its project risks as fully as possible and capture the knowledge gained through its project risk experiences in order to develop greater organisational 'wisdom' about risk.

An important part of strategic risk management is the development, testing, and verification of risk management tools and models, including ensuring that, through practice and rehearsal, relevant staff are familiarised with their use.

All these strategies clearly relate to the process stages in the systematic cycle of PRM presented in Figure 4.2 and in the ensuing chapters of this book that were devoted to each of these stages. Much of the content of this chapter therefore relates directly to that material. However, our aim here is to consider the stages of the RMS from a strategic perspective, exploring the best approaches for each perceived situation. Before that, however, several issues surrounding strategic management must be addressed.

Where necessary, reference will be made to the content, tables, and figures of earlier chapters and to the case studies in the Appendices.

16.2 Strategic Issues for Project Risk Management

It is tempting to place the main focus of this chapter upon organisational strategies to deal with the risks that arise on projects (e.g. how should an organisation respond to them?). For many organisations, however, wider strategic thinking should be carried out well before this occurs. Table 16.1 lists issues for consideration, together with possible strategy options that may be available to the organisation, and factors that could guide strategic thinking and decision making.

Table 16.1 Strategic PRM issues.

Strategic PRMS issue	Sub issues/options	Guidance factors
a) Implementing formal PRMS	• Do not implement • Partial implementation • Full implementation • Specialist help	Motivation and risk; resources; costs; awareness requirements; organisational culture
b) System separation	• Separate systems • Separate department • Project focused • Organisation wide (+ hybrids)	Organisational structure and capacity; compliance requirements; operational processes
c) System inception	• Single-project trial • Multiple-project trial • Full implementation	Urgency; capacity; project suitability; resources
d) Initial system application	• Selected RMS stages • All RMS stages • Selected project phases • All project phases	Tools and methods availability; team capability; urgency; project suitability
e) Roles and responsibilities	• Fixed allocation • Organic development • Recruitment • Agency situations	Management style; organisational culture; staff availability; training resources; employment market; project locations; project organisational structures
f) PRMS process approach	• Individual or team based • Workshop frequency • Workshop participation • Tools and methods • Risk registers • PRMS stages	Project contexts and schedules; PRMS scope; staff availability and capability; legal compliance; legal guidance
g) Risk knowledge management	• Knowledge purpose and content • Silo-based? • Organisational diffusion • External sources	Risk knowledge needs; management responsibility; KM specialist staff availability
h) PRMS maintenance and development	• Review frequency • Benchmarking • Refreshment	RM intensity; risk experiences; RMS decay rate; staff turnover
i) Disaster preparedness	• Disaster fixation • Separated responsibility • Integrated responsibility • Planning and practice	Motivation; resources; costs; organisational capacity; staff attitudes; availability

KM, Knowledge management; PRM, project risk management; PRMS, project risk management system; RM, risk manager; RMS, risk management system.

The table is not a complete list and simply provides an indication of the type and direction of strategic thinking that should be employed. This thinking requires the involvement of senior management in the organisation.

16.2.1 PRMS Implementation

If an organisation has no formal system of PRM in place, the key strategic issue is whether or not to implement one. If the issue is not even considered, then the default situation is that project risks will not be formally managed but dealt with on an ad hoc basis or reactively as risk events occur. Case Study D (an engineering train mock-up project) is an example of the default position, although it is perhaps fairer to say that some risks for this project (e.g. safety and the environment) are formally managed as part of legal compliance requirements. Case Study E (a hot-rod car project) is another example. Here, the engineer carrying out his 'passion' project did not consider formal risk management, relying instead on his engineering ability and ingenuity to deal with uncertainties. However, here too observance of legal requirements can be found in terms of compliance with national automotive design rules (albeit with sanctioned exceptions) and with safe welding practice.

Case Study C Part 1 (an international aid-funded project consultant) reveals an instance of a considered decision not to adopt a formal project risk management system (PRMS). The decision, although informed by a comprehensive understanding of formal risk management, was based upon the excessively high premium costs of professional indemnity insurance for a small consulting practice, and the comprehensive understanding held by the consultant of his own abilities and practices. The strategy he decided upon was to contain the likelihood of occurrence of liability by limiting the growth of the consultancy and the type of projects he undertakes. Wisely, the consultant keeps the issue of implementing a formal risk management system under regular review.

Many organisations employ formal risk-based management systems aimed at the organisational level rather than specifically for projects. Typically, these too would cover aspects such as occupational health and safety, environmental harm, and even quality assurance. An organisation might therefore decide upon a strategy of relying on these existing arrangements to embrace project risks, at least in part and in perhaps their more important compliance aspects.

Increasingly, however, distinct PRMSs are implemented. This is due partly to the 'projectification' of many business endeavours, partly to the expectations and stipulations of project clients, and partly to the more effective and efficient management of project risks that can be achieved.

The strategic options are therefore: no PRMS implementation (default position), partial implementation, or full implementation. Partial PRMS implementation is the introduction of a stand-alone system dealing only with projects. Full implementation seeks to go further and integrate organisational risk management and PRM.

Whether or not to engage specialist help is an important factor to consider in terms of partial or full PRMS implementation. To a large extent, this strategy will be determined by the amount of risk management experience existing in the organisation, the nature of its activities, and the level of confidence that can be placed in the specialists available, bearing in mind that there are no prescribed qualifications for this skill and no restrictions on offering it as a service. Due diligence should therefore be undertaken to assess and reduce the uncertainty associated with a decision to engage consultants, and any decision to do so must acknowledge the magnitude and implications of the organisational change that will inevitably follow. Consultants are not necessarily equipped to deal with this, as it may be out of their remit.

Factors that can influence strategic decision making about PRMS implementation include: motivation and risk awareness, organisational culture, and resource and cost constraints.

Motivation for formal PRM often arises from subjective (and occasionally objective) perceptions that project risks are not being well-handled – or at least not on some projects. The perceptions may arise anywhere in the organisation but are eventually brought to the attention of someone with sufficient power (director or senior manager) to initiate a proposal for system change. Such an individual is then likely to become the 'champion' for formal PRM. This is a desirable stratagem.

The organisational culture may itself be a positive or negative factor in deciding upon PRMS implementation. Positively, the culture may be derived from increasing awareness of project risks which then translates into a movement for improving management of them. Negatively, the culture may be stubbornly resistant to change, relying instead upon traditional approaches that are possibly becoming increasingly ineffective.

Perceived resource requirements and costs are often the overt constraints associated with PRMS implementation decision making, especially if specialist assistance is contemplated. These factors are rarely investigated in any great depth, mainly because of the difficulty of measuring the tangible benefits of a PRMS. Instead, resource and cost issues are used to cloak the influence of a covert negative organisational culture towards formal PRM.

16.2.2   System Separation

We have alluded in this chapter to situations where some formal risk management is already in place in an organisation. The strategic issue here is whether or not a PRMS should be implemented as a separate system or, perhaps more logically, if existing systems should remain separate from the PRMS.

An existing system for managing risks may reside within any department in the organisation that deals with a particular set of processes. It may even constitute a stand-alone unit. It may have a specific project focus, have organisation-wide application, or be used in a hybrid manner for both. In contemporary organisations, such system separations tend to deal mainly with occupational health and safety concerns, quality assurance, and financial management, but sometimes also with environmental management. More rarely, they are found in human resource management.

Factors influencing strategic thinking about system separation include: the organisational structure, organisational capacity, compliance requirements, and operational processes.

Where management of a particular type of risk is already established in a separate unit or department within an organisation, the best strategy may be to leave that in place, at least for the time being (especially if it is regarded as effective), and ensure that strong direct communication links are established with the PRMS. For example, ISO 45001 (2018) which deals with occupational health and safety management is a recent addition to the suite of international standards (after many years of collaborative development with several national standards organisations). It covers all kinds of work safety hazards (including psychological ill health) and contains prescriptive requirements for compliance and certification. There is little point (at least logistically) in duplicating administration of the standard for the organisation and separately for projects. On the other hand, ISO 14001 (2015), one of a suite of standards developed from an earlier issue that dealt with environmental management systems, is less mandatory and more of a guidance document. The environmental conditions for the organisation might be substantially different from those encountered physically on projects, and the organisation–project link may be less evident. While a separation strategy for environmental risk management might be appropriate, a more convincing supporting argument may be needed. Standards used in other fields may also warrant strategic consideration in terms of system separation or combination.

Case examples where the PRMS is a stand-alone system that embraces all types of project risks are found in Case Studies A (a correctional facility PPP project) and B (a rail improvement programme). In each case, risks pertaining to the organisation itself (as distinct from those arising within or crossing the actual project system boundary) are dealt with separately in each organisation. In both cases, the separate organisational risk management relates mainly to safety and environmental concerns that are different in nature or magnitude from those encountered on the projects. The project tasks and activities are also more vulnerable to public and third-party liability issues, and thus warrant their separated risk treatment.

16.2.3   System Inception

The strategic issue of PRMS inception relates to how quickly and comprehensively the system is introduced. Options available include trialling the proposed PRMS on a single project, on several projects, or on all current projects. Factors guiding the inception strategy decision may include the urgency associated with implementation, the organisational capacity to implement, the types of projects, and the resources required.

The urgency of PRMS implementation will be influenced by perceptions of the inadequacies of existing PRM processes, and this in turn will be informed by the views and opinions of those directly involved in them. Senior management in the organisation therefore need to engage in some form of data collection or fact finding, preferably through appropriate consultation with project managers.

Implementing PRMS across all projects straightaway may result in a degree of change that severely taxes the organisation's resilience. The potential flow-on implications of this should be carefully considered.

Applying the PRMS to one project may appear to be the most appropriate strategy, but it is likely to lead to over-reliance on the findings of what is essentially a single case study. This presents a threat risk in itself, given the unique nature claimed for all projects, but especially if the trial project has been selected solely on the grounds of simplicity or convenience.

A better strategy, organisational capacity and resources permitting, is to deploy the PRMS on several projects, not necessarily simultaneously but as closely synchronised as possible. It is not essential that the target projects are all at the same stage and progressing at the same rate, although that would be desirable. What is more important is that, as far as possible, the projects selected for trialling PRMS implementation should be heterogeneous. Heterogeneity in this context might mean different types of projects, different types of clients, different project values, different locations or environments, and different project personnel. The findings of multiple trials would yield a richer picture for further strategic considerations.

16.2.4   Initial System Application

By initial system application strategy, we refer to incremental or complete application of PRMS processes to trial projects (or indeed to all projects if that was a prior inception strategy decision). Figures 3.3 (Project Elements, in Chapter 3) and 4.2 (The Dynamic Cycle of Project Risk Management, in Chapter 4) provide the context for strategic thinking here. Generally, an incremental system application strategy is likely to work best.

In practice, the capacity to trial PRMS on different project phases of several projects is unlikely to be feasible unless the projects are short-term endeavours such as the information technology (IT) project illustrated in Figures 4.5 and 4.6. It is usually more practicable to trial a PRMS incrementally on each stage of the PRM cycle, although the project contextualising and risk identification (RID) processes could be trialled at the same time.

Factors guiding strategic thinking here will include the availability of appropriate tools and methods, risk management team capability, any urgency attached to the trials, and the suitability of the projects for the application. While progressive and incremental application of risk management processes on trial projects may take more time, the benefits gained include the opportunity to adjust the application to more suitably (or more fully) embrace particular project conditions, and the greater risk management experience acquired by the participants. Radical attempts at full and immediate application of a PRMS carry their own failure risks.

16.2.5   Roles and Responsibilities

Roles and responsibilities in PRM should be assigned as part of the strategic management of an organisation. It is important to recognise, however, that these are rarely fixed in the medium term, let alone the long term, for any organisation. They may also be subject to organisational change and to greater risk management maturity in organisations (see Chapter 17).

A risk management role does not necessarily include 'risk' in its title, as some role titles are synonymous with risk or implicitly denote some responsibility for managing particular risks. 'Safety officer' and 'finance officer', for example, are intrinsically associated with risks pertaining to their discrete areas of management.

Typically, these titles also denote broader responsibility than just for projects. Strategic risk management should recognise this. If the project risk responsibility must be distinguished from wider organisational responsibility, then an appropriate tactic would be to add 'Project' at the front of each role label. However, a strategy of separating risk management responsibility in this way must be supported by specific references in the respective job descriptions.

Perhaps the worst role responsibility in any organisation, in any industry, would be that of 'risk manager'(RM). The title seems explicit – indeed, could anything be clearer?

However, the problem is that in almost every case there is misunderstanding about expectations for the role and the level and nature of the responsibility associated with it.

An organisation that expects its RM to manage all the uncertainties associated with its activities is misleading itself. Only in a very small organisation with a narrow range of project activities would that be likely to be carried out effectively. In all other instances, the expected role responsibility would be just too great.

A more appropriate role responsibility would be to expect the RM to manage the risk management activities of others in the organisation. Case Study B (a rail improvement programme) is a good example of this, whereby the rail crossing removal authority's RM is responsible for reviewing the risk management activities of the project managers for each crossing project and reporting the findings to the board of the authority.

This approach accords with a general principle in risk management of locating management responsibility where the most direct and greatest control can be exerted over a risk. The principle is shown in the risk severity/responsibility triangles in Figure 10.1. If any doubt (i.e. more uncertainty) arises about the level of risk, an appropriate tactic would be to escalate responsibility up the chain of management, but at the same time ensure that the general strategy is not neglected. There is little point in including someone in that chain who does not understand the nature and parameters of a particular risk, and effectively the RM becomes a manager of managers (as in Case Study B).

In this chapter's introduction, we stated that strategic risk management aims to achieve convergent risk thinking across the organisation. In an organisation displaying good risk management maturity (Chapter 17), risk awareness is diffused across the whole organisation, and risk management responsibility should follow suit. A good strategy is to adopt a team approach to this.

While PRM roles and responsibilities may be assigned on a fixed basis in the early implementation of a PRMS in an organisation, we suggest that this should be considered a temporary strategy at this point. As a PRMS 'beds down' in the organisation, it may emerge that some staff are not suited for the tasks of risk management. This is not necessarily because they lack risk awareness, but rather because they are unable to develop a reliable emotional 'feel' for, or appreciation of, the uncertainties associated with risk. Other project personnel may 'grow' quickly and easily into their risk management responsibilities. The uncertainties of human nature are themselves a risk here.

The development of risk awareness and its consonance with risk management skills and ability is likely to be a dynamic, but uneven, process in any organisation. Role and responsibility assignment should recognise this. Patience, especially during trial PRMS implementation, is a good stratagem, but sometimes recruitment may be the only solution eventually; and for this, clear criteria must be established at the senior management level.

We cannot leave the issue of risk management responsibility without referring again to agency situations in project management. This was discussed in the introductory section to Chapter 4.

An organisation or individual acting in an agency capacity (e.g. as a professional consultant) for a project client must be clear about the role and responsibility involved in managing project risks on the client's behalf. A clear and direct communication strategy is essential, and assuming that an implicit role boundary exists is actually poor project management. At the same time, the agent should develop an internal strategy to deal with the uncertainties associated with the agency's own risks in the client–consultant relationship.

16.2.6   PRMS Process Approach

Strategic issues relating to the PRMS process include matters such as: individual/team-based processes, risk management workshops, tools and methods, and project risk registers. They arise in the planned and staged cycle of PRM activity indicated in Figure 4.2.

Factors likely to influence strategic thinking here include: the general project contexts for the organisation, typical project schedules, the planned scope of the PRMS, and staff availability and capability. Case Study F (an aquatic theme park project) is a good example of strategic risk-based thinking applied to the earliest stage of project involvement, where a list of criteria for organisation–project alignment (Table F.2) informs a management decision as to whether or not to respond to a call for expressions of interest to tender for a project. The list addresses internal and external context issues.

However, neither the strategic issues nor the factors that influence them are likely to remain static in any organisation, and a good commencing strategy would be to conduct periodic reviews of the organisation's strategic use of its PRMS.

Throughout this book, we have advocated a team approach to PRM, as being carried out individually renders it too vulnerable to bias. It constrains risk awareness and an appropriately balanced appreciation of what can be achieved. However, the nature of a particular organisation and its projects may mean that PRM must be carried out by individuals. In such cases, a strategy for consistent and frequent review should be implemented. Continual self-reliance (such as that indicated in Part 1 of Case Study C) may be inadvisable in the longer term, hence the regular review strategy of the consultant in this case study.

To some extent, the same issue will inform a strategy for risk management workshops. A team-based risk management strategy more easily accommodates a project risk workshop strategy.

Figures 4.3–4.6 in Chapter 4 indicate the timing of risk workshops proposed for two different types of projects, but these are only suggestions. An appropriate workshop strategy would be to hold them as frequently as possible subject to the schedule of each project, the existing scope of the PRMS, the availability of staff, and their individual risk management capabilities. Generally, the less experienced and the less capable the staff, the more frequent should be the project risk management workshops, at least in the early stages of a project.

The project scope, complexity, and packaged procurement system for Case Study B (a rail improvement programme) necessitated more extensive risk management workshop arrangements. In the early stages, these were held at the removal authority level with other stakeholder authorities participating. Later, as design and construction work progressed, each package contractor was required to undertake its own risk management workshops in order to comply with the ISO 31000 requirements. Interestingly, in Case Study B, the engineers and managers at various levels (project, design, authority, operations, etc.) organised their own risk forums (as focus groups or 'communities of practice') to share risk issues, strategies, and experiences.

As a general strategy, participation in PRM workshops should be confined to people directly involved in risk management. This limits the capacity to combine PRM matters with project meetings where the agenda usually incorporates a range of other items and other people. Project progress meetings, or the regular site meetings typical of construction projects, are examples of this. Convenience is often cited as a reason for combining PRM workshops with other meetings, especially when some participants may be involved with both purposes or when travel becomes a complicating factor.

However, there is seldom anything worse than having 'passengers' in a meeting, and there is always the danger (risk) that project risk matters will not receive sufficient attention. This is not to say that risk-related items should not be dealt with in other meetings or workshops, but where they constitute a large proportion of the agenda they should be addressed on a separate occasion. Flexibility is a good strategy, with the project RM taking responsibility for determining whether or not a separate meeting or workshop is warranted. Convenience might be satisfied by arranging a separate risk-focused meeting immediately before or after the other one. If this stratagem is adopted, our preference would be for a post-meeting risk workshop. In this way, any risk-related decisions arising from the preceding meeting can be addressed in a more focused manner. In the earliest stages of a project, stand-alone risk workshops are essential.

Tools and methods used in risk management should always be subject to strategic thinking beforehand. Even unstructured brainstorming should be considered from a strategic perspective. Strategies must be developed for initiating, guiding, controlling, and sustaining brainstorming in terms of encouraging participation, steering the direction of the topic, 'gatekeeping' for reluctant or dominating contributors, and knowing when to prolong or discontinue what is often an intellectually and physically tiring exercise.

Armed with suitable experimentation and trials, strategic management must think about the tools and methods needed to service the PRMS. Chapters 6 and 7 dealt with these in greater detail, and there we emphasised the benefits that can be derived by using tools primarily intended for other management purposes. While this strategy can deliver extra value from such tools, this should not be the primary argument for using them in risk management. Project risk management process strategies should always be based upon what delivers the best risk management outcomes in practice for projects. Good strategies will identify preferred tools and methods, but leave the way open to further adaptation or developing new ones as more PRM experience is gained.

Project risk registers, risk register item records, and an organisational risk register together form an important element of PRM and have been more extensively discussed in Chapters 4, 7, and 11. From a strategic perspective, they represent multi-level documentary evidence of risk decision making and action for particular projects. Organisational risk registers gather not only risk information from each of the projects undertaken but also that from the wider organisation itself. At an empirical level, they present an archival record of risk management and a repository of risk knowledge. Strategically, an organisation must decide how such information and knowledge are to be gathered, arranged, and used.

Strategies developed about the content and format of the registers must recognise that, in certain circumstances, there might be an obligation to produce them (or at least relevant parts) as evidence in legal or quasi-legal proceeding and enquiries.

16.2.7 Risk Knowledge Management

Strategies for risk knowledge management (KM) go beyond the 'what' of risk information and knowledge, to then focus upon 'how' the knowledge can be used. Strategic thinking will consider not only the essential purposes of the knowledge but also what value additional uses of such knowledge might deliver. A critical strategy would be directed towards increasing project risk 'wisdom' through cleverer, more appropriate, more effective, and more efficient risk management. For example, could the risk information and knowledge be used to 'map' the organisation's project risks from various dimensions in order to identify predominant types, frequencies, impacts, and consequences, or more vulnerable project stages? Could it be used to improve compliance with statutory requirements? Would the collection of risk information from external sources add further value to the organisation's project risk knowledge and wisdom?

Much of this was discussed more fully in Chapter 11, largely from a perspective of interactive risk knowledge generation, but factors that could influence strategic thinking here go back to the intended purposes of the knowledge and the extent of the need for it in the organisation. KM requirements, responsibility, and staff capability are pertinent sub-issues. In many circumstances, specialist librarian skills may be needed. In terms of any statutory compliance in risk knowledge matters, legal guidance may be required.

16.2.8 PRMS Maintenance and Development

The strategic issues of implementing a formal PRMS were presented earlier in this chapter, and the processes of building and maintaining the PRMS will be considered in Chapter 17. Assuming that a PRMS is in place, strategies for system maintenance and development should be considered.

Just as we have noted that neither projects nor project risks are static, so too is it necessary to ensure that the PRMS is not allowed to stagnate (or decay). Many organisations with such systems in place have policies that require their application to most, if not all, areas of the projects that are undertaken. Case Studies A (a PPP correctional facility) and B (a rail improvement programme) are examples. None of the case studies, however, provide clear evidence of PRMS maintenance or continuing system development.

An initial strategy might be to ensure that the PRMS is reviewed on a regular basis, preferably at intervals not greater than two years, regardless of the duration of any projects. Not all projects would require review, and specimen projects could be selected according to desired PRMS parameters such as project size, scope, complexity, and staff capability. System quality assurance and benchmarking are important sub-issues to be considered in strategic thinking.

System reviews should cover each of the stages in the PRM cycle (Figure 4.2) in terms of the processes followed and their resource requirements. These are considered more fully, from a strategic thinking perspective, later in this chapter, but it is important to recognise that, besides the actual processes of risk management, staffing (in terms of capability and turnover) is likely to be an influential factor. System and staff refreshment should not be overlooked, and strategies should be devised to deal with them.

16.2.9 Disaster Preparedness

Two pertinent matters must be considered before we can discuss disaster preparedness in strategic risk management. These are the roles of emotion and catastrophising. They are both capable of leading to what might be termed a fixation with disaster.

In Chapter 2, we noted that risk is a psychosocial construct. Flowing from this, we argue that strategic risk management should have an emotional engagement for those who participate in it. In many organisations, a 'management' view may preside whereby emotion is deliberately excluded or rejected in strategic thinking, on the grounds of avoiding subjectivity or of ensuring greater effectiveness and efficiency in making and implementing strategic decisions. While we agree that decision making should be carried out dispassionately as far as possible, we do advocate that emotion be welcomed into the discussion and allowed to make a controlled contribution to it. Passion has its place in any human endeavour, and totally excluding it in a vital area of management then risks creating a 'mechanistic' approach that, in the long term, benefits neither the organisation's risk management nor the people involved in it. Used constructively, emotion has much to offer to strategic risk management. However, caution must be exercised so that it does not develop into excessive negativity, particularly since threat risks dominate much of an organisation's risk management effort.

'Catastrophising' represents just such negativity and is exhibited in constantly imagining how a situation can get progressively worse and worse. It is tantamount to a mental disorder whereby personal thinking may become progressively less and less rational and more and more distorted. It may lead to incapacitation in terms of further rational thinking and action. The disorder may be sourced in anxiety or lead to more acute sensations of this fearful state. It can also be found in depression.

Catastrophising can manifest in several ways including: completely rejecting hope or positive attitudes, or recasting them into an even more negative frame; displaying highly biased or selective thinking that is based upon too few aspects of a situation; exaggerating selected threat issues and their anticipated outcomes; drawing highly adverse conclusions from inadequate evidence; or overgeneralising in a negative manner. In some circumstances, it may also be accompanied by severe feelings of self-blame.

The condition is a form of cognitive distortion, and we can all become prone to it to some degree. Irrational over-optimism, its counterpart, is less often encountered.

Catastrophising differs from anxiety in that anxiety, in its milder forms, may have cognitive benefits that can lead to positive solution finding, whereas catastrophising is always severe and yields no such benefits. Its significance for disaster preparedness lies in this difference and in the masking effect of rationalisations such as 'We must always prepare for the worst'. Catastrophising should not be confused with scenario testing (Section 7.7, Chapter 7).

Where catastrophising is encountered in strategic thinking, mitigating actions might include:

- Openly confronting it and encouraging self-recognition of its presence
- Avoiding or disallowing over-exaggeration
- Avoiding over-extrapolation of the past into the future
- Sticking with the strategy, if everyone has already agreed that it is a good one
- Breaking irrational links in the catastrophic thinking chain
- Emphasising that it represents thoughts and anxieties that are rarely ever realities or certainties
- Encouraging greater focus upon hope and resilience.

Given this caution, disaster preparedness is an important aspect of strategic risk management and can be considered in terms of *locus* (where it is dealt with in the organisation) and how it is managed.

Disasters are not restricted to physical events alone (e.g. total or substantial loss of data would be a disaster in an IT data migration project for a major bank). An organisation should first carry out high-level strategic thinking about the type of disaster scenarios that might be encountered. It is less important that *every* possible type of disaster scenario should be imagined, than it is that a few significant ones (in terms of consequences) should be proposed. The main purpose of this conceptualising strategy is to induce a preparedness frame of mind amongst senior management. The location of strategic consideration of disaster preparedness should therefore be at the highest level of decision making in an organisation.

Once appropriate scenarios have been identified, strategic thinking for each one should then move towards preparedness strategies associated with factors such as: forecasting imminence, avoidance, impact minimisation, and post-disaster recovery.

The nature of each disaster scenario will determine what strategies will best deal with each factor, but it is quite likely that common strategies will begin to emerge for issues such as: the need for more detailed planning, emergency services requirements and notification to relevant authorities, staff and resource allocation and prioritisation, backup resource acquisition, containment, staff training, scenario rehearsal routines and frequency, and public relations and media attention.

16.3 PRMS Process Strategies

In Case Study B (a rail improvement programme), strategic objectives were established for the PRMS itself. These included:

- A simple process compliant with ISO 31000
- A system relevant to the methods of procurement used which could be applied consistently across the alternatives
- A system that would be defendable for the removal authority
- A system that would demonstrate a good risk culture from a top-down basis.

These are good objectives to bear in mind when discussing PRMS process strategies.

As has been the case for much of this book, the cycle of PRM (Figure 4.2) continues to provide the structural frame for discussion in this section. Here we look at more detailed strategic risk management relating to project contexts, project RID, risk analysis and assessment, response and treatment, risk monitoring and control, and knowledge elicitation.

16.3.1 Project Contextualisation

Strategic thinking is important for the internal contexts of projects, particularly in terms of the strategic fit of a project for the stakeholder organisation. An appropriate question here is 'How well does this project fit our organisation?' The question can be framed to address several 'fit' perspectives including the organisation's mission, aims, and objectives; its organisational capacity and existing project load (particularly issues relating to finance, resources, and technologies); and its initial risk appetite for the project. As we noted, Case Study F (an aquatic theme park project) is a good example of how this question can be addressed.

There should also be a strategy for setting the project involvement agenda and considering its implications. Strategic questions here relate to organisational involvement in any or all of the project stages: inception/design, delivery, operation, and termination. This thinking is important for clarifying the extent of PRM that will be required. Yet again, Case Study F exemplifies this. The civil engineering company's decision to take an equity share in the aquatic theme park project (a novel role for the organisation) influenced its risk management processes.

For external project contexts, strategies should be established for addressing each of the four risk 'drivers' shown in Figure 5.2: physical, technical, economic, and social.

An appropriate organisational stratagem would be that, whatever the organisation's experience and familiarity with particular types of projects, no new project proposal should escape high-level consideration of its internal and external contexts.

16.3.2   Project Risk Identification Strategies

A wide array of RID tools and techniques were presented in Chapters 6 and 7. In establishing RID tool preference strategies, an organisation should consider aspects such as: tool adaptability from any primary alternative use, user-friendliness, comparative reliability, timing, and the availability and adequacy of information needed to service the tool.

For an organisation relatively inexperienced in PRM and with a low level of risk management maturity, an appropriate RID preferment strategy might be to commence with checklists (see e.g. Chapter 7, Section 7.4, Associated Representative Tools), and then to move as quickly as possible to more activity-based methods such as a work breakdown schedule or project programme schedule.

We strongly advocate a robust 'project risk statement strategy' (see the argument presented in Chapter 7, Section 7.9), since in our view it is critical for effective risk management. An appropriate strategy should mandate consultative preparation of each risk statement; formulation according to a predetermined format dealing with event likelihood, consequences, duration, and associated uncertainties; and the inclusion of the agreed statements in the project risk register. Whatever abbreviations may ensue for project risks, they should be underpinned by precise original statements.

Devising a 'classification system' for project risks is also an essential task of strategic risk management. The need for appropriate classification was considered in Chapter 2 (Section 2.8). A fallback strategy might be to adopt a generic source-event approach (Tables 2.2–2.4). Alternatively, categorisation according to organisational governance and structure might be preferred (Table 2.5), or a more customised version (Table 2.6).

We also remind readers that, for many organisations, multiple risk categorisation approaches (Chapter 2) can be adopted if deemed useful for different purposes, but such a strategy might give rise to difficulties in achieving common understanding within and beyond the project stakeholder organisation. Shared meaning is considered more fully in Chapter 19 (Communicating Risk).

16.3.3   Quantitative and Qualitative Risk Analysis Strategies

Case Studies A (a PPP correctional facility) and F (an aquatic theme park project) indicate where contractors take a strategic view towards quantitative risk analysis. It is preferred where very high probability combined with very high financial impact is associated with particular risks, and where sufficient and reliable risk data can be sourced.

Qualitative risk analysis (Chapter 8) is not without its strategic concerns. Among these are its comparative simplicity to understand and use, and the speed with which it can be used in project environments where time is at a premium. While these are benefits, they can be illusory, deceptive, and vulnerable to personal bias. Strategic thinking should be aware of this.

In Chapter 8, we described several scale measures for estimating uncertainties in the likelihood and impacts of project risk events, and we advocated five-point scales rather than the simpler but cruder three-point low/medium/high scales. As we noted in that chapter, the implementation of qualitative scale measures for project risk analysis is a strategic management decision, and it is essential that the scales are known and understood across the entire organisation (i.e. not just among project teams). The scale intervals must be relevant and logically progressive. They should be meaningful in an organisational context. Where multiple measures of impact are devised, concordance across each scale interval for each type of impact can be difficult to determine. If multiple impact measures are used in analysis, a strategy should be adopted whereby the assessment for each risk is based upon the highest impact level measure encountered for it. In addition to the scale design, strategies must be implemented to ensure that common understanding for *each* interval measure is achieved.

With qualitative risk analysis, a good additional strategy is to ensure that the analysis is revisited for project risks where high levels of probability are also found to be matched by high levels of impact, especially for risk exposures over long periods.

Testing and periodically recalibrating the adequacy and reliability of qualitative scale measures for risk analysis are vital.

Beyond this, a further strategic management issue might be whether or not project risk analysis is required at all. While such a proposition might seem to fly in the face of the old adage that 'What cannot be measured, cannot be managed', as well as much of the intention of this book, a little thought will show that we are not actually saying that. Instead, we are proposing that, although risk analysis could be carried out, some project circumstances mean that such analysis is not really necessary. The three main circumstances that might influence a decision to omit risk analysis are: absence of choice for risk response, risks falling below the intolerance level, and the value of common sense.

Absence of choice typically arises in situations such as that for safety threat risks. In most jurisdictions, occupational health and safety compliance legislation requires some formal assessment of potential hazards and the response treatments mandated for them. The work 'Safety Hazard Analysis' record shown in Table 7.7 is a template example of typical safety compliance requirements. In reality, the template does not actually need to indicate probability or impact estimates. If the hazard is real, something must be done about it. For example, work in confined spaces (e.g. inside an empty storage tank or cylinder) presents safety hazards relating to access, carrying out the work, and egress. Each of these might present sub-risks that contribute to the overall work hazard. While analysis and measurement of each sub-risk might be possible, in practice this would serve to complicate a threat risk situation where something has to be done in terms of legal compliance and duty-of-care responsibility. In most such instances, the assessment measures included in the template form are not really needed and may just provide an auditable indication of the seriousness the organisation has accorded to the hazard.

The identification process is likely to surface a large number of risks (hundreds, in some cases). Many of these might be considered trivial in magnitude. They would normally fall into the lowest 'as low as reasonably practicable' (ALARP) category (Figure 9.1), that is, far below an organisation's risk intolerance level. The risk response judgement could be made on the grounds of common sense without the need for risk analysis.

Common sense prevails over more methodical risk analysis when the risk taker *intuitively* understands the probability and impact parameters for particular risks and responds accordingly. As a risk analysis strategy, this runs counter to the key strategy of implementing *systematic* PRM. It can be seen in Case Study E (a hot-rod car project) and to some extent for the consultant in Case Study C Part 1 (an aid-funded project). Where common sense is conceded as an alternative to quantitative or qualitative risk analysis strategies, this should only be after reliable validity has been demonstrated through sufficient experience in a project environment. Clearly, it will work best in a very small organisation or in a sole practitioner situation. Relying upon intuition or common sense alone also means that a valuable audit trail is lost. That said, a strategy of *aligning* common sense with a quantitative or qualitative risk analysis strategy (or both) will yield valuable benefits for PRM.

16.3.4 Risk Response and Treatment Strategies

While risk assessment is a guide to selecting appropriate responses and treatments for project risks, this stage in the PRM cycle (Figure 4.2, Stage D) is actually a key decision point in the whole cycle.

The first strategy here should be that, even where an active response is prescribed in terms of compliance requirements, no response decision should be made until at least some degree of prior risk assessment has been carried out. Uninformed responses to risks come with their own threat risks. That said, our comment in the previous section about intuitive assessment still stands.

A second strategy is to ensure that, where mitigating action is proposed, the mitigated risk is reanalysed and reassessed to ensure that it now complies with the organisation's ALARP principles (Figure 9.1). This strategy is represented by the dotted feedback arrow from Stage D to Stage C_1 in Figure 4.2.

Yet another response strategy is found among construction contractors, and Case Studies A (a correctional facility) and F (an aquatic theme park project) provide examples. A contractor may decide to include in its bid the full impact cost of any threat risk showing characteristics of high probability allied to high financial impact. This is tantamount to retaining the risk but fully rewarding oneself for doing so. In most instances, the contractor is also aware that by doing this, its bid is likely to be less competitive. Should the bid be successful, however, and the risk does not eventuate, the contractor makes a windfall profit. In other contexts, the risk taker might elect to spend up to the risk exposure amount (probability × financial impact) on risk mitigation or transfer costs for risks that have been quantitatively assessed.

Risk response and treatment options should be practically feasible for the particular project context. The project stakeholder organisation must have the capacity to undertake its preferred response and treatment for any risk. In considering this, the organisation should give heed to responses and treatments already decided for other project risks to ensure that the capacity is sustainable.

16.3.5   Risk Monitoring and Control Strategies

For the monitoring and control of risks during the project life cycle, key strategies revolve around the 'who', 'what', and 'when' questions.

A clear strategy is that the monitoring and control process should be undertaken by people (preferably a team) closest to the potential risk situation. This is not always easy to achieve, and some staff required for this process may not have been involved in earlier risk management stages for a project. While compromises may have to be made, the relevant staff must be properly briefed, and there should always be a clear 'chain of command' in terms of responsibility and reporting.

It may be impracticable to monitor constantly all risks included in the project risk register. Prioritising decisions may have to be made. The nature and extent of risk mitigation treatments could inform this strategy. Where more resources and money have been used to mitigate a particular risk, it makes sense to accord it greater priority for monitoring and control purposes, and it may be possible to create a 'top 10' list of risks.

The monitoring and control process sometimes indicates the possibility that new project risks could emerge or that existing risk parameters are changing. This will necessitate a return to earlier stages in the PRM cycle, and this too is indicated by the dotted arrows from Stage E to Stage B and Stage C_1 in Figure 4.2.

16.3.6   Risk Knowledge Capture Strategies

For any organisation that is relatively new to PRM, establishing strategies for risk knowledge capture can be quite difficult. Chapter 11 provided a much fuller treatment of this topic, and the brief discussion earlier in this chapter emphasised the importance of the 'what' and 'how' questions ('What knowledge is needed?', 'How should it be collected?', and 'How should it be used?'). Organisational project risk knowledge strategies should address these questions carefully, particularly in terms of strategy differences in capturing explicit and tacit project risk knowledge.

A good risk knowledge capture strategy, also applicable to risk control and monitoring procedures, is to discourage 'copy and paste' processes for reporting risk. This short-cut approach may often result in too much irrelevant information (thereby losing the key risk focus), missing information, or erratic reporting frequency.

'Silo-ing' of information can also occur, with valuable risk knowledge kept back on the erroneous assumption that it is not pertinent to other professional disciplines, other parts of the project, or other projects. Unless a careful and experienced eye is maintained on the knowledge capture process, valuable risk insights may be distorted or lost.

16.4   Summary

After completing this chapter, as authors we have to admit that we underestimated the extent of strategic management needed in dealing with project risks. Nor did we fully appreciate the value of an approach to strategic thinking along the lines of 'This is the plan – let's own it'. Strategies must be understood and accepted across the project stakeholder organisation.

Strategic management is needed for every stage and every process in the systematic cycle of PRM. However, it is important to remember that strategies are *plans*. They are not set in stone and, as the case study examples have shown, some flexibility is desirable.

Having emphasised the importance of strategic management, it is appropriate to explore the ramifications of building a PRMS.

References

International Organisation for Standardisation (ISO) (2015). *Environmental Management Systems* (ISO 14001). Geneva: International Organisation for Standardisation.

International Organisation for Standardisation (ISO) (2018). *Risk Management – Principles and Guidelines.* (ISO 31000). Geneva: International Organisation for Standardisation.

International Organisation for Standardisation (ISO) (2018). *Occupational Health and Safety Management Systems* (ISO 45001). Geneva: International Organisation for Standardisation.

17

Planning, Building, and Maturing a Project Risk Management System

17.1 Introduction

This chapter is primarily intended for students and for readers working in organisations or environments where little or no systematic project risk management (PRM) takes place. However, we think that some topics will be of interest to those who have experience in sophisticated and quite elaborate risk management procedures and system documentation.

Readers who are unfamiliar with project risk management systems (PRMSs) are recommended to first do fast refresher-reading of Chapters 5–10 and 16. Much of the topic content of Chapter 11 is also pertinent.

While we all deal with risk in some way, either as individuals or in groups and organisations, questions arise about how we do it and how well we do it.

As individuals, we manage many day-to-day risks intuitively. Other risks are dealt with on a more formal basis, and there we tend to rely on specialist help, such as insurance advice for our potential medical expenses, our cars, our property, and our personal possessions. We may consult stockbrokers about our investments, or financial planners about arrangements for our future retirement income. We go to doctors about our health risks.

We may also deal with risk intuitively in the projects that we undertake. Case Study E (a hot-rod car project) is an example of this. Even quite small projects involve processes where risk arises and must be managed. The renovation project schedule described in Chapter 7 (Table 7.4) reveals instances where risks are addressed. In Table 7.4, Items 2, 4, 5, 8, 14, 32, and 33 all indicate avoidance or reduction measures for perceived risks.

Risk management may be carried out informally in small groups and associations. However, the duty of care now legally imposed in many jurisdictions means that informal approaches must now be replaced by more systematic procedures in most circumstances.

Many organisations have been forced into formal and systematic risk management by the prescriptive requirements of occupational health and safety legislation. In the United Kingdom, the Construction (Design and Management) regulations (CDM; HSE 2015) have extended this reach beyond construction contractors and into the professional disciplines involved in the construction industry. Similar legislation has been enacted in Australia (Lingard *et al.* 2014; Safe Work Australia 2016). The latter document is actually a set of model guidelines and has no legal effect in itself. It is intended

Managing Project Risks, First Edition. Peter J Edwards, Paulo Vaz Serra and Michael Edwards.
© 2020 John Wiley & Sons Ltd. Published 2020 by John Wiley & Sons Ltd.

to promote nationally consistent regulations federally and across the states and territories of Australia. This has now been achieved. Other countries have taken similar approaches. Through developments such as these, all aimed at improving health and safety risk management beyond the purview of projects, many organisations have also implemented risk management systems designed to embrace other types of risks.

Expectations of sound risk management are increasing. It is now not unusual for a key stakeholder (particularly a public sector client) to call for evidence of systematic risk management capability as part of its qualification criteria for selecting potential participants and bidders for its projects.

Responsibility is a key issue in managing risk, and it is now generally accepted that risk management is more a people issue than a mathematical exercise. For many stakeholders, the PRM process actually revolves around managing the people who are dealing operationally with risks. Encouraging them to have an informed understanding and awareness of risk in general is one important aspect; getting them to adopt formal approaches to identifying and dealing with specific risks affecting their organisation is another. The implications of this were discussed in Chapter 16.

Implementing PRM in an organisation, or expanding an existing system, *is a project in itself*, and readers should bear this in mind throughout this chapter.

The topics to be considered include establishing the risk management requirements of the organisation, planning and designing a PRMS, organisational maturity, building the system, practising risk management, operating the system, PRMS benchmarking and improvement, and dealing with system decay. Our approach will be that of designing, planning, and implementing a new PRMS in an organisation.

17.2 PRMS Objectives

A critical first task is to find a champion in the organisation: someone who is enthusiastic about the PRMS implementation project, is willing to take it on, and has sufficient authority to drive it through to operational readiness. In many instances, this person is actually the initiator for PRMS implementation anyway; the champion has already agitated for this development in the organisation.

The champion should immediately seek to surround herself or himself with a small, like-minded, and committed team. People invited to join the team should be familiar with the project environment of the organisation, eager to improve its project management performance, and unafraid of change.

The team should establish objectives for the PRMS, develop a business case if this is required, formulate a budget, and prepare a delivery schedule.

Among others, the team is likely to set objectives for the PRMS such as:

- To systematise risk management in the organisation for all the projects that it undertakes
- To establish consistent and reliable PRM processes
- To improve PRM performance
- To raise risk awareness in the organisation
- To harmonise the PRMS with other areas of risk management in the organisation
- To establish a PRMS that will comply with international standards for risk management.

The objectives are not listed in any order of priority or importance.

Every member of the team should understand risk concepts and terms, and have a reasonable grasp of what the PRMS should comprise (at least to the extent discussed in Chapter 4). Familiarity with the cycle of PRM (Chapter 4, Figure 4.2) is essential. Ideally, the team should be able to perceive the value added at each stage of the PRMS.

The support of senior management is essential throughout the PRMS implementation process, but it is particularly important in the early start-up period. Part of the champion's role is to ensure that this support, together with considered approval for the PRMS plans and proposals, is forthcoming in a timely way. Good communication skills are required here as the champion has to act as a go-between, encouraging the implementation team and helping them with their work, taking the PRMS plans and proposals to senior management, and ensuring that appropriate feedback is given to the team.

It is also reasonable to expect senior management to review material for the PRMS and its implementation at regular intervals. Briefings and team presentation seminars can be used for this purpose.

Early in the planning process, a scan should be carried out to assess the current risk management maturity in the organisation. PRMS maturity is discussed later in this chapter. The scan will help to determine what level of maturity is required and achievable in the immediate, short, medium, and long-term future. It will also inform the business case and system delivery schedule.

A short one-day or half-day workshop should suffice for this investigative exercise. The PRMS team should participate, together with invited representatives from key areas of the organisation. If thought necessary, the workshop could be facilitated by a professional consultant, but in most circumstances a team member (ideally the PRMS champion) should be able to fill this role.

17.3 Planning and Designing the PRMS

If the organisation is new to PRM, then the start-up should be kept as simple as possible. Our advice would be to defer (at least for the short term) any intention to integrate a PRMS with any existing systems (such as health and safety management) already in place. This will avoid any dilution of those systems and any potential domination of project risk priorities simply because they are already there. Furthermore, complete integration at this point might represent indigestible change for the organisation and is better left as a more manageable change later on.

17.3.1 Planning the PRMS

PRMS planning starts with the determination of system requirements. If the organisational scan has not already gathered this information, and the PRMS objectives do not fully encapsulate it, then answers to further questions are needed:

- Which projects (or project activities) clearly require formal risk management?
- How are decisions made about them?
- What risk attitudes are evident in this decision making?
- What *informal* PRM activity already takes place?

- How effective is it?
- What are the obvious gaps in the organisation's current risk management practice?
- How could the gaps be filled?
- Who should be involved in that?

Inevitably, the picture produced by an organisational scan will be blurred in places and lack a consistently clear focus. There will be evidence of activity resembling risk management (but not necessarily labelled as such) in some parts of the organisation, but blank spaces for others. Of particular interest will be any gaps or misalignments relating to the management of the projects undertaken by the organisation. Aspects concerning project decision making should be given special attention. The scan is also likely to uncover evidence of existing organisational risk cultures and risk attitudes. While some of the information gathered by the scan will be factual, most will subjectively represent the perceptions and opinions of contributors and observers.

Overall, the scan should provide a richer picture of the state of risk management practice within the organisation and guide the directions in which it can be more formally systematised and improved. It should now be possible to formulate a coherent PRM policy which can be presented, together with the business case and delivery schedule, to senior management for approval. The policy should be specific rather than general. It should state why the organisation should adopt a more formal approach to managing risk and indicate if the PRMS is to be targeted at particular projects, to all projects, or even across all the activities of the organisation. Where possible, risk management responsibilities should be defined, using an organisational structure diagram if necessary.

17.3.2   Designing the System

System design should first develop an appropriate PRMS framework. A good basis for a PRMS framework design would be the cycle of PRM presented in Figure 4.2 (Chapter 4). The design should then be expanded to accommodate the processes and techniques envisaged for each stage in the cycle. Table 17.1 proposes an expanded PRMS design framework, incorporating references to changes proposed for other parts of the organisation and implications for resource requirements and training. Guidance for this expansion can be derived from earlier chapters in this book (Chapters 4–11). The international risk management standard ISO 31000 (2018) and its companion handbook SA/SNZ HB 436 (2013) also offer guidance about risk management system design frameworks.

The content of Table 17.1 is suggested as an indicator of parameters that should be considered in PRMS design. Tables 4.1 and 4.2 (Chapter 4) can also be used as an early design template, but nothing should be set in stone at this point. The design should be kept as flexible as possible throughout this process.

Important design principles for PRMS operation are that the system should be capable of functioning consistently across all types of projects and be user-friendly. It should encapsulate all the organisation's key objectives for PRM, but not necessarily all objectives at once.

Table 17.1 PRMS design framework.

Processes, tools, and documents

PRMS stage	Workshops	Team/individual responsibility	Tool/technique	Project risk register entry	Organisational risk register entry	Comments	Implications
A. Establish project contexts	Yes	Team	Structured brainstorming	Concise	Concise transfer	Consistency? Commitment?	RM culture
B. Identify risks	Yes	Team	Checklists? WBS? Other? Brainstorming	Risk statements	Concise transfer	Policy? Strategies? Existing?	Training? Practice?
B. Classify risks	?	Individual	Generic? Other?	Multiple categories?	Concise transfer	Policy? Existing?	Rationale?
C$_1$. Analyse risks	?	Team	Quantitative? Qualitative (measures)?	Concise	Concise transfer	Policy? Existing?	RM culture Training? Practice?
C$_2$. Evaluate risks	?	Team	Qualitative (measures?)	Concise	Concise transfer	Policy? Strategies? Existing?	RM culture training? Practice?
D. Response options	?	Team	Guided brainstorming	Concise	Concise transfer	Policy? Strategies? Existing?	Feasibility?
D. Response treatments	?	Team	Guided brainstorming (ALARP?)	Concise	Concise transfer	Policy? Strategies? Existing?	Resources?

(Continued)

Table 17.1 (Continued)

Processes, tools, and documents

PRMS stage	Workshops	Team/individual responsibility	Tool/technique	Project risk register entry	Organisational risk register entry	Comments	Implications
E. Risk monitoring	No	?	Programmed frequency	Concise	Concise transfer	Policy? Strategies? Existing?	Resources?
E. Risk controls	No	?	Risk specific	Concise	Concise transfer	Policy? Strategies? Existing?	Resources?
F. Risk knowledge capture	Yes	Team	Feedback (structured?)	No	Concise transfer	Policy? Strategies? Existing?	Structure? Use? Access?

ALARP, As low as reasonably practicable; PRMS, project risk management system; RM, risk management; WBS, work breakdown structure.

Other important considerations in early PRMS design are the fundamental operational characteristics required. Is the system intended to be passive, reflecting 'historical' decision making about project risks and their management? Or should it be active at appropriate points? By *active*, we infer system characteristics such as automatic prompting for system data entry, automatic recalculation or sorting of cell data, and calendar flagging for risk monitoring and control activities. How should interactions between the PRMS and its users, and with other relevant risk-related systems in the organisation, be accommodated? What interconnectivity and interoperability are feasible and worthwhile?

In our view, and depending upon the risk management maturity of the organisation and its resource capacity, a relatively passive and straightforward PRMS design may be best as the initial start-up aim, particularly for a small organisation with limited resources. Added sophistication can be introduced as maturity increases and as other organisational changes become necessary. However, in Chapter 11, we noted how quickly knowledge management is developing, particularly in the areas of artificial intelligence (AI) and machine learning, so we concede that our advice here may be out of date.

Many of the question-marked entries in Table 17.1 relate to uncertainties about how processes will be carried out and how risk information will be dealt with operationally. A more comprehensive discussion and treatment of risk knowledge management is given in Chapter 11, and any comment in this chapter will be brief and specific to particular issues. The same caveat applies to philosophies, explanations, and arguments relating to PRMS processes, so that undue repetition of earlier material is avoided.

The extent of system building needed will be influenced by the level of PRM maturity existing in the organisation.

17.4   Risk Management Maturity

The maturity of PRM in any organisation can be related to any one of four levels:

1) Mostly unaware
2) Starting
3) Growing
4) Maturing.

The maturity concept draws upon the work of Hillson (2002) and is shown in Figure 17.1.

Importantly, organisational risk management maturity is always seen against a background of change occurring in the organisation. It is also important to note that there is a threat risk of system decay, at least for Levels 2 and 3. Beyond Level 3, the system should be fundamentally stable and in a cycle of continuous improvement.

17.4.1   Level 1 PRMS Maturity (Mostly Unaware)

Although Level 1 organisational PRM maturity is labelled 'mostly unaware', this cannot be taken as meaning that no PRM takes place. Every organisation does something about some project risks. The immaturity arises because such management is not systematic. It is likely to be sporadic, reactive rather than proactive, ad hoc, fragmented, and inconsistent across projects. Management responsibility will be on an 'as needs' and 'availability' basis, and any experiential risk knowledge will reside tacitly in individuals in the organisation.

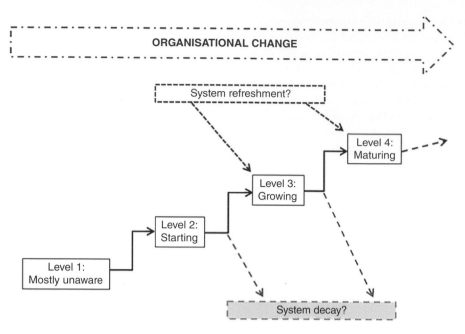

Figure 17.1 Risk management maturity levels.

At this level, risk awareness may be generally weak, and any organisational risk culture covert and negative rather than overt and positive. The 'as low as reasonably practical' (ALARP) principle (Chapter 9, Figure 9.1) is almost certainly not observable, and the number of threat risks simply retained without mitigating treatment is likely to be high because they will not have been properly identified and assessed.

17.4.2   Level 2 PRMS Maturity (Starting)

Figure 17.2 illustrates what might be a maturity Level 2 (Starting) PRM structure for a project stakeholder organisation. This level might be observed soon after the organisation has recognised the need for a more formal and systematic approach to managing its projects risks and has implemented a PRMS. The type of system is not an indicator of maturity here, since a simple PRMS might be working well but a more sophisticated system could be presenting a learning curve for staff that is too steep.

The diagram proposes an organisation which has a stakeholder role in at least three projects, and for Project A there are at least two other stakeholders.

The organisation has a typical 'departmental' functional structure, and we have deliberately omitted a governance hierarchy in order to avoid unnecessary graphic complexity.

At this level, at least some semblance of systematic PRM will be observable. What is probably a separate 'Project Risk Management' department (Figure 17.2, grey shading) conducts risk management activities for each project by communicating through the organisation's internal project management unit. It may even try to coordinate the risk-related activities of other departments for each project, although this is unlikely in practice during the early period of this maturity level. Establishing a separate PRM department treats it in the same way as any other in the organisation and facilitates a start-up approach under the 'champion's' leadership. It is an obvious response to a situation where a project client might stipulate a requirement for a compliant PRMS as a condition precedent for awarding a project contract. Practically, of course, it also presents an easy route for dismantling the unit if it does not prove successful in terms of the project stakeholder's expectations.

LEVEL 2 Risk Management Maturity

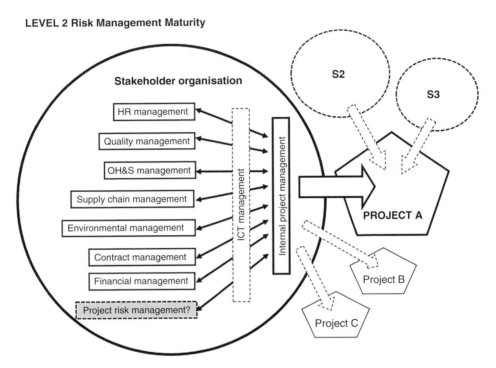

Figure 17.2 Level 2 organisational project risk management maturity.

At Level 2 PRMS maturity, some evidence of greater risk awareness will be found (compared to Level 1 maturity), but this is likely to be patchy and concentrated in the PRM department and the internal project management unit. However, a strong risk management culture is unlikely to be evident in the organisation.

The 'champion' or head of the PRM department may be undertaking this responsibility either on a secondment basis or in addition to other duties. If start-up limits have not been set, the department team will be expected to manage risks for most, if not all, of the organisation's projects. If consultants have been hired to help with PRMS design and implementation, they will still be present at times in the department, and a help-line facility is likely to be operating. Depending upon the size of the organisation, and the number and scope of its projects, the team is unlikely to exceed five or six staff. One or two of these may have been recruited for the role, and one or two more may have been encouraged to participate in some type of external risk management training. The type, level, and quality of this training may not be fully known.

Across the organisation, there will be a continuing reliance on reactive rather than proactive approaches to managing risk. Project risk knowledge capture is likely to be sparse, and any risk learning not yet widely disseminated within the organisation.

Transfer or part-transfer (by insurance or subcontracting) may be the predominant risk response in Level 2 organisations, since they will still be largely unfamiliar with robust mitigation techniques. Opportunity risk exploitation is rarely found at this level.

17.4.3 Level 3 PRMS Maturity (Growing)

Growing maturity in PRM is represented by Level 3 and shown in Figure 17.3. The organisational context is similar to that of Figure 17.2, but now a risk management department sits between the organisation's projects and other departments and may more closely resemble the situation for managing information and computer technology (ICT) systems across the organisation. It is shown adjacent to the internal project management unit, but in some cases may have been subsumed into this unit or department and function more or less within a project management office (PMO).

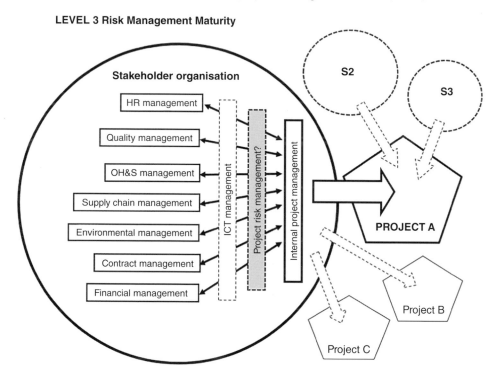

Figure 17.3 Level 3 organisational project risk management maturity.

Within a Level 3 maturity organisation, risk management activity will be seen to be growing, but it is likely to still be almost exclusively project-based. Risk awareness will be evident, and a stronger culture of risk management may be detected, or at least an awareness of the need for it, but generally only in the parts of the organisation directly involved in projects. The 'champion' may have returned to a former role in the organisation, and more specialised risk management staff may have been recruited. If the organisation is wholly project-driven, change management may have placed responsibility for occupational health and safety (OH & S) within the PRM unit, although this is unlikely if a PMO is in place. The rationale for this change would be to get OH & S responsibility closer to the project environment and to bring what was likely to have been a small OH & S team under better management scrutiny.

Most of the PRMS stages and processes at Level 3 PRMS maturity will have been standardised and applied to all projects, although some unique or highly complex projects may test the capacity and effectiveness of the system. Pro-forma approaches to documentation will be evident. A cycle of improvement to some risk management procedures may exist, and the PRMS itself may have been upgraded. Some system benchmarking may have been carried out, with simple key performance indicators established.

The stakeholder organisation may be using its PRMS as a marketing feature on its website.

Projects will each have an individual project risk register (PRR). They will be subjected to risk debriefing at completion, and an organisational risk register (ORR) will exist. There may be evidence of fairly simple analysis of selected data from the registers, and risk information (particularly for observable risk clusters or trends) will be communicated at least across the operational parts of the organisation. Selected risk occurrences will be treated as learning experiences, as a means of internal training for staff, and as the basis for procedural improvements.

Risk transfer responses will be subjected to increasing levels of benefit–cost analysis, and no risk will be retained without being subjected to at least a brainstorming attempt at reducing (mitigating) it. The 'retain' level ALARP principle (Chapter 9, Figure 9.1) may have been set at a lower level of risk severity.

More opportunity risks will have been recognised, and limited exploitation of them may have occurred.

17.4.4 Level 4 RM Maturity (Maturing)

At Level 4 maturity, it is possible that the stakeholder organisation has been radically changed with respect to managing risks, not only in terms of its projects but also for the more general organisational risks it faces. Figure 17.4 depicts this, but the precise structure of the organisation will be more uncertain now, given that structural changes may have been implemented.

Alert readers will have noticed that the heading to this subsection has changed from 'PRMS' to 'RM'. The maturity focus has necessarily widened to embrace general, as well as project, risk management needs.

In conjunction with growing compliance requirements, staff capability needs, and the type of business it is in, major restructuring of the whole organisation may be contemplated or occurring. Amalgamation of the PMO with departments responsible for other functions (such as quality assurance, environmental management, OH & S, and contract management) is possible, although the last of these may be left as a smaller legally focused unit. Any changes of this magnitude, however, will have brought further risks (i.e. more uncertainty) with them.

The wider perspective means that Level 4 risk management maturity should still be seen as an aim rather than a finished state. While an active 'champion' may no longer be needed, someone in the organisation at a senior level (e.g. a director or partner) will be accorded responsibility for general oversight of all risk management activity.

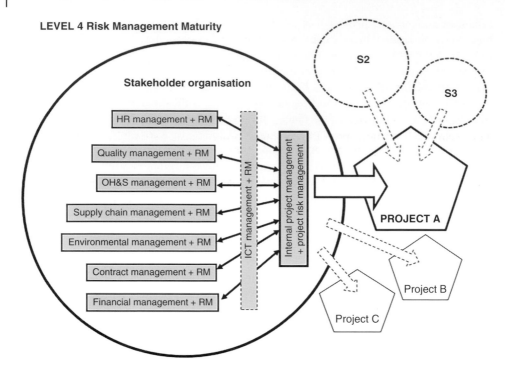

Figure 17.4 Level 4 organisational project risk management maturity.

A high level of risk awareness should be discernible across the organisation, accompanied by a robust risk culture. A dedicated and fully resourced PMO may be in place, filtering risk information from other departments in terms of the implications for projects. The PMO will directly manage all project risks and report directly to senior management. No project is allowed to proceed unless acceptable initial risk management procedures have been implemented for it, and no project will lie beyond the organisation's capacity to assess its riskiness. Risk knowledge capture will occur across the organisation, with relevant information fed into the ORR. Externally sourced risk information will be collected and similarly filtered. All PRRs will be archived in the ORR, which will be fully searchable through multiple parameters and terms. The whole system is likely to be fully interactive and highly automated. A small specialised data analysis and modelling unit may exist, undertaking risk modelling for new projects as well as carrying out historical data analysis and interpretation, thus enabling risk policy development to proceed on an informed basis.

The PRMS may have gone through further cycles of improvement. System benchmarking will be more sophisticated. A learning attitude will be taken towards project risk failures, and project case studies will be used for risk management training in the organisation.

Increasing attention to opportunity risks may mean that a specialised business development unit may be needed in order to exploit them more thoroughly.

Diffusing risk management across the organisation (as shown in Figure 17.4) may well have brought additional benefit in developing more innovative ways to reduce threat risks. This in itself may open up more opportunities.

The overall effect of Level 4 risk management maturity may be confirmed by the organisation's greater ability to improve or maintain its competitive advantage.

17.5 Building the PRMS

Building a PRMS comprises doing everything necessary to bring the design into reality. The preliminary framework of PRMS stages, processes, and strategies is already developed, at least conceptually. It is like having the drawings, plans, and perhaps even models for a physical project on your desk and now going onto the factory floor or out to the site to get your hands dirty! Armed with all the necessary materials, resources, and technologies, you begin to methodically build the actual structure or basic shell, add the necessary coverings and finishes, and install the requisite services and operating mechanisms and controls, to the point where you feel confident in switching on and starting up the whole system. Before that point is reached, however, you will probably want to test the integrity of some project components along the way. You will also want to ensure that you have the correct operating fuel available.

A myriad of details must be settled, and many small and large decisions made. The major 'what?', 'why?', and 'how?' questions will have already been addressed in the system design architecture (and were dealt with in Chapters 4–11) according to the strategic needs of the organisation (Chapter 16). System building looks more closely at the 'who?', 'where?', and 'when?' questions.

Implementing a PRMS in an organisation thus requires organising and building skills, knowing when specialist help is needed, component testing capability, trialling arrangements, and a system roll-out schedule. All of this must be planned and resourced in advance.

17.5.1 Organising the PRMS Project

A project leader is needed for system building, but this does not automatically mean that the PRMS 'champion' should take this role. Such champions often have wonderful 'starter' energy to build up momentum, and high motivating skills, but are not necessarily good 'finishers'. A PRMS building project leader needs good interpersonal relationship and communication skills, a methodological manner, and attention to detail.

A simple schedule of system building activities should be prepared, with a feasible calendar of delivery milestones. This, together with the accumulation of system and process design information, will determine if the team has the requisite skills and resources for the system building project or if specialist help is needed.

17.5.2 PRMS Specialists

A dilemma arises when considering whether or not specialist help is needed for PRMS implementation. While the in-house project team may lack confidence in undertaking the system building task, it does have the priceless advantage of intimate knowledge of the host organisation and hopefully its existing risk culture. Even experienced PRM consultants will be largely unaware of the organisation in terms of its structure and business processes. They are likely to be completely ignorant of the prevailing organisational culture.

For their contribution to be effective, consultants must be properly briefed. An issue then arises as to when consultants should be engaged. Subsection 17.5.1 partly reflects our view about this, which may be considered somewhat radical, and we accept that consultants may not agree.

Our concern is about system 'ownership'. An organisation lacking any confidence in its own ability to establish even a preliminary framework for PRM represents, to some extent, a contradiction in terms. We might be justified in suspecting that it could be an organisation lacking vibrancy and business acumen in other areas. Given the whole content of this book, and the guidance it offers, hiring consultants to carry out system design and lead strategic thinking in the organisation may actually create a fertile environment for an adverse organisational risk culture to grow that reflects 'their system' (i.e. the consultants') ownership perceptions for the PRMS. The host organisation must adopt full ownership of its PRM from the very beginning of its awareness of the need to be more systematic in this area of management.

Instead of ultra-early consultant engagement, we advocate that thorough PRMS building effort by the host organisation is first employed, until it becomes obvious that specialist help may be needed with particular tasks. PRMS consultants can always be brought in later to review system performance since this requires dispassionate scrutiny.

To some extent, this also applies to assistance with information technology (IT) requirements for the PRMS, but here early involvement of the organisation's internal IT resource staff should be arranged, and their advice heeded about commissioning external assistance with PRMS computerisation.

17.5.3 System Building Tasks

The various tasks involved in building a PRMS tend to be practical rather than theoretical, once the guiding framework has been established. They involve activities such as:

- Establishing preferences for PRM workshops in terms of when they are held, where they can be held, and who should attend.
- Creating checklists for project contextualisation (Figure 4.2, Stage A). This was discussed in Chapter 5.
- Adaptation of other management tools for risk identification (Figure 4.2, Stage B). Chapters 6 and 7 dealt with this topic.
- Creating template formats for precise risk statements and preferred risk classification approaches. Risk statements were considered in Chapter 7, while risk classification was discussed in Chapter 2.
- Clarifying justifications for quantitative risk analysis (Figure 4.2, Stage C_1) and checking the organisational capacity to undertake quantitative analysis. Chapter 8 provided guidance about this. If it is thought that external help will be needed, then early contact with the mathematics department of a local university could be a solution.
- Confirming the design parameters for qualitative risk analysis (Figure 4.2, Stage C_1) and for risk evaluation (Figure 4.2, Stage C_2), especially where linguistic scale measures are to be used (Chapter 8).
- Specifying treatment procedures for risk response decisions (Figure 4.2, Stage D) and indicating how an ALARP strategy is to be applied (Chapter 9).
- Establishing the type, nature, and frequency of risk monitoring and control procedures (Figure 4.2, Stage E; and Chapter 10).

- Determining the nature, extent, frequency, and preferred environments for project debriefing, and the formats and methods for information gathering and knowledge capture (Figure 4.2, Stage F; and Chapter 11).

For several system building activities, the adequacy of the component outcomes should be tested.

17.5.4   Component Testing

Each link in the PRMS chain of processes will involve components that can be tested individually before system trials are conducted.

Generally, focus groups will be the most appropriate method of testing. For example, a project contextualisation checklist can be tested in a focus group comprising representatives from senior management and from the operational level of the organisation. The group is simply asked to consider if the list is sufficiently representative of the organisation, its business environment, and the projects it typically undertakes. Particular emphasis should be placed upon external project contexts.

In testing adaptations of other management tools for use in a PRM setting, the focus group should be more homogeneous and comprise staff with expertise in the tool's original purpose.

Testing risk identification statement formats and proposed risk classifications is best done with a series of examples, asking the focus group (now comprising project managers) if each statement sufficiently captures the particular risk situation and if the proposed classification is suitable.

The subjectively derived scale measures intended for qualitative risk analysis and evaluation should be thoroughly tested, since their outputs will substantially inform subsequent risk response and treatment decisions. Preferred testing here would be through a series of focus group meetings, with each group comprising different representatives from the organisation, at senior and operational levels. The aim is to achieve common understanding for each interval definition for each scale measure.

Risk response and treatment mechanisms are best tested by recourse to actual risks experienced on previous projects, to relevant organisational policies and procedural mechanisms, and to scenarios.

Monitoring and control measures should be tested by operational-level staff, and also by anyone involved in compliance requirements for risks where these are stipulated.

Testing PRMS components in this way can take quite a lot of time, and this should be considered carefully in system building planning. The benefits of testing are that component modification can then be undertaken without unduly disrupting the whole building process.

When the majority of component testing has been completed, attention can turn towards trialling at least parts of the PRMS.

17.5.5   PRMS Trials

Once testing of the separate components has been completed, the PRMS should be trialled progressively with several consecutive stages/processes at a time, before the whole system is trialled. Focus groups can be used again or even just a few key operational project staff.

Using the cycle of PRM (Chapter 4, Figure 4.2), the best approach is probably to examine two stages at a time, followed by a single stage, and culminating in a full system trial. This arrangement would comprise:

1) Stages A (project contextualisation) and B (project risk identification)
2) Stages C_1 (risk analysis) and C_2 (risk evaluation)
3) Stages D (risk response and treatment) and E (risk monitoring and control)
4) Stage F (risk knowledge capture)
5) Full PRMS.

While a 'live' project could be used as the vehicle for the first three trials, it is probably better to use one or two historic projects for the fourth and fifth trials, as it will not be possible to apply them completely to current projects.

The trials are intended to explore the adequacy of stage application and user-friendliness for each part of the PRMS. The benefit of using focus groups for much of this is that the system building team can observe proceedings and also gather impressions of the time needed for each trial, as this will inform real-life scheduling for PRM activities. Difficulties encountered with particular parts of the PRMS can also be recorded. For example, preparing precise risk statements can be daunting to anyone who has never done it before. Even brainstorming might require some practice among staff unfamiliar with this technique (especially with its limitations). Lessons in dealing with group dynamics may be necessary. It is also not unusual to encounter further confusion among staff in an organisation about particular concepts of risk, such as uncertainty and probability, so system trialling may need to be followed up with additional education and training.

When reasonable satisfaction with the trials has been achieved, and any further system modifications completed, the PRMS should be ready for implementation.

17.5.6   PRMS Roll-Out

In Chapter 16, PRMS implementation was considered from a strategic management perspective. We advocated a low-key roll-out of a stand-alone system (i.e. the PRMS is not integrated with other systems in the organisation) on a few selected live projects, preferably those that are heterogeneous. The target projects do not all have to be at the same state of progress, as 'catch-up' risk management activities should be possible. Staff availability for each roll-out will determine the number of projects for initial PRMS deployment, but the target should not exceed three or four projects.

Responsibility for administering the PRMS on each project should be allocated to suitable staff selected (or volunteered) from each project team. Appropriate training may be necessary, but judicious choice of projects in the preceding system trials would mitigate the extent of this. Ideally, the project risk managers (RMs) should be people who are flexible and resilient, have extensive experience and knowledge of the projects undertaken by the organisation, and possess good communication skills. Given a start-up situation such as that envisaged by Figure 17.2, the project RMs may have to be seconded from their project teams (and sometimes from other parts of the organisation) to work in a newly formed and separate PRM unit. The new unit members should be mentored by the founder members of the system building team.

In the early stages of PRMS implementation, and depending upon the type of projects involved, some thought should be given to the need for emergency rostering of project

RMs, and even for out-of-workhours communication facilities. However, such arrangements should be considered only a short to medium-term measure while the PRMS is 'bedded in' on the guinea-pig projects.

Comprehensive and consistent feedback about all aspects of the PRMS from each project RM is a key element for evaluating the system roll-out. Any system inadequacies, or even partial failures, should be treated as opportunities for learning and improvement.

Again, depending upon the nature of the projects involved in PRMS implementation, early arrangements should be made for undertaking system performance reviews and, after essential modifications have been made, extending the roll-out to all projects in the organisation. This should also represent the beginning of a continuous cycle of performance improvement.

We again remind readers of the 'two hats' risk management issue for professionals, especially project managers, who are engaged in a consultancy role in projects.

17.6 PRMS Performance Review and Improvement Cycle

One of the most difficult things to deal with in maintaining a PRMS in an organisation is evaluating its performance. After all, if risks are successfully managed so that they either do not eventuate or do not have the impact they might have had if they were left unmanaged, how is the organisation to measure what might be an invisible benefit? What are the key performance indicators and critical success factors for a risk management system in a project stakeholder organisation? Obviously, these must be determined by each organisation and based upon the objectives established for the PRMS as a whole and the functions of the constituent parts. Given this, however, difficulty will almost certainly still be encountered in attempting to measure system costs and system value outcomes. Even if this can be accomplished, how can it be benchmarked against similar systems in other organisations?

We attempted some consideration of these issues from a strategic risk management perspective in Chapter 16, but the whole area of PRMS performance is considerably under-researched, hence the multitude of questions in the paragraph above!

The great variety encountered in project types, scopes, procurement arrangements, and environments complicates the performance review dilemma and our treatment of it here. Discussion will therefore have to be largely general.

For the most part, a PRMS performance review relies upon the critical assessment of feedback evidence from staff involved in using the system. Much of this evidence will comprise subjective opinion. Provided this is gathered in a methodical way, and validated wherever possible by triangulation and comparison, the results should be reliable.

PRMS performance review issues to be dealt with here include: review criteria, benchmarking, system decay, and review frequency.

17.6.1 Review Criteria

Some indicators for PRMS performance review are suggested in Table 17.2. They are ordered according to the stages of PRM. As an organisation becomes more experienced (i.e. matures) in its PRM, additional system performance indicators will emerge, together with more concrete measures for existing indicators.

Table 17.2 PRMS performance review criteria.

PRMS stage	Performance focus	Possible performance criteria
Project contextualisation	• Comprehensiveness	• Internal and external contexts covered consistently? • Internal context parameters dealt with project and organisational factors? • External contexts updated? • Does project contextualisation contribute significantly to identifying risks? • Does it significantly influence project risk assessment?
Risk identification	• Effectiveness of threat risk identification techniques and processes • Sufficiency of risks identified • Adequacy of risk descriptions	• What difficulties did project team members experience in using techniques? • Sufficient coverage of most project areas and phases? • Was brainstorming effective? • How many foreseeable risks were missed and then discovered later in a project? • Precision of risk statements? • Good risk classification?
Risk analysis	• Effectiveness of threat risk analysis techniques and processes • Interpretation of outcomes	• Appropriateness and reliability of quantitative analysis techniques? • Proper interpretation of modelling outcomes? • How consistent were the subjective qualitative analyses? • Any issues experienced with scale measures?
Risk evaluation	• Appropriateness of threat risk assessment level	• How accurate and reliable were any quantitative assessments? • What anomalies can be detected in assessment levels? • Any issues experienced with scale measures?
Risk response and treatment	• Appropriateness and effectiveness of threat risk response and treatment decisions	• Were any obvious risk mitigation treatments missed? • Were any straight risk retention decisions mistaken? • How effective were risk transfer/part-transfer actions? • Could more risks have been avoided? • What comparisons can be made between before/after treatment risk severity scores or cluster maps?

(Continued)

Table 17.2 (Continued)

PRMS stage	Performance focus	Possible performance criteria
Risk monitoring and control	• Effectiveness of threat risk monitoring and control procedures	• Did any procedures overlap with other management actions (e.g. value management, quality management, or safety management)? • Were any risks proactively averted because of the PRMS?
Risk documentation and knowledge capture	• Adequacy and effectiveness of the project risk register • Usefulness of the organisational risk register	• How is the PRR used, and how frequently? • What PRR information is most useful? • What PRR information is least useful? • How is information transferred from the PRR to the ORR? • How is post-project risk knowledge captured and documented? • How is the ORR used, and how frequently? • What ORR information is most useful? • What ORR information is least useful? • To what extent is ORR information mined and analysed to explore risk trends, create case studies, develop better disaster preparedness, etc.?
General	• Value-adding	• In what ways is the PRMS adding value to the organisation and its business operations? • Has the contingency spend rate on projects decreased? • How much attention is given to opportunity risks?

ORR, Organisational risk register; PRMS, project risk management system; PRR, project risk register.

Performance indicators for the process of establishing project contexts are often difficult to assess. This is due partly to this stage being the first in the PRMS cycle and thus being furthest back in post-project memories. It may also be the least well-documented stage, simply because it comprises descriptive material that originally relied heavily on subjective opinion.

In reviewing risk identification performance, it can be seen from Table 17.2 that a multi-focus approach is needed, covering the use of techniques, the number of risks identified, and the descriptive recording of them. Particular attention should be paid to the brainstorming process where it has been used, as it is highly susceptible to group dynamics. If the dynamics of a workshop team are not working successfully, the quality of its outputs will be threatened.

Risk identification tools adapted from other project management purposes should be closely examined in terms of their practicability and user-friendliness, and for the best timing for their use. Trying to use a risk identification technique too early in the project management process, when there may not be sufficient information available to service it, can be counterproductive.

In our view, the completeness and precision of identified risk statements are critical for good PRM. Single- or two-word risk descriptions can lead to communication errors; they are not unequivocal. Good risk statements not only inform the processes of risk analysis and evaluation, but also contain the seed for appropriate response decisions and treatments. They can even guide the monitoring and control procedures.

For project risk analysis and evaluation, the PRMS performance review should explore how robust the techniques are and how reliable the outcomes have been.

The decision making associated with risk responses and treatments should be reviewed with a view to exposing instances where, from a post-project perspective, wrong or inadequate decisions have been made. From a system performance viewpoint, it is more important to identify decision errors than to try to assess how correct the decisions were.

PRMS performance reviews for risk monitoring and control processes should always seek to match the identified risks with their subsequent management and how effectively that was carried out. In the interests of project management efficiency, a lookout should also be kept for the extent to which the PRMS is duplicating, or contributing effectively to, other aspects such as safety, quality, value, and environmental management.

In assessing the performance of project risk documentation and post-project knowledge capture, the PRRs and the ORR will come under close scrutiny. This can be a lengthy process. For an organisation with Level 2 risk management maturity (Figure 17.2), this may involve a relatively small amount of computerised information and a much larger body of paper-based evidence. One purpose of a continuous improvement cycle would be to reverse this situation. As it improves, a quantitative system performance measure comes into play. It should be possible to record the number of 'strikes' (i.e. access occasions) for the registers and even the staff who are accessing them. While this may be a rather crude measure, it is an indicator of increasing risk awareness in the organisation. For an organisation with Level 4 risk management maturity, the extent to which register data are mined for analytical and modelling purposes is a further indicator of system performance. A higher performance is achieved when the system data are used to improve the capacity of the organisation to recover from potential disaster situations.

A PRMS performance review should end with consideration of the contribution the PRMS is making to the organisation and its management of projects. While this is likely to be largely subjective, for some organisations it might be possible to find quantitative evidence through the rate at which any financial contingency amounts included in project budgets are spent. If the spend rate is decreasing on projects, the PRMS is making a solid contribution. This leads us to consider PRMS benchmarking.

17.6.2 System Benchmarking

Benchmarking a PRMS requires something to benchmark against. Although this seems like stating the obvious, it does mean that careful thought should be given to this approach to system performance review.

With internal benchmarking, it is usually only possible to observe (and in some way measure) how the organisation's PRMS performs across some or all of the projects it undertakes. While this exercise will yield valuable insights, it will not necessarily expose parts of the system that tend to perform poorly if the same level of performance occurs on all projects. Nor will such comparisons reveal missing or partly missing PRM activities, as long as some activity is happening in all projects at each stage of the system cycle.

While internal benchmarking has weaknesses, external benchmarking is often just too difficult to arrange. The secrecy associated with perceived 'competitive advantage' in almost every industry or business militates against the sharing of information, even if it is to each party's mutual longer term advantage. Yet external benchmarking is more robust than its internal counterpart.

Where several companies are combined in a larger group, semi-external PRMS benchmarking becomes possible, but even here a culture of inter-company rivalry may prevail, and in such conglomerates the business purposes of each company may be different so that comparison between similar projects, although possible, is not straightforward. However, useful insights may be gained, especially if the companies use a similar PRMS but their actual risk management practices are distinctly different.

Case Study D (a high-capacity metropolitan train [HCMT] mock-up project) revealed evidence of knowledge 'silo-ing' among grouped companies, with the risk of losing intellectual property (IP) offered as a rationale.

Semi-external system benchmarking is also feasible for large organisations which operate autonomous regional or international branches.

Completely external PRMS benchmarking is probably best initiated at an industry level, whereby industry and professional associations promote the need for it and offer to be the broker for member organisations willing to participate in benchmarking activities. A valuable outcome, apart from the obvious incentive for system improvement, is the more systematic detection of PRMS decay that would ensue from inter-organisational benchmarking.

17.6.3 Addressing System Decay

PRMS decay is usually insidious in its progress and therefore also difficult to detect. Generally, the higher the level of risk management maturity, the more gradual is the degradation of the system, although at Level 4 PRM maturity (Figure 17.4) the PRMS should be close to a self-sustaining level. Decay is unlikely to occur slowly in an organisation at Level 1 risk management maturity, since it will be largely unaware of its maturity level anyway. Spectacular collapse is a more likely outcome, as luck is trusted once too often. Curiously, organisations at Level 2 or 3 PRM maturity (Figures 17.2 and 17.3) may be more susceptible to PRMS decay if they have completed several projects without any clearly obvious benefit being derived from their application of the system.

Some signs of decay to look for in system performance reviews include:

- Neglect of project context establishment or indifference to the significant influence of contextual risk drivers.
- Evidence of rote activity in the risk identification stage.
- Perfunctory analysis and evaluation of identified risks (other than for risks where analysis is not required).
- Evidence that proposed risk response decisions and treatments are not actually implemented.

- Planned risk monitoring and control procedures are not carried out diligently.
- Risk knowledge capture is either not carried out at all or not done thoroughly.
- Project and organisational risk registers show signs of excessive copy and paste activity.
- The organisational risk register is stagnant in terms of content or knowledge mining activity.
- Little or no evidence of risk data analysis in terms of patterns or clusters.
- Little or no evidence of opportunity risk exploration or exploitation.

Contributory factors in PRMS decay include issues such as loss of the system champion (moves on to other roles), high staff churn, priority given to other staff responsibilities, negative and covert organisational culture allowed to prevail, changes in risk attitudes and appetites in the project organisation, or advice from external specialists that proves to be impractical or ineffective. Occasionally, an organisation will consider that its PRMS has become too cumbersome or costly to operate and maintain. The ORR can be a vulnerable component here. The system cost argument is rarely promoted on the basis of factual and convincing evidence.

Addressing PRMS decay should be undertaken as quickly as possible after one or two signs have been detected. For the most part, it is likely to be a matter of system refreshment or more realistically the refreshment of PRMS operators, managers, and owners. Motivation must be re-energised, adverse organisational cultures eradicated (a difficult task), and strategic thinking revisited.

Given the possibility of system decay, how often should PRMS performance reviews take place?

17.6.4   Review Frequency

To a limited extent, the frequency of PRMS performance reviews is guided by an organisation's level of risk management maturity. For a Level 2 organisation, the frequency should be annually or biannually. At Level 3, the review interval could be stretched to every three to four years; at Level 4, once every four to five years may be appropriate. Five years should probably be the maximum interval between PRMS performance reviews. This might also correspond with the time usually required to plan, develop, and implement technological and organisational change in most industries. It could even match the typical duration of some projects. However, watchfulness for signs of system decay will trigger the need for more frequent review.

17.7   Summary

The risk management maturity model should enable organisations to assess the extent to which they currently engage in formal (and informal) risk management practices, and it provides target points for change.

Organisations must establish a clear risk management policy, objectives, and an implementation strategy for a PRMS, as creating such a system is a project in itself.

Careful design of the system framework is needed. In the early stages of risk management maturity, a simple spreadsheet approach may suffice. Issues of responsibility for the PRMS must be resolved, and responsibility should be clearly seen to extend to the highest levels within the organisation.

System building and performance will be enhanced if due attention is paid to communication, trialling techniques, and training staff.

The effectiveness of each part of the PRMS should be tested and the whole system trialled before it is made operational. Appropriate periodic performance review mechanisms for the PRMS should be put in place. Signs of system decay should not be ignored, but addressed as a matter of urgency.

Before completing our discussion of systematic PRM, it will be useful to consider the topic of PRMS computerisation. We do this in the next chapter.

References

Health & Safety Executive (HSE) (2015). *Construction (Design & Management) Regulations.* London: Health & Safety Executive (HSE). ISBN: 9780717666263.

Hillson, D. (2002). *Risk Management Maturity Level Development.* Newtown Square, PA: Risk Management Specific Interest Group, Project Management Institute.

International Organisation for Standardisation (ISO) (2018). *Risk Management: Principles and Guidelines.* (ISO 31000.) Geneva: International Organisation for Standardisation.

Lingard, H., Pirzadeh, P., Harley, J. *et al.* (2014). *Safety in Design.* Melbourne, Australia: Centre for Construction Work Health and Safety, RMIT University.

PMI (2013). *A guide to the Project Management Book of Knowledge*, 5e. Project Management Institute. Newtown Square, PA: PMI. ISBN: 9781935589679.

Safe Work Australia (2016). *Model Work Health and Safety Regulations.* Canberra, Australia: Parliamentary Counsel's Committee.

Standards Australia/Standards New Zealand (SA/SNZ) (2013). *Risk Management Guidelines – Companion to AS/NZS ISO 31000:2009* (SA/SNZ HB 436). Sydney, Australia: SAI Global Ltd. ISBN: 9781743426333.

18

Computer Applications

18.1 Introduction

In planning the structure of this book, it was tempting to put a chapter about computerised applications for project risk management immediately after Chapter 4 (on risk management systems [RMSs]), as that seemed to be a natural sequence. Indeed, there was a possible case to be made for including the material directly in that chapter. However, we were worried that doing so might infer a mechanistic understanding of project risk management that would negate the psychosocial underpinning that we have argued is an appropriate conceptual frame for risk (Chapter 2), and which we believe you must engage with for good risk management. Even if computerised risk management incorporated artificial intelligence (AI) elements, it could not replace the large elements of human emotions and societal values, beliefs, and assumptions (the social culture of risk) that drive project risk management in practice.

In structuring the content of Chapter 11 (Risk Knowledge Management), we again thought of including a section about computer applications there, but rejected this notion on similar grounds. Tacit knowledge and emotional intelligence transcend AI and machine learning.

Then, given our reluctance to rely on information with a notoriously short shelf life, we debated whether or not to omit the chapter completely. Our fear was based upon personal experience. In the late 1990s, the terms 'risk' and 'risk management' were entered in an internet search engine, yielding about 13 000 hits. Many of these were framed in the actuarial context of the insurance industry. A few related to awakening awareness of the need for risk management among civic authorities, as they began to realise, and respond to, the enormity of their public liability burdens. Less than 10 of the 1990s websites included any content about computerisation. Fast-forward two decades, and the search was repeated, this time just using 'risk management software' as an entry term. This produced a staggering 220+ million hits! It was not possible to access all of them, and the first few hundred were inspected on a random basis. Several sites offered comparisons of 'top 10' systems carried out by 'independent reviewers'. No explanations were given for the selection of the systems nor of the reviewers. Most sites described their systems as having a 'compliance' focus relating to many different areas, such as governance, safety, security, finance, foreign exchange, quality, and the environment. No explanations of the compliance needs were offered. Many sellers offered 'free trials' of their commercial software application products.

Managing Project Risks, First Edition. Peter J Edwards, Paulo Vaz Serra and Michael Edwards.
© 2020 John Wiley & Sons Ltd. Published 2020 by John Wiley & Sons Ltd.

While it is impractical for us, as authors, to deal with the whole array of computer software applications available for risk management, our dilemma is that we cannot ignore it entirely.

Eventually, we decided to place a separate chapter here and to frame the topics around the types and capabilities of computer applications rather than attempt to deal comprehensively with particular systems on offer. We also thought it important to include brief consideration of other information technologies with computer connectivity that have a place in project risk management.

Readers who may regard our rationale as a shallow avoidance compromise and who seek more information on this topic are advised to undertake physical strengthening exercises, and develop high levels of patience and resilience, before embarking on their own internet journeys in search of risk management software!

In terms of employing computerised software applications for project risk management, every project organisation must decide:

- What it needs (scope of functional requirements)
- What it wants (i.e. over and above the needs, what else might be desirable?)
- What importance the outputs may represent for the organisation
- What data can be reliably and safely input into the application
- What default application values available within the application are realistic
- What trust can be placed in the application outputs
- What transparency exists in the application (avoiding the 'black box' syndrome)
- What flexibility is contained in the software and how easily it accommodates changes in the project risk environment
- What level of application audit capability is needed and available.

There are thus strategic issues involved.

We first explore the basic types of software applications available and discuss their capabilities. This section adopts a critically reflective 'consumer beware' approach. Following this, speculative consideration is given to other information technologies and devices that potentially have a role in project risk management.

The chapter also comes with a product warning! It is not a catalogue. Nor is it completely comprehensive. Neither the chapter nor its content includes any guarantees relating to fitness for purpose of any product or for product quality, availability, costs, and service maintenance or upgrading. Caveat emptor!

18.2 Project Risk Management System (PRMS) Software Applications

Computerised software applications for project risk management broadly comprise nine different types. These are shown in Table 18.1 with some of their functional attributes. The types include simple word-processing tables and matrices, user-defined standard spreadsheets, commercially available semi-customised spreadsheets, commercial customised spreadsheets, commercial standard (basic) project management applications, commercial advanced project management systems, commercial customised advanced project management systems, and completely bespoke knowledge management systems.

Table 18.1 PRMS computer software application types.

RMS computerisation type		System attributes							
	Label format	Text data entry	Numerical data entry	Calculation capacity	Modelling function	Charting options	Export text	Export data	Export graphics
User-defined table or matrix	Manually in cells	Manually in cells	Manually in cells	User-defined (limited)	None	None	Copy and paste	Copy and paste	None
Commercial semi-customised matrix	Limited user-defined	Limited user-defined	Manually in cells	Standard options (limited)	None	None	Standard options	Standard options	None
User-defined standard spreadsheet	Manually in cells	Manually in cells	Manually in cells	Standard options	Standard options + add-on modules	Standard options	Standard options	Standard options	Standard options
Commercial semi-customised template spreadsheet	Limited user-defined	Limited user-defined	Manually in cells	Standard options	Standard options + add-on modules	Standard options	Standard options	Standard options	Standard options
Commercial customised spreadsheet	Limited user-defined	User-defined	User-defined	Customised	Customised	Customised	Customised	Customised	Customised
Commercial standard basic PM application	Limited user-defined	User-defined	User-defined	Standard options	Standard options	Standard options	Standard options	Standard options	Standard options

(Continued)

Table 18.1 (Continued)

RMS computerisation type	Label format	Text data entry	Numerical data entry	Calculation capacity	Modelling function	Charting options	Export text	Export data	Export graphics
					System attributes				
Commercial advanced PM application	Limited user-defined	User-defined	User-defined	Standard options	Standard options + add-on modules	Advanced options	Advanced options	Advanced options	Advanced options
Commercial advanced customised PM application	Advanced options	Advanced options	Advanced options	Advanced options	Advanced options	Advanced options	Advanced options	Advanced options	Advanced options
Bespoke RKMS	User-defined	User-defined	User-defined	User-defined	User-defined	User-defined	User-defined	User-defined	User-defined

PM = Project management; RKMS = risk knowledge management system; RMS = risk management system.

When exploring computerised risk management applications, readers should bear in mind the decision issues listed in Section 18.1. Since the issues may be unique to particular project organisations, they are not considered further in this section.

18.2.1 Tables and Matrices

Most word-processing software enables users to create tables or matrices similar to that used for Table 18.1 (and for all the tables in this book). User-defined formats for the number of rows and columns, with variable column widths and row heights, allow basic matrices to be designed. A wide selection of font styles and sizes is available for data entry. The tabular format can be used to insert user-defined column labels across the top row(s) and similar row labels down the first column. The size of the table is determined by the extent and nature of the data inserted into the cells of the matrix and the page layout determined by the number and width of the table columns and rows. The size of the table is also constrained by paper size if the finished table is to be printed off on a single page. The general format is deliberately designed to be simple for users.

Somewhat surprisingly, commercially available semi-customised versions of simple matrices for risk management can be found on the internet (usually accompanied by free limited-period trial offers). The level of customisation offered is not determined by purchasers' requirements, but by the commercial designers' understanding of particular compliance needs. Commercial matrices are available for risk management compliance in areas such as governance, financial administration, and environmental management. Their origins almost certainly lie in the proliferation of specialised national and international standards now available (Chapter 4).

When considered across the system attributes in Table 18.1, the commercial variants offer little more than user-created tables, and in some respects they provide less, since column and row label definitions and captions are likely to be fixed. Also, calculation and output options may be more restricted. In practice, the commercial variants serve as brightly captioned and coloured 'memory joggers'. An important consideration is that compliance requirements differ from country to country, and a partially customised matrix may be inappropriate for a particular jurisdiction and impossible to adapt. Few commercially available matrix-based risk management applications reach the level of coverage such as that shown in Table 7.13 (Chapter 7), where a resourced work breakdown structure (WBS) array and a risk category array are combined to create a customised risk identification tool.

Tables and matrices used in the way we have presented most of them in this book are a useful means of recording information about a wide array of factors and criteria for dealing with them. They provide a basis for ensuing discussion, as we have done, and exploration of their implications for projects. Generally, they do not in themselves add meaning to a topic, and that task is left to authors or individual readers. The usefulness of their application for active project risk management is therefore limited.

In a similar way, some commercial risk management applications incorporate 'fish bone' or 'bow tie' diagrams relating to existing control systems. These are simply graphical devices through which project risk information can be presented and arranged so as to guide risk decision making. The diagrams themselves do not create meaning: that still has to be done by the user. Their value lies in providing a visually stimulating structure for users to populate.

18.2.2 Spreadsheets

Standard spread-sheeting software applications are a management 'workhorse'. They follow a tabular formatting approach similar to tables, but offer greater flexibility in terms of labelling, cell data entry, data import and export options, cell calculation options, and output report formats. The PRMS templates shown in Tables 4.1 and 4.2 (Chapter 4) are typical, simple spreadsheet examples created for project risk management application.

These applications are 'semi-passive' in that they require users to enter data and then select formulae-driven processing routines, although automatic recalculation is possible for new data entry values. In some instances, more 'active' applications are available which can be user-programmed to create alerts, or links to other documents or systems, such as a calendar/time reminder for a risk monitoring and control activity to be updated.

Customised spreadsheet applications are tailored to individual user requirements. Most of these incorporate modules for quantitative modelling. Many have 'active' features.

Spreadsheet-based PRMSs can usually deal with text as well as numeric data entry. Data processing capacity and capability are quite comprehensive and often only limited by the extent of the user's imagination and willingness to 'play' with the system. The latter will determine how much meaningful output is built into the system, as distinct from passive information which must then be interpreted completely by the user.

A spreadsheet PRMS can be used quite easily for quantitative risk modelling (especially if appropriate add-on modules for computation involving Monte Carlo analysis, Bayesian analysis, and fuzzy logic are available), but their ability to deal with text 'string' variables requires a high level of manipulation skill on the part of users. 'Off the shelf' spreadsheet versions may not accommodate this, but where this facility is available the PRMS can be programmed to accept linguistic scale measure data for risk probability and impact, and will then deliver a similar linguistic measure of threat risk severity.

We can demonstrate the latter concept through a series of conditional logic statements for a three-point risk severity assignment table (Chapter 8, Table 8.6). For this basic low/medium/high severity assessment, nine conditional statements are needed, each containing 'IF', 'AND', and 'THEN' logic. These are shown in Table 18.2.

Table 18.2 Conditional statements for three-point risk severity rating.

Statement no.	Likelihood	Impact	Severity
1	IF Likelihood = Low	AND Impact = Low	THEN Severity = Low
2	IF Likelihood = Low	AND Impact = Medium	THEN Severity = Medium
3	IF Likelihood = Medium	AND Impact = Low	THEN Severity = Medium
4	IF Likelihood = Medium	AND Impact = Medium	THEN Severity = Medium
5	IF Likelihood = Low	AND Impact = High	THEN Severity = High
6	IF Likelihood = Medium	AND Impact = High	THEN Severity = High
7	IF Likelihood = High	AND Impact = Low	THEN Severity = High
8	IF Likelihood = High	AND Impact = Medium	THEN Severity = High
9	IF Likelihood = High	AND Impact = High	THEN Severity = High

Note that Table 18.2 assumes that the basic assignment of severity levels has been established through strategic risk management thinking (Chapter 16). The rationale of that severity distribution needs to be thoroughly tested. For example, in some circumstances an organisation might wish to assign a 'High' severity rating to condition 4 in Table 18.2 or a 'Medium' severity to condition 7.

Given our preference for the more fine-grained risk assessment ratings provided by the five-point scale shown in Table 8.7, 25 conditional severity statements would be required. Bear in mind that for spreadsheet-based systems, this might seem that all the statements would have to be combined in order to incorporate the result into one cell. Additional 'OR' and 'ELSE' logic conditions might also be needed. Practically, given a simple RMS template such as that presented in Tables 4.1 and 4.2 (Chapter 4), it might be easier to insert additional columns into the spreadsheet so that each severity level has a separate column. The text string variables might also be represented by nominal values to further simplify statement formatting, as long as it is accepted that they are nominal and not continuous data to be used arithmetically. A customised commercial spreadsheet application would probably be a minimum system requirement to accommodate the automated capability.

Risk severity assessments can then be mapped or charted in several ways for risk management purposes (see Chapter 8, Figure 8.4). Cell formatting allows coloured shading (e.g. red, amber, and green) to be assigned to different levels of severity. Additionally, project risks can be ranked according to their severity, allowing the 'top 10', for example, to be surfaced for closer attention (see Case Study A, Table A.3). Such system features contribute substantially to the quality of system outputs, especially when reporting formats are carefully defined.

One of the benefits of building an in-house computerised PRMS based on a spreadsheet format (e.g. Chapter 4, Tables 4.1 and 4.2) is that, for each new project, a copy of the project risk register (PRR) for a previously completed project can be retrieved from the organisational risk register (ORR), used as a template, and resaved with a new project file identifier label. The pre-populated risk data cells can be colour-shaded and then used as a checklist to explore and enter new risks pertaining to the new project. In this way, the new PRR itself acts as a risk identification tool. Later, cell data relating to the historic project that is no longer relevant is simply deleted. However, this practice can lead to 'project tunnel vision', and it is inadvisable to place sole reliance on it. Partially pre-populated PRR templates were considered in Chapter 11.

An important caveat for potential buyers of highly customised applications is that the more customised such systems are, the more they tend to be 'black box', lacking transparency in terms of background calculation routines and output format flexibility.

18.2.3 Project Management Systems

Computerised project management systems (CPMSs) are available in a relatively wide range of capacities, capabilities, and prices. Purchase costs range from a few hundred dollars to many thousands for more advanced and commercially customised versions.

Since system capacity is usually defined in terms of the number of separate project activities that can be accommodated, almost all CPMSs adopt a WBS (Chapter 7, Tables 7.1 and 7.2) as a foundation system design structure.

Good CPMSs will incorporate the calculation, analytical, and modelling features noted here for risk management purposes. User entry includes determined or estimated values for defined model parameters. User-defined or default values (e.g. beta distributions of probability) are possible for many of the variables in the calculation algorithms. However, the range of analytical tools is often limited, even with advanced customised systems. Greater variation is found in outputs, with multiple options available and data displayed or exported as reports, tables, histograms, torpedo diagrams, and spider charts.

A user caveat applies to all CPMSs. While they are designed to be user-friendly (hence their popularity), they do not come with automatic interpretation of the meaning of outputs. Users must have a reasonable understanding of the theory and principles underlying the analyses, modelling routines, and outputs delivered by the system. Many actually over-deliver information in some respects: they offer more features than most users actually require or can probably comprehend. Lack of system transparency can aggravate this issue.

Generally, CPMSs are aimed at the design and procurement/delivery phases of projects. The inception phase has to be well advanced before they are capable of yielding meaningful project management benefits, but project risk management has to start from the very outset of a project. At the other end of the project time continuum, integrating the operational and disposal phases of a project into a single CPMS may be difficult and impractical. Separate systems might be required or desirable for this.

It is also important to remember that the primary application purpose for these systems is general project management. While more advanced systems are currently available which integrate other project-related systems, such as computer-aided design (CAD), energy efficiency and environmental management, and building information modelling (BIM), most CPMSs are not designed or intended to act as specific PRRs. Customisation is possible to achieve this, but is likely to be expensive. For Case Study A, the construction contractor employed a customised combined PRMS. In Case Study B (a rail improvement project), participants in each of the two different packages (design and construct [D & C] or alliance) were expected to use systems with an acceptable level of reporting interoperability with the rail crossing removal authority's system.

None of the CPMSs that we know of incorporate automatic recognition and management of project risks (i.e. they are not based upon high-level AI processes). On the other hand, it must also be acknowledged that a variety of user-training opportunities are normally available for CPMSs. For the more basic system applications, courses can be found at local technical colleges and universities. For advanced customised systems, the provider will initially offer free specialised training for a purchaser's nominated staff, and then subsequently on a fee-for-service basis.

18.2.4   Bespoke Risk Knowledge Management Systems (RKMSs)

Completely bespoke project RKMSs address many of the issues and shortcomings of other applications, albeit at high cost to the commissioning organisation.

The principles and processes of knowledge management (KM) were discussed more fully in Chapter 11, together with suggestions for KM system design architecture. Besides having the capacity to incorporate the features of a CPMS (as described in

Section 18.2.3), a bespoke RKMS should be able to alert users to particular risks, provide and pre-populate template PRRs for new projects with information from the ORR, and automatically update PRRs and the ORR upon user entry of new data. A bespoke system should easily accommodate quantitative risk modelling and the automatic calculation of risk severity ratings as described in this chapter for spreadsheet applications, as well as other automated functions.

A strategic issue with RKMS that must be resolved at senior management level is the extent of RKMS integration with other knowledge management requirements for the organisation. Obviously, the higher the level of integration sought, the higher the capital, training, and maintenance costs the organisation will incur.

Beyond these computerised applications for project and project risk management, other information technologies and tools can make a useful contribution to managing project risks.

18.3 Other Information Technologies and Tools

Our discussion of other information technologies and tools for project risk management must necessarily be limited to those for which we have at least some experience. It is tempting to be more speculative, but that would probably take us beyond the 'comfort zone' of our personal risk profiles!

We can broadly categorise other information technologies and tools as simulations, smart sensors, and aerial drone systems, although some are found in combination.

18.3.1 Simulation Systems

Keeping abreast of developments in project simulation systems is difficult.

Virtual reality applications have progressed from keyboard-entered static scene progressions, through jerky movement renderings of process operations, to fluid representations of complete activities. Virtual constructed environments (VCEs), for example, can portray materials being delivered to a site, unloaded, part-assembled at the ground level, and then being craned into their required positions and installed. Clever algorithms programmed into the application will alert screen viewers to vehicle access limitations, storage area demarcations, safety zones around crane jib arcs, obstructions to crane operations, and many other practical construction work issues. The VCE renderings can also be rotated for three-dimensional (3D) effects or manipulated to create internal 'tours' in a similar manner to modern CAD systems. Increasingly, AI-based interpretation features are being developed for VCEs, using pattern recognition routines to deliver alerts to users. Harnessed to user-wearable 3D viewing devices, these simulation systems can deliver powerful visual (and often audio) stimuli for project risk identification, particularly for safety risks.

18.3.2 Smart Sensors

Smart sensor technologies are developing at a rate matching that of simulation systems. These sensors measure a particular aspect of their related environment and either log data for passive retrieval or transmit it to convenient computerised receiver systems.

Remote wireless sensors can be located in situations where human access is difficult or impossible. For example, devices can be installed to monitor movements in temporary engineering support structures (e.g. shores, sheet piling, or cofferdams) and alert when safe movement limits are exceeded. Radio frequency identification devices (RFIDs) can be attached to components, and their identity and position checked by a nearby radio monitor. An example of this is the use of RFID to check the component type and placement of steel bar reinforcement in reinforced concrete structures, whereby the reinforcement can be checked immediately after installation and again after concrete has been placed and cured.

Wearable smart sensors can serve a variety of purposes but are especially aimed at safety risk management. Wristwatch sensors are now available, for example, for workers operating alone or in confined spaces. These AI-based devices measure several health-related factors such as body temperature, blood pressure, and pulse rate, as well as external thermal conditions. The devices have Bluetooth connectivity to smartphones capable of alerting the nearest emergency services.

In a similar manner, electronic screening 'gates' (similar to those installed for airport security) can detect if the correct personal protective equipment (PPE) is being worn by workers and site visitors, and alert for any missing, incorrect, or incorrectly fitted equipment.

18.3.3   Aerial Drones

As with other IT developments, aerial drone systems seem to have no limits to their application potential, other than deliberate restrictions imposed upon their use in terms of the environments and circumstances in which they can be operated. Besides simple observation, they can now accommodate pattern recognition, including colours, tags, markers, and shapes, through the incorporation of AI-based algorithms in the operating and recording system software. Still in its infancy in many ways, aerial drone technology is set to make a significant contribution to risk management for physical outdoor projects, and even for those involving large indoor spaces.

18.4   Summary

Three requirements for computerised PRMSs are: appropriate information input capacity, sufficient information processing capability, and meaningful output deliverables.

Tabular or matrix system formats deal more or less adequately with the first requirement, and poorly or not at all with the second; and users have to find their own meaning in any outputs. Commercially available systems of this type offer limited value.

Spreadsheet systems cope well with the first and second requirements, but require careful programming to deliver the third. Ubiquitous in nature, they can be customised by enthusiastic users. Commercially customised versions serve their purpose, but tend to be 'black box' systems for users. Importantly, they may 'lock' users into a particular level of project risk management maturity, and may be difficult and expensive to re-engineer as the nature and extent of application of risk management change in an organisation.

CPMSs adapted to deal with project risks are just that – adaptations. While they are usually quantitatively powerful, their lack of transparency and limited capacity for subjective qualitative risk analysis are weaknesses, even in expensive customised versions. In our view, the increasing multisystem integrations found in these applications are unlikely to deliver comprehensive and acceptable project risk management capability in the short term, and possibly not in the medium term, unless more attention is given to the psychosocial characteristics of risk and less to mechanistic treatments.

Bespoke RKMSs can be designed to address most of the shortcomings of other computer applications, but may lie beyond the financial capacity of smaller organisations. Strategic thinking about the organisation's knowledge requirements is necessary, and for staff training and capability. High initial and system maintenance costs will be incurred.

We have looked at only three other types of information technologies that are finding a place in project risk management. Simulations, smart sensors, and aerial drone technologies have probably barely even completed the first lap in their development cycle. Their future potential is huge, and we look forward to the exciting technological discoveries yet to be made.

19

Communicating Risk

19.1   Introduction

Research into the factors that are critical for project success (however that is defined) invariably ranks communication highest in importance.

Effective communication is an essential ingredient in project risk management, since risks can only be managed if they are known and understood, and such knowledge and understanding can only be gained if the participants in the project risk management process are able to find shared meaning about the nature and extent of each of the risks that they must deal with.

Risk communication must be effective intra-organisationally (i.e. within the project stakeholder organisation), and communication effectiveness should also be sought inter-organisationally (i.e. between project stakeholder organisations). Before this, however, there should be intrapersonal congruency in risk communication: that what is perceived or experienced as risk is translated intellectually, emotionally, and behaviourally into responses capable of dealing with it. Beyond the inter-organisational perspective lies extra-organisational communication, whereby information may need to be passed on to the general public.

When we talk about risk communication, we may refer to anything in the whole gamut of information, knowledge, opinions, advice, suggestions, and warnings about risks in general and project risks in particular that may have to be shared with relevant people within an organisation, between organisations, and/or between organisations and authorities, or distributed to the public at large. Put simply, risk communication comprises messages about risk that are aimed at targeted audiences.

Project risk communication is grounded within the general theory and principles of communication and follows much of that larger field's precepts and models. It provides the starting topic for this chapter, where we also consider components in the process. After that, we consider aspects of communication in the project risk management cycle, communicating risk beyond the project organisation, and evaluating project risk communication.

Managing Project Risks, First Edition. Peter J Edwards, Paulo Vaz Serra and Michael Edwards.
© 2020 John Wiley & Sons Ltd. Published 2020 by John Wiley & Sons Ltd.

19.2 Communication Theory and Models

Models of human communication abound, and we do not attempt to deal with all of them. Instead, we propose to 'cherry-pick' from this array and offer a hybrid communication multi-model that we think better serves the purposes of communicating risk.

The aim of human communication is to achieve shared meaning of a message between the parties involved. Underlying this simple purpose, however, is a wealth of theoretical frameworks and processes. Figure 19.1 presents a proposed model for risk communication that draws upon selected theories and processes.

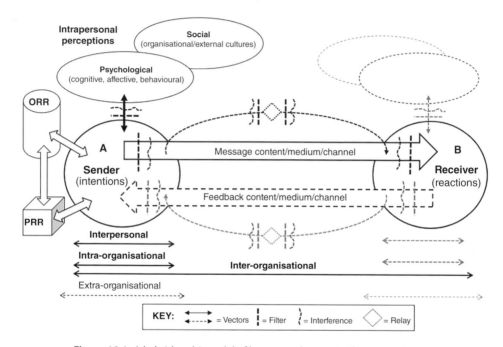

Figure 19.1 A hybrid multi-model of human and project risk communication.

This is a 'busy' and complex model and we need to unpack it if, as authors, we aim to share its meaning more fully with you, our readers. Discussion focuses upon underlying theories for the model and the communication processes involved.

19.2.1 Communication Theories in the Model

In this section, we focus briefly on five theories of communication reflected in our model: mechanistic approaches, psychological theory, technical theory, sociological perspectives, and a transactional view.

A mechanistic theory of communication relates to vectors in terms of the intended direction of messages, to channels as the means of conveying them, to filters as devices for encoding and decoding messages, and to senders and receivers as the start and end points in a linear process. This theory also considers forces and stressors as factors that can cause interference and strain in the communication process. Furthermore, space and time are important to the mechanistic viewpoint, implying movement and a dynamic quality for communication.

A psychological theory of communication offers cognitive, affective, and behavioural dimensions to the model by considering the role of human perceptions and emotions. Intellectual and emotional intelligence come into play. Organisational psychology adds an important dimension for project stakeholders.

A technological approach to communication embraces system delivery, transmission, and reception. Technologies also deal with message amplification (volume or intensity) and attenuation (quietness or dispersion).

Sociological perspectives of communication relate to the acceptability of the messages and the delivery media. From the intrapersonal psychological dimension at the start of all communication, sociological considerations determine the parties subsequently involved. As we noted in the introduction to this chapter, for risk communication these may extend from interpersonal to intra-organisational vectors, and from there to inter-organisational destinations. Beyond this may lie extra-organisational audience targets.

A transactional view of communication looks at the value or benefit derived by message senders and receivers in their communication. It also emphasises the importance of feedback (message confirmation) in the achievement of shared meaning. This suggests that, while risk communication can be represented as a closed-loop system, it is not necessarily a seamless, continuous process but is likely to be staged in some manner. This is particularly so for risk management communication, since the project risk management cycle is staged (Chapter 4, Figure 4.2).

19.2.2 Other Theory Elements of the Model

Other theory elements of this communication model relate to perspectives of project risk management that have been discussed in previous chapters. They include:

- Risk psychology (Chapter 2)
- Parties in risk communication (Chapter 2)
- Decision making and uncertainty (Chapter 3)[a]
- Linking the stages of the project risk management system (PRMS) cycle (Chapter 4)[a]
- The internal and external contexts for projects (Chapter 5)
- Each of the stage processes in the PRMS cycle (Chapters 6–10)[a]
- The management of risk management knowledge (Chapter 11)
- Organisational and societal cultures (Chapter 12)
- Project complexity (Chapter 13)[a]
- Political risk (Chapter 14)[a]
- Opportunity risk (Chapter 15)[a]
- Strategic risk management (Chapter 16)[a]
- Planning and building a PRMS (Chapter 17)[a]
- PRMS computerisation (Chapter 18)[a].

a Implicit and not shown in Figure 19.1.

19.2.3 Processes in the Model

The communication model in Figure 19.1 commences with individually separate intrapersonal communication being enacted in one or more people involved in managing the risks on a project for a project stakeholder organisation ('A' in the diagram). This will involve cognitive, affective, and behavioural subprocesses relating to each individual, together with his or her views about the social and organisational cultures that relate to the project. A limited amount of intrapersonal feedback may occur to provide internal conformational validity of perceptions, emotional reactions, and decision making relating to the various project risk situations. The fluency with which this information processing occurs may be influenced by experiential characteristics in the human processor (Slovic 2010, p. 110). Essentially, the more risk management experience the person has, the more fluent will be the intrapersonal processes. However, such fluency may also be influenced by latency and recency effects. The perceived wisdom drawn from older experiences may be obscured by personal views about more recent ones.

The outcomes of this intrapersonal process then feed into the various stages of the project risk management cycle. They are shared at each stage with other participants in the cycle, on an interpersonal basis. This subprocess is portrayed in the model diagram by the solid double-ended arrow at bottom left, which denotes transmission and feedback communication vectors. A similar subprocess then directs the project-related matters to other parts of the stakeholder organisation on an intra-organisational basis (again represented by a double-ended arrow denoting dual vectors). The organisation structures depicted in Figures 3.11 and 3.12 (Chapter 3) and in Figures 17.2–17.4 (Chapter 17) indicate typical target receivers for the project risk messages. These intra-organisational message receivers would deal with the project risk messages in a similar way to the intrapersonal and interpersonal communication processes described here.

Beyond the stakeholder organisation itself, inter-organisational communication (represented by the long double-ended arrow at the bottom of the diagram) may be necessary with other project stakeholders, and thus the subprocess routines would be replicated in the receiver organisation 'B' (shown with grey dashed ovals and vectors and with less detail). Message feedback/confirmation from 'B' would be expected by 'A'. The broad single-ended arrows in the middle of Figure 19.1 indicate this, and we have also added the possibility of relays occurring (i.e. the process may not be direct but may be relayed through another party or communication device).

To avoid overcomplication of the model, we have not included other project stakeholders, nor have we attempted to represent communication links from 'A' or 'B' to other projects being undertaken by those stakeholders. However, this structural complexity must always be considered when dealing with risk communication, in addition to that arising within the communication process itself.

In our proposed hybrid multi-model of communication in project risk management, components can be found that are common to all the theoretical models.

19.3 Components in the Communication Process

Figure 19.1 shows some of the components in the communication process. Others are not shown simply because their inclusion, as with the complexity noted in Section 19.2.3, would render the diagram largely incomprehensible.

The components of communication to be considered here include: senders, receivers, messages, media, channels, filters, interference, and feedback.

19.3.1   Senders

Senders initiate messages in the project risk communication process. They may choose between different options in the process or be guided or bound by the policies and requirements of their host organisations. Primary senders actually become secondary receivers when they receive feedback from their original message targets, but for our purposes we will ignore such distinctions.

19.3.2   Receivers

Receivers are the intended (or sometimes unintended) audiences for messages. Intended audiences are targeted. They may be individuals, selected groups, or even mass audiences. Unintended audiences for messages may be unharmed by the experience, or they can become 'collateral damage' in a media 'war'.

Message receivers may be positively receptive or negatively unreceptive. Less interference is found with the former.

19.3.3   Messages

Messages express the sender's intentions. These may be personal or impersonal, direct or indirect, physical, verbal, visual, aural, textual, graphical, numerical, symbolic, plain or enciphered, whole or fragmented, abbreviated or extensive, explicit or implicit, and human or system-generated. They may also be found in facial expressions and in body language and can be sensory or insensible. Messages may be amplified (made 'louder' in terms of aural volume or emphasis) or attenuated (softened, quietened, and dissipated). Contemporary 'social networking' has brought many of these message characteristics into play and has invented many more (e.g. emojis). The range of message characteristics is vast, and we can only cherry-pick a few relevant examples from it.

Weather advisory messages are intended to inform target audiences about potentially adverse weather conditions likely to arise in the short-term future. The target audience is usually restricted to a local population (whether on land, at sea, or in the air), but may have wider regional or national relevance. The messages may be text-based, visual, or graphical, depending upon the transmission and receiving systems. Message content may comprise system-generated data (by remote sensors) or human interpretations of sensor data and data modelling outputs. Weather messages may be received aurally from other people ('Weather's not looking so good right now, is it?') or from radio and TV presenters. The messages can be transmitted electronically to 'weather apps' on smartphones and similar handheld devices. They are intentionally brief and direct. On the other hand, risk messages about the climate change effects of global warming are rarely brief and are more likely to be delivered through longer interviews and essays.

In crane operations, wind sensors detect prevailing wind speeds and gusty conditions and send audible and visual warnings to operators. They may also have a control function for automatically shutting down or limiting crane jib movement capacity.

Just as geo-seismic instruments measure the epicentres and magnitude of earthquakes, remote ocean sensors now measure wave heights and motion in order to detect early signs of tsunamis and then transmit data to shore-based receivers. Strategic location of the sensors enables the data analyses to provide early-warning indications of the likely direction and magnitude of a potential tsunami.

Modern smartphones provide message alerts that may be aural (variable ringtones to suit personal preferences and to distinguish between types of messages) or physical (vibrations).

Risk-related messages are associated with every stage in the cycle of project risk management, but we have emphasised the importance and benefits of preparing precise risk identification statements (Chapter 7). If these statements are not sufficiently precise or if they are not communicated effectively interpersonally, intra-organisationally, inter-organisationally, or even extra-organisationally, then the whole process of project risk management may be compromised.

By way of example, consider the identification statement (message) 'There is a safety risk on the project'. We know what type of risk this is, but what is the actual risk; what are its circumstances, and what uncertainty is associated with it? Does it mean that project conditions are unsafe for workers? Or is there a likelihood that workers will not observe safety precautions? Are the materials unsafe for use? Is the completed project likely to be unsafe for operational use? Is it possible that, eventually, project disposal cannot be carried out safely? Note that any of these alternative explanations will entail different assessment, decision making, and subsequent control action for the remaining stages of the project risk management cycle.

19.3.4   Media

Media are the vehicles by which messages are communicated. 'Multimedia' refer to multiple methods of message transmission. Many of the message characteristics described in Section 19.3.3 are actually associated with the message medium, but contemporary language use has left them synonymous with the message rather than the medium. A 'tweet' is a typical example. We tend to accept these shifts in meaning in communication without great resistance, despite their obvious contradiction with the principles of communication, and media flexibility is now regarded as a contemporary employment skill.

In mass communication, we might construe multimedia as newspapers, radio, and TV broadcasts distributed to large audiences but actually targeted at individual receivers. However, we should note that 'spam' email also fits this description! It is also worth noting that, despite many prophets of gloom, the human voice is still probably the most popular means of human communication. Social media networks have a somewhat murky relationship with project risk management, since legislation to control them in many jurisdictions substantially lags behind their use in practice. In contrast, mass transmission of text messages alerting cell phone users to imminent dangers, such as bush fires and storm flooding, has proved to be a valuable resource in public emergencies. In some countries. this service is offered to authorities at no charge by cellular service providers.

Some communication media have largely fallen into disuse (e.g. mechanical semaphore signalling) other than for a few exceptional circumstances. Electronic facsimile transmission (fax) may be the most recent and short-lived example of such obsolescence.

Media for risk messages must be consistent and reliable.

19.3.5 Channels

Just as media are treated synonymously with messages, so are communication channels often confused with media. For most purposes, this does not matter. It arises because, for some communication theorists, messages have connotations relating to content, direction, medium, and transmission. An email message has text content intended for a select audience, using electronic transmission over the internet.

Communication channels relate to the method (i.e. the technology) of message delivery. To some extent, this may also include the status of the sending and receiving mechanisms. We sometimes refer to 'dedicated' channels, as transmission mechanisms set aside for specific types of messages between specific senders and specific receivers.

Channels may be human (e.g. brain synapses, voice, hand delivery, or hand signals), physical (e.g. copper wire), or electronic (e.g. wireless). They may be passive (e.g. billboards and posters) or dynamically active (e.g. videos). All channels are vulnerable to interference, including brain synapses through chemical effects in the blood system or through mental disease.

19.3.6 Relays

Communication relays come into play as intended or unintended, known or unknown, diversions from direct sender–receiver communication channel vectors. They may constitute communication channels that exist alongside other channels.

Email communication provides a good example of intended but largely unknown message relays. Messages are typed in text onto a screen by means of an appropriate software application. They are then digitally transformed by mail servers in the system and authenticated in terms of the validity of sender and receiver addresses. The message is then uploaded (or relayed to a local 'backbone' system and then uploaded) to a dedicated communication satellite located in space and orbiting the earth. The satellite system seeks the fastest and most efficient transmission link available, and, after several rerouting links, the message is downloaded to the most convenient backbone or incoming mail server and directed through the internal mail system to the receiver. We have deliberately ignored additional procedures such as message encryption and decryption (communication filters) and file attachments (additional messages), but, importantly here, the digitised message is also broken up into fragments ('packets'), and each packet is relayed wirelessly and electronically via the communication satellites. Every packet may thus pass through different transmission routes until it is eventually reunited with its fellow arrivals for delivery as a complete and coherent message to the recipient.

Other than this example, communication relays can result in some loss of message content or quality. The family game of 'Chinese whispers' is a good example. Depending upon the channels used, relays should be carefully monitored in risk communication.

19.3.7 Filters

Communication filters modify messages or their intended meaning in some way. Intrapersonal filters may be influenced by knowledge and experience, and by physical, mental, and emotional states that may give rise to message content selectivity, in both sender and receiver modes. We might 'tune out' or modify unwanted or undesirable message content. This is a form of 'gatekeeping' that is also found in interpersonal and other levels of communication. At these wider levels, organisational cultures, policies, procedures, and even politics act as communication filters.

Communication filters may also comprise message encoding and decoding processes. These may be intuitive or fully rationalised and defined. Message encryption and decryption is a similar filtering process.

19.3.8   Interference

Although interference was originally associated with the mechanistic theory of communication, it arises (in various guises) in all other theory perspectives and constitutes a barrier to effective communication. 'Noise' is another term for this effect in communication.

Intrapersonal message interference is most often attributable to personal bias. At the interpersonal and intra-organisational levels, it may be similar in nature and origin to the barriers to effective decision making described in Chapter 4.

Inter-organisational message interference can arise from conflicting organisational cultures and the excessive intrusion of competition in the communication process. Beyond this, extra-organisational 'noise' probably emanates most often from the effect of politics upon the message or communication channel. Community expectations and attitudes also play a part. Lack of congruency in values also causes interference in risk communication.

19.3.9   Feedback

Feedback may be the most neglected component of communication. It is the confirmation that the sent message is shared and understood by the receiver. Mutual meaning has been achieved. That is not the same as saying that the receiver agrees with the sender. Intrapersonal communication feedback is not immune from neglect; we can choose to ignore what our brains and emotions are telling us.

Feedback confirmation does not necessarily use the same media or channels as the original message. An email message receiver may opt to give feedback verbally in person, by telephone, by text message, or by letter, or may relay it through a third party.

The importance of feedback should not be underrated in project risk management. Relying on 'sign-off' procedures may not be sufficient if the possibility exists that a risk (and any proposed treatment) is not fully and mutually understood by the parties involved. That possibility constitutes a communication threat risk for a project. It is why contemporary Safety Hazard Analysis (SHA) records must include evidence of consultation with workers and not rely solely on 'sign-offs'.

19.4   Communicating Risk in the PRMS Cycle

Communication messages are required in every stage of the PRMS cycle for a project. They have been demonstrated throughout this book. The Australian/New Zealand standard for managing risks in projects (AS/NZS IEC 62198; Standards Australia/

Standards New Zealand 2015) makes specific reference to internal project communication and reporting mechanisms and offers additional guidance for communicating with other project stakeholders.

In the context establishment stage, formal recognition of the contextual drivers is needed, particularly in terms of how they 'shape' project risks for the stakeholder organisation.

In risk identification, messages deal with not only precise risk statements but also their classification and possible reallocation.

During risk analysis, communication is often necessary for the methods of analysis proposed, the data inputs, and the modelling outcomes and their effect upon project decision making.

Risk evaluation will entail communication about the relative severity of threat risks and the exploitability of opportunity risks.

Extensive communication is likely to be needed in deciding upon response options and treatment alternatives for each risk, and for the monitoring and control activities needed for managing them.

The processes of risk knowledge capture inevitably entail considerable communication, both formal and informal, and this was discussed in Chapter 11 (Project Risk Knowledge Management).

Given the qualified definition we have adopted for risk (based upon ISO 31000 [ISO 2018], and see Chapter 2), communicating uncertainty is particularly important in risk management. Uncertainty content in project risk messages should be explicit and not left to implied recognition.

It is important here to acknowledge that communication, especially in projects, may occur at two levels: formal and informal. Case Study C (an aid-funded project) demonstrates this. Figure C.1 shows the senior advisor for the project in an advisory communication role for the donor and recipient governments. These communication channels and vectors may be regarded as formal – the advisor is appointed for that role, to give and receive information and advice to and from either government. Those communications will be 'coded' and 'filtered' by the advisor in terms of clarifying their meaning and augmenting them with the advisor's professional judgement. In reality, of course, the advisor maintains informal communications with all the project stakeholders with a view to facilitating their decision making. His relationship with the two governments is formal, but by virtue of his appointment (and for all practical purposes), his relationships with other stakeholders are more informal (but backed by his formal status). In this way, he is able to interact with the project management and design team, the main contractor, and most of the subcontractors. Much of this interaction is risk-based, since it largely relates to uncertainties associated with decision making.

The communication in Case Study C is within the remit of the project, but for some project circumstances it will be necessary to engage in extra-organisational risk communication beyond the immediate project stakeholders. The rail crossing removal authority in Case Study B (a rail improvement project) has conducted invitation seminars and public information sessions with a wide range of audiences (e.g. universities, professional associations, and local communities) on several topics including safety risks, technical issues, and station upgrading proposals.

19.5 Communicating Project Risk Beyond the Project Stakeholder Organisations

Extra-organisational risk communication in projects tends to arise in one of three circumstances: as promotional announcements; in adverse, challenging environments; or in extensive advisory inclusion loops.

19.5.1 Promotional Announcements

Promotional announcements about projects tend to deal with opportunity risks, although they are rarely explicit about this. More often than not, they are marketing messages. For example, a developer might announce that a further 500 residential building sites in a development project have been made available for public purchase, thus inviting people to exploit the opportunity. Potential threat risks in terms of poor location or inadequate public services either are not mentioned in the messages, or they are minimised or camouflaged in some way. These announcements are not considered further here, since *caveat emptor* ('let the buyer beware') should be the message receivers' response.

19.5.2 Communicating Risk in Adverse or Challenging Environments

In more negative project risk environments, extra-organisational communication frequently targets mass audiences, and this will influence all aspects of such communication.

The perceived concerns and opinions of people, particularly in terms of their informed understanding of a situation, have to be carefully addressed. Rumours and misinformation can be intractable to manage effectively, and contemporary social networks (now the archetype of mass communication) do not make it easier to do so. In fact, they operate so rapidly that message control over communication is almost impossible. Nor does it help that the circumstances are often controversial and can lead to high levels of concern and stress.

In such situations, risk messages should state the context simply and concisely using plain language(s) wherever possible. Euphemisms, hyperbole, and grotesque forms of language must be avoided. Any sector of the mass audience that is particularly at risk, such as a small community, should be identified and, where this is practical, addressed as a separate audience.

Public participation in the communication process should be encouraged, with as much transparency as possible accorded to any risk decision making processes. Fairness is also an important ingredient. The audience must be able to trust the message source.

Inadequate or ineffective communication of project risk messages beyond the stakeholder organisation can lead to a phenomenon known as the 'social amplification' of risk, whereby Kasperson *et al.* (1988) observed that quite minor threat risks attract strong and concerned reactions from public audiences. The risks are socially amplified. Conversely, severe risks may elicit little or only mild attention; they are attenuated (diluted). The risk amplification or attenuation is influenced by the meanings (accidental or deliberate) contained in the messages sent and received about those risks. While message content, language, and tone are the main influences, message timing, media, and delivery channels also have an effect.

The researchers have theorised that the social amplification phenomenon can be explained by interactions between the various theoretical constructs of communication that we dealt with earlier in this chapter. While their framing was in terms of a public audience, we can regard this as a cultural 'shaping' of risk (Chapter 12) that may also be encountered at the project level.

The distortions of the social amplification of risk can affect the management of risk at every stage in the risk management cycle. Excessive amplification ('turning up the volume') can lead to overspending or misallocation of management resources. On the other hand, unwarranted attenuation ('audience softening') of risk messages may give rise to poor understanding of the situation and lack of responsive attention to it.

Stakeholder organisations must not only manage their project risks effectively but also *be seen to be* managing them. This gives rise to another extra-organisational risk communication concept: the 'social licence' for risk, through which communities set unwritten public limits on the levels of risks beyond which they are not prepared to grant risk leeway (community tolerance) to an organisation.

Both of these phenomena are strongly associated with risk communication filters and interference. Good project risk management must be prepared to deal with them.

Risk communication in crisis or disaster situations raises the level of required and expected communication standards in every aspect, mainly by increasing the pressure on those involved in the process. Often, there is little by way of previous project experience to guide the communicators.

In such cases, the content and delivery of the risk messages must clearly distinguish between the purposes of explanation, persuasion, and prescription.

19.5.3   Communication in Extensive Advisory Loops

Probably we have all been victims of the 'shotgun' approach to the dissemination of email messages. We are included in an advisory loop, even though we may have little or no involvement in the matter being dealt with. The sender wants to make sure, perhaps as a perceived threat risk response, that anyone who could be even remotely involved should be kept informed by including them in the loop. Extracting oneself from such loops (i.e. distribution lists) can be difficult and time-consuming. Many of us just resort to deleting email with certain subject headers, without reading the messages contained in them.

To some extent, this phenomenon is demonstrated in the organogram for the hospital refurbishment project described in Figure 3.7 of Chapter 3. The large array of organisations involved, with their disparate power relationships and communication demands, means that some of them will not consider themselves as direct stakeholders in the project, but any communication that does take place with them will inevitably include matters relating to project risks. The problem with this is that the message receivers in this instance may have insufficient capability to understand the full meaning and implications of the intended risk information or its relevance for them. If the same receiver is on the remote fringes of many such projects, information overload may occur and give rise to communication threat risks. Alternatively, a receiver could accept the message as an opportunity to exert additional influence (often inappropriately) over project decision making.

The phenomenon of extensive advisory loops in project risk communication can and should be addressed in the early stages of any project. Creating an accurate and comprehensive organisational structure diagram for a project will help if the place and role of peripheral organisations are clearly indicated and defined. The 'people finder' tool described in Chapter 11 is useful in this regard.

19.6 Evaluating Risk Communication

Although the evaluation of risk communication in project management relies heavily on the availability of confirmatory feedback from message receivers, senders can conduct independent audits.

Communication effectiveness evaluation criteria may include questions such as:

- Were the messages succinct yet sufficient?
- Were receivers properly identified?
- Were the message media appropriate and sufficient?
- What communication channels were used; were they appropriate; were they sufficient?
- What communication filters and relays were in operation; were they appropriate and sufficient?
- What interference could be detected at any point in the communication process; what was its effect on the risk messages, and how was it dealt with?
- Were feedback confirmation expectations made sufficiently explicit; did the feedback show evidence of shared meaning for the risk messages?
- Has social amplification shaped the communication of project risks in any way; if so, how was it dealt with?
- Are there any 'social licence' implications involved?

Thorough evaluation should seek to apply these criteria to every stage and every level of the project risk management process.

19.7 Summary

Just as projects are cloaked in risk, so are project risks enveloped in communication. To add a further metaphor, communication is the lubricant for project risk management. Essentially, risk communication seeks to achieve shared understanding and meaning in the risk messages that are sent and received.

Communication is one of the most widely published topics in contemporary business management and is ranked first in many indicators of business and project success. It is theoretically underpinned from several perspectives, and in project risk management it is necessarily practised at different levels.

The perspectives may be defined as mechanistic, psychological, technological, sociological, and transactional. The levels of risk communication may be intrapersonal, interpersonal, intra-organisational, inter-organisational, and extra-organisational. All are likely to be encountered in a single project, regardless of its type, scope, cost, or value.

The hybrid multi-model of communication in project risk management (Figure 19.1) provides a prism through which the theoretical perspectives and levels can be framed and understood, and then fitted into the cycle of project risk management.

The components in project risk communication are not unique. They comprise senders, receivers, messages, media, channels, filters, relays, interference, and feedback.

Precision and concision are of paramount importance in risk communication, in our view particularly with respect to risk identification statements. If a risk you have identified is not properly described, the potential for 'noise' to enter the communication process and distort the meaning of the risk for others is greatly increased. Feedback confirmation is a vital, but often neglected, precaution against this.

Risk communication beyond the project system boundary to external audiences is difficult to undertake effectively. More often than not, it has to take place in circumstances that are pressured and challenging. Transparency, fairness, and participation should be the watchwords for messages in such situations.

Prudent project risk management will seek to audit and evaluate risk communication effectiveness at all stages and levels in the risk management cycle.

In the following and final chapter of this book (Chapter 20), we offer some conclusions about managing project risks.

References

International Organisation for Standardisation (ISO) (2018). *Risk Management – Principles and Guidelines*. ISO 31000. Geneva: International Organisation for Standardisation.

Kasperson, R.E., Renn, O., Slovic, P. *et al.* (1988). The social amplification of risk: A conceptual framework. *Risk Analysis* 8 (2): 177–187.

Slovic, P. (2010). *The Feeling of Risk: New Perspectives on Risk Perception*. London: Earthscan Ltd.

Standards Australia/Standards New Zealand (2015). *Managing Risk in Projects – Application Guidelines*. AS/NZS IEC 62198. Sydney, NSW 2001: SAI Global Ltd.

20

Conclusions

20.1   Introduction

In concluding this book, we hope first of all that you have enjoyed reading it. It is not a blockbuster novel with a huge page-turning attraction leading to a thrilling completion, so we do not mind if you have read selectively. On any second reading, however, please include the chapters that you missed!

Our aim has been to provide an introduction to project risk management that would guide and assist people tasked with dealing with those risks. Our objectives were to:

1) Communicate a conceptual and philosophical understanding of risk.
2) Consider the nature of projects and the stakeholders involved in them.
3) Present a systematic and logically progressive approach to the processes of project risk management.
4) Explore the drivers of risk and the factors which shape them.
5) Emphasise the importance of capturing and exploiting project risk knowledge.
6) Provide guidance about designing, building, and implementing project risk management systems (PRMSs).

We think we have achieved our aim and fulfilled our objectives, and hope that we have actually offered more.

In this final chapter, we offer some thoughts about the contemporary state of project risk management and attempt to look forwards to its future. We also suggest some 'take-away' checklist synopses for the major topics which may help you to measure your reading satisfaction.

Throughout this book, we have consistently referred back to statements, tables, and diagrams included in earlier chapters. Some readers may have found the reminders irritating, but repetition is learning reinforcement, and each time we have attempted to relate the item to new aspects of project risk management. We certainly offer no apology for the repeated references to Figure 4.2 (Chapter 4). It works! In addition to representing a staged cycle for project risk management, this diagram demonstrates an experiential learning loop and can be used to identify the areas in an organisation where it is important to strengthen practice by developing a good risk management culture. It also provides a basis for performance review and benchmarking for each part of the PRMS.

Managing Project Risks, First Edition. Peter J Edwards, Paulo Vaz Serra and Michael Edwards.
© 2020 John Wiley & Sons Ltd. Published 2020 by John Wiley & Sons Ltd.

For several reasons, Chapter 11 (Project Risk Knowledge Management) has been the most difficult to write. The chapter is really a fulcrum or hinge for the whole book, but the rate of development in knowledge management systems almost outstrips the capacity to incorporate the developments into system explanations. After a brief knowledge theory presentation, the chapter shifts to risk-based knowledge management systems and suggests further development possibilities (i.e. mechanisms that are not yet evident in the decade of publication for this book but which can reasonably be expected to appear in the short-term future). This chapter also summarises the importance of knowledge gained through the practical processes of project risk management described in the preceding chapters and then points forwards to the chapters that show where that knowledge can be applied and augmented. An organisation should not just want to identify the risks it faces on projects, but ideally should also want to know enough about those risks so as to develop organisational wisdom about managing them on future projects. The processes of capturing and using risk knowledge are therefore critical for developing a fuller understanding about risk and risk management, for better diffusing risk awareness and culture throughout the organisation, and for developing greater wisdom in decision making and greater ability and capacity to manage project risks. The three models offered in this chapter (Figures 11.1–11.3) should help readers to more fully understand the importance of the knowledge transformation sequence, the knowledge creation cycle, and the interactive process of using knowledge in the project risk management cycle.

Chapter 19 (Computer Applications) could have presented similar but even greater difficulties. Here, however, to avoid troublesome references to proprietary computer applications (and all the necessary permissions processes that would have involved), we preferred to present what we think are some durable principles for system selection, and to point out their merits and demerits. The short shelf life of many computer applications also influenced our decisions about chapter content.

Project risk management is not just for mega- and major projects. Without wishing to place definitions on any of these labels, it also delivers value for small and mini- or even micro-projects. Applying it systematically improves the capacity in an organisation to manage its projects effectively. Experience and skill in using a PRMS can offer potential career shifts for people.

Some of these benefits may become clearer as we consider the current state of project risk management and explore its future development.

20.2   Current State of Project Risk Management

Asking anyone to discuss the current state of project risk management is rather like asking about the weather. Everywhere is slightly different, and globally it probably displays the full gamut of possible conditions. Pursuing a contiguous climate change analogy, however, subtle changes may also have occurred over time.

From an overall perspective, risk management has developed from an ad hoc or intuitive approach to dealing with project risks to one that is generally much more systematic. Less than 50 years ago, almost all risks were managed reactively: an event occurred, and we would then try to cope with the consequences for our project. Now (at least for many projects), the approach is more proactive and systematic, although it has to be said that many of the old ad hoc and reactive practices remain.

This situation is not confined to particular project types or to specific industries. Even construction, which came late to the party, now has a relatively strong grasp of what is required to manage project risks more effectively. Any reluctance to change has almost certainly been due to hardened attitudes and preferences, whether corporately or individually held.

The shift towards more systematic project risk management can be ascribed to five sources: changes in business conditions, more serious impacts of risk events, public expectations and regulations, publication of dedicated standards and texts, and curriculum changes in tertiary institutions. Issues still remain, however.

20.2.1   Changes in Business Conditions

The continuing growth in privatisation, as a policy of governments, has led to more competitive business conditions whereby protection of the 'bottom line' has become a major priority. The global financial crisis of 2008 has also contributed to this by exposing poor risk management practices in the corporate world paralleled by similar inefficiencies in public sector risk management and governance. Improvements in threat risk management have occurred almost exponentially in the decade since then.

20.2.2   More Serious Risk Impacts and Consequences

A trend towards bigger and bigger projects has implications in terms of the greater consequences of any failure in them. Mega-project failures may incur costs associated with factors such as unrealistic demand forecasts, project stakeholder insolvencies, environmental clean-ups, and an ever-increasing recourse to litigation. The 'too big to fail' syndrome has exerted its own cautionary effect.

While the monetary amounts involved may be far higher for larger projects, their cumulative influence on many smaller projects is not insignificant.

20.2.3   Public Expectations and Regulations

Partly driven by the more serious consequences for projects noted so far in this section, public expectations of good risk management have increased markedly over the past three decades. This has resulted in more focused public policy development in this area and more extensive regulatory regimes. Most public agencies now have prescriptive frameworks and procedures in place for the management of their perceived risks. Occupational health and safety has been at the forefront of this shift but, in the private and commercial world, due diligence, ethical compliance, and corporate social responsibility expectations have also played their part. Environmental risks have become more prominent through public concerns about global warming and climate change effects.

20.2.4   Publication of Standards and Texts

Over the past 20 years, there has been a veritable avalanche in terms of official and semi-official publications about risk and risk management. Each has tended to focus upon particular aspects associated with its interest domain. Few have attempted more comprehensive treatment, although national and international standards for risk management now proliferate. While these publications present necessary and admirable frameworks, none bring the 'skeleton' to life with flesh, sinews, muscles, nerves, and brain.

Increasing attention has been paid to the psychological cognitive, affective, and behavioural aspects of risk and the social contexts associated with them. This has informed a less mechanistic understanding of risk, moving it beyond a preoccupation with numerical expressions of uncertainty.

20.2.5   Tertiary Curriculum Changes

More and more tertiary curricula now incorporate courses that deal with risk and risk management. The shift has also been from postgraduate to undergraduate degrees as a response by tertiary institutions to perceived market demands. The same demand has widened the course delivery from narrow mathematically based engineering and similar cognate disciplines to management and sociology degree programmes.

A danger here, however, is that the additional pressure on curriculum timetabling, through the inclusion of more and more content, tends to result in less time being devoted for each topic – even those traditionally regarded as 'core' subjects. Ideally, risk and risk management learning should be delivered as a formal stand-alone basic course, and the range of applications then dealt with separately in each relevant discipline subject. In our view, it is also desirable to separate risk understanding and management from purely quantitative mathematical risk modelling. Since it is not essential for every area of risk management application, quantitative risk analysis could be offered as an elective subject.

20.2.6   Continuing Issues with Contemporary PRMSs

Despite the growth observed in the application of systematic project risk management, it is also possible to discern continuing issues with its use.

Risk awareness and appreciation of cultural contexts are patchy, as evidenced in Case Studies C, D, and E. This is not just about developing 'risk smart' organisations, but more about diffusing genuine risk awareness throughout them.

There is an over-reliance on 'black box' techniques, and system misuse may occur through excessive 'copy and paste' entry of risk information. Associated with this is a concern that the outputs of risk-based models are not sufficiently understood, nor are they always correctly and consistently interpreted in terms of the inherent uncertainty in the values expressed in them.

Lack of continuing commitment to the PRMS (system decay) is observable, and some stages of the project risk management cycle are ignored or bypassed. Risk knowledge capture and management are often weak, or the processes are sacrificed in the rush to engage in the next project. In some organisations, a PRMS is only applied to projects where it is stipulated as a contract requirement by clients.

Given this brief overview of contemporary project risk management practice, what lies in the future?

20.3   Future Project Risk Management

A recent statement by a global accountancy and business management company attracted our attention: '[The] future of risk management is not about precision but about leadership and management' (http://www.youtube.com/watch?v=rcMZrOdeDAo). In an otherwise useful publication about the topic, this claim unfortunately stands out as being less than accurate on several counts.

Firstly, the issue is not about a choice (i.e. between precision on the one hand and leadership and management on the other). Secondly, precision does enter the debate, since it is always appropriate to strive for it. We suspect the claim was referring to precision in the context of numerical risk analysis, and our views about this were largely presented in Chapter 8. Quantitative and qualitative approaches to risk analysis each present unique demands about precision in terms of the analytical modelling techniques, the data used to service them, the language of expression, and the attitudes of the analysts. We have consistently pointed out the need for precise risk statements and precise descriptors for subjective scale measure interval labels for risk assessments.

In the same statement, 'leadership and management' is a tautology in two respects: risk management does require *management* (why else would it be so named?), and by extension management requires leadership. It is neither an impersonal nor a mechanised process. The only danger arising here is if the development and implementation of artificial intelligence (AI)-based 'smart' PRMSs stray too far into the territory of 'black box' systems that discourage human reflection upon, and wise management of, project risks.

Realistically, the future of risk management, particularly in a project context, lies in its capacity, through the commitment and capability of the people involved, to demonstrate effectiveness efficiently and consistently. This is the important challenge.

Systematic project risk management is growing and will continue to grow, both in the extent of its application and in the development of the techniques that support it. However, the driver of growth is important.

Growth in the extent of project risk management application will be driven by the persuasive arguments of a sound business case and by the increasingly prescriptive expectations of project clients. Case Study A demonstrates the former, and Case Study B to some extent reveals the latter. Both can be regarded as effective instances of using a PRMS, but the challenge is for the first argument to prevail over the second – carrots are more appetising than sticks! The point here is that, for future risk management, organisations should embrace PRMSs willingly and proactively and not in response to prescriptive contract clauses.

Techniques development is likely to include more auto-detection of project risks by creating and using the relational databases associated with computerised project management systems, so as to generate risk checklists for particular project types, scopes, values, and contexts. Since these systems largely depend upon the disaggregation of projects into identifiable requirements, and from there into project activities (in the manner of a work breakdown structure), it would be entirely possible to develop customised computer applications to represent the whole array of project risk management requirements and system content for most of the projects that an organisation undertakes. Even untypical projects would be largely covered by the system umbrella, with anomalies flagged and dealt with separately. The relational project risk database would be populated with knowledge mined from the organisational risk register and associated historic project risk registers.

Virtual reality simulation techniques, harnessed to AI algorithms, will generate situational risks in 3D renderings of physical project processes based upon pattern recognition, marker flagging, and other detection devices.

We think that, using a decade as a long-term horizon, the development period for most of these techniques from design to widespread practice will be in the short to medium term rather than in the long term.

In the same way, exploration and modelling of less tangible risks will become more sophisticated. Currently, for example, financial feasibility analysis for many projects uses discounted cash flow modelling to provide outcomes variously expressed as net present value, annual equivalent value, and yield rate or internal rate of return. While the calculating algorithms in such models are automatic, many user-determined data inputs are necessary, and sensitivity testing is usually carried out in a stepwise manner for one variable at a time – the model user either accepting default options or defining the step intervals and the target variable each time. Future financial models of this nature are likely to more fully harness the power of relational databases and use AI and pattern recognition, applied through more sophisticated and much greater modelling power, to enhance modelling capacity and usefulness.

For example, financial modelling (using contemporary applications) for a high-rise commercial, office, and residential development project might indicate that the internal rate of return on the investment over a 30-year time horizon is below the hurdle rate for the developer. The model does not indicate why this is so and provides only a limited expression of the uncertainty associated with this outcome. The developer may assume that it is attributable to excessively high costs of construction (a favourite target for blame) and order a reduction in this area of capital cost by sacrificing building quality. The project may then be remodelled iteratively, adjusting various input factors each time, until the hurdle rate is passed. This is a crude but widespread approach to the financial analysis of projects. Modelling improvements such as those identified here could transform such analyses. A cleverer system would automatically assess the relative strength of the financial contribution of each of the factors in the model, compare them to historic factors patterns in the database, and deliver value outcomes for stepwise increases and decreases in various factor combinations. Currently, this remodelling is done for one factor at a time without due regard for any interactive effects. Construction costs are reduced without properly considering the implications of lower building quality over the project life cycle and its effect on future maintenance costs. Nor is the likelihood considered that rental income and tenant churn forecasts will be adversely affected by a cheaper, less prestigious, and thus less attractive building. More sophisticated and more powerful financial modelling (which is essentially a part of risk management) will embrace such issues in a comprehensive yet still transparent way.

However, all these advances in project risk management must come at a cost that is not only financial (the enhanced applications will be expensive to install, operate, and maintain) but also psychosocial. They will expose organisations to the threat risk of becoming more 'detached' from the project risks they are seeking to manage. Much care will be needed to ensure that organisational risk awareness is not dissipated or diluted by the prospects of greater sophistication and automation in future systems of project risk management. Besides being better assessed, project risks must still be mutually understood and properly communicated across the organisation and beyond.

We hope to see a far greater deliberate, systematic, and proactive search for, and exploitation of, opportunity risks on projects in future.

Whether or not you agree with our prognostications will, to a large extent, depend upon the value you have extracted from the contents of this book.

20.4 Checking Your Reading Satisfaction

Sales are a publisher's raw measure of reader satisfaction for a book, but this is only partially true for authors. For us, your intellectual satisfaction is important and, for a book such as this, the value it delivers for your practice of project risk management.

As a framework for your individual assessments, we conclude with a list of possible 'take-aways' for consideration. These are arranged according to the chapter topic sequence.

20.4.1 Risk

Project risk is the effect of uncertainty upon project objectives. If project objectives are not clearly framed and understood, then project risk will not be properly perceived or effectively managed.

Risk uncertainty arises out of events and their consequences. If neither generates uncertainty, there is no project risk.

The psychosocial characteristics of risk may cloud its perception, analysis, treatment, and control in projects. They colour its interpretation and the received wisdom arising from project risk experiences. Project risks are closely associated with project decision making, which in itself is a psychosocial process.

The classification of risks not only assists with their management (and the resources needed to manage them) but also may provide a risk typology that is useful beyond project, stakeholder organisation, and industry contexts. Our preference is for a generic universal classification, but this is still a distant target in contemporary risk management. A classification based upon risk source events is best, but some systems use risk consequences as the basis (e.g. 'Financial Impacts' or 'Production Capacity'). Other classifications may confuse both events and consequences, and this limits their usefulness. Customised 'in-house' risk classification systems are common but may not communicate well beyond the stakeholder organisation's project system boundary. Their 'shelf life' is easily affected by organisational change.

20.4.2 Projects

We hope we have provided some transformational thinking about projects. In many cases, they are not finite endeavours marked by an inception and an ending following immediately after delivery, but can also include operational and eventual disposal phases. Different risks may be associated with each phase, and the magnitudes of similar risks occurring in each phase may vary considerably.

Projects are purposeful endeavours. However, they do not always fulfil all the objectives set for them. Few exceed expectations. In most instances, they consume resources that are expensive and hard to replace, hence the importance of good project risk management.

Few projects proceed smoothly. Some stutter or go ahead jerkily; others stagnate for considerable periods. The 'fluidity' of projects can affect the effectiveness of project risk management. So too can the interactions with other stakeholders in projects.

20.4.3 PRMSs

PRMSs are systematic approaches to dealing with project risks, with an emphasis on proactive management.

Risk management is not about preventing risks on projects, and our distinction between threat and opportunity risk affirms this. Nor does risk management guarantee project success. Project risks can be mitigated, but, for various reasons, the project itself may still fail. Effective project risk management should either enhance the probability of project success or reduce the probability of failure. At best it will do both, but, as with the nature of risk, there is no certainty to this outcome.

We do not support the notion of a universal PRMS to manage the risks of all stakeholders participating in a project. Differences between the project objectives of each stakeholder preclude this, and each stakeholder should take separate responsibility for managing its own project risks. Those acting in an agency capacity (i.e. as a consultant) on behalf of another project stakeholder should be particularly aware of this, as a dual risk management approach is almost certainly required: helping the client organisation to manage its project risks, and applying good risk management processes to the consultant's own agency risks.

National and international standards provide useful frameworks for risk management, based upon the stages identifiable in the process. We prefer to represent this framework as a dynamic cycle of project risk management activities, roughly corresponding to the well-known experiential learning cycle. At certain points in the cycle, iterative return loops may be needed as risk conditions change or as new project risks are identified.

An early deliverable in the PRMS cycle should be a project risk register capable of acting as a document of record and as a plan for subsequent risk management action.

We strongly advocate team responsibility and workshop settings for almost the whole of the project risk management cycle. Over-reliance on individuals in this management process itself constitutes a threat risk to an organisation.

Effective and systematic project risk management raises risk awareness in an organisation.

20.4.4 Risk Contexts

Time spent in establishing the project contexts for risk is never wasted. The contextualising process frames project participants' perceptions of the nature of the project, its implications for the stakeholder, and its setting in the wider world.

Overfamiliarity with some types of projects may lead to project context establishment being bypassed in risk management. Given that every project is claimed to be unique in some way, this is a false economy.

Perceptions of external project risk contexts can be grouped under four perspective lenses: physical, economic, technical, and social. Each may involve smaller prisms for focusing on specific matters. Together, they form the drivers and shapers of project risks. Adopting a project system boundary facilitates distinction between internal and external contexts and project risk issues. It better informs risk treatment options.

20.4.5   Risk Identification

The importance of project risk identification is demonstrated by the two chapters we have devoted to this stage in project risk management (Chapters 6 and 7). Risks that are not identified cannot be managed proactively, but at best are dealt with only reactively after they have occurred.

Brainstorming lies at the heart of the risk identification process, thus reinforcing our support for team-based workshops as risk identification environments. While brainstorming can stand alone (and often does), it is usually more effective to incorporate it into a more structured approach through the use of appropriate techniques and tools. In principle, these should follow the processes of project decision making.

A 'narrowing' approach is usually best. This may involve using multiple identification techniques that match as far as possible the progressive life-cycle sequence of the project and the concomitant availability of more project information. Risk uncertainty is always mitigated by greater information.

Checklists are still probably the most common tool for project risk identification, but it is worth exploring the practicality of adopting (and if necessary adapting) other project management tools for the purpose. We have typified these as: activity-related tools, analytical tools, associated and representative techniques, functional value tools, matrices, simulation and visualisation techniques, speculation techniques, and structural and management tools. Of these, scenario testing (speculation) is possibly the least used. Some analytical tools are more appropriate for subsequent risk assessment, but if they can be used earlier for risk identification they can deliver additional risk management value.

Willingness to practise with them is vitally important for the use of risk identification tools and techniques. Even more important, in our view, are precise and concise risk identification statements. Preparing them needs practice, too.

20.4.6   Risk Assessment

Some risks are susceptible to precise measurement of their associated uncertainty; many can be assessed only subjectively and approximately. Other risks may not be worth assessing in any detail, either because they are too trivial in effect or, at the other extreme, because they are so serious something will have to be done about them regardless of any assessment outcomes.

Quantitative and qualitative approaches to risk assessment each have their place in project risk management. They are not necessarily mutually exclusive. Qualitative assessment generally, but not invariably, precedes quantitative assessment. Knowing which approach is more appropriate, and when, is a hallmark of good project risk management.

With any form of risk assessment, proper understanding of the inputs, analytical processes, and outputs must be demonstrated and communicated. The less 'black box' and the more transparent the processes, the more easily this understanding will be achieved.

In a good project risk register for a PRMS, threat risks will be ranked according to their assessed severity (highest first), so that required risk treatment resources can be prioritised.

Risk mapping is a useful tool for recognising patterns among project risks.

20.4.7   Risk Response

Ideally, something should be done about all project risks, subject only to resource constraints and management capability. In some circumstances, however, project risks may be retained without further risk management treatment being envisaged. This should only happen on an informed basis.

Other responses (besides retention) include: reduction and retention of the residual risk, transfer, and avoidance. Response combinations are also found for risk reduction and transfer.

Risk responses and treatments are often a matter of balancing needs and resources. They are also influenced by the risk profiles of the stakeholder organisation. Profiles are based upon attitudes towards, and appetites for, project risks. In organisations, these are established by the project decision makers. Risk attitudes and appetites are subject to personal (and thence corporate) biases and may shift over time.

Decisions about project risks should be preceded by investigation into any measures already in place to manage them, as treatment may then focus more appropriately on bridging any gaps. Care is needed so that risk treatments do not thereby introduce other, more severe risks. All risk treatments have costs associated with them, and proposed risk reduction measures entail a feedback return loop to risk assessment to determine if the post-treatment risk now meets the organisation's criteria for acceptability.

Reducing the likelihood of occurrence for a risk event generally means doing things better or doing them differently. Reducing the consequences might entail ensuring the availability of backup resources or routines, or setting aside money in budgets to deal with uncertainties (contingencies).

Risk response decisions and treatments should be recorded in the project risk register, always bearing in mind that the register may be accorded evidential status in the proceedings of any formal inquiry.

20.4.8   Risk Monitoring and Control

The planned action dimension of the project risk register comes into play with the processes of monitoring and controlling project risks. If this aspect is allowed to remain passive, and is not enacted in tangible ways throughout the remaining phases of a project, then nearly all of the potential benefits of the PRMS will be lost for that project.

Assignment of responsibility is of paramount importance, as are the frequency and nature of monitoring and control activities. The greater the severity of the risk, the more extensive and expensive these processes are likely to be, and the more senior the responsibility accorded to them, especially where 'due diligence' and 'duty of care' expectations come into play. Recording risk monitoring and control outcomes is usually essential, and it is inevitably required where disaster recovery is involved. Creating a transparent audit trail may be vital.

20.4.9 Risk Knowledge Management

No contemporary organisation can afford to ignore the knowledge generated through carrying out its activities. It is a vital resource that is capable of adding value and contributing to an organisation's competitive advantage. It is particularly important in project risk management.

Project risk information from each stage in the PRMS cycle can be summarised initially in a 'risk register item' record for each risk. Salient information is then transferred to the project risk register as a working risk management process document. At appropriate times as the project proceeds, this information is also captured in the organisational risk register, together with the tacit knowledge gained from actual project risk experiences and captured through a 'lessons learned' process. The capture process should be structured and systematic, as ad hoc approaches leave it vulnerable to individual memory loss and the priorities of the next project.

The product of a systematic capture process becomes a repository of organisational risk knowledge which, when suitably arranged and augmented by risk information from other sources, then becomes a knowledge resource which can be mined, interpreted, understood, and reflected upon, thus leading to risk wisdom that can be applied in future project decision making.

20.4.10 Risk and Culture

Our perceptions and assessments of risk are culturally shaped. This may be through the intrapersonal world views we hold, the assumptions we make, and the values and beliefs that are dear to us. However, risk shaping is also influenced by the organisational culture of the project stakeholder, together with the cultures of the external environments in which the stakeholder operates, and including the organisational cultures of other stakeholders involved in the project.

Culture in a project risk context is the influence brought to bear through the assumptions, beliefs, and values of an organisation, and those of general society. The influence is directed towards project decision making, and it surfaces sharply in risk management.

The effects of cultural shaping of risk may be positive or negative. Organisational cultures may exist across the whole organisation or as subcultures. They may be overt or covert. The latter are difficult to detect, are often negative, and are usually hard to shift, particularly if they exist as subcultures.

20.4.11 Complexity

'Complex' is a term often found in close juxtaposition to 'project', yet its meaning is not universally defined and understood. Nor are criteria established for distinguishing it. It infers some state whereby a project comprises an intricate and complicated combination of parts or attributes that could prove more difficult to deal with than the components of another project not similarly burdened. To that extent, complexity is always relative.

A useful initial concept for complexity is to examine a project in terms of the extent of uncertainty it presents, compared to the solution space available to it. The higher the potential uncertainty, and the greater the potential solution space available, the more complex the project is likely to be. Then, in the context of project risks, complexity can be further understood as the uncertainty associated with conditions of differentiation and interdependency inherent in a project. This takes us out of the cognitive bind of relating complexity only to large projects.

'Differentiation' refers to the number of constituent parts which *should* be distinguished in a project in terms of our capacity to manage them. Management is the key complexity qualifier here. A project might comprise many other separate parts that we *could* distinguish, but, if they do not present a problem for managing them, they do not necessarily constitute project complexity.

'Interdependency' concerns relationships between the differentiated parts of a project. Pooled interdependency occurs when project activities are carried out individually but not in any strict sequence. This is rare for most projects. In sequential interdependency, conditional sequencing relationships exist between two or more parts of the project. One activity cannot commence until another is completed or at least partially complete to a predefined point. Reciprocal interdependency occurs when any change or volatility in the nature or conditions of one part of a project entails consequential change or volatility in one or more of the other parts.

Uncertainty exacerbates complexity situations, and, because projects take time, the complexity may be dynamic. Mapping complexity situations and uncertainties can help to provide a richer picture for the purposes of project risk management, but such mapping may itself be a complicated process.

20.4.12   Political Risk

Few projects are entirely free from political risk, especially as the word 'politics' has expanded from the confines of its original connotation as the affairs of public government. Politics can exert influence intra-organisationally, inter-organisationally, and extra-organisationally. Its influences may be mild or drastic. At one extreme, the slender fingers of politics may simply stroke project egos. At the other extreme, the hard shoulder of politics can be struck by the fist of corruption.

Political risks in projects tend to be subversive and disruptive in effect. They cover a wide range of possible circumstances and events. Their true sources are usually hard to detect, and analysis and assessment are also difficult.

Managing political risks can be a tortuous process with few 'winners'. Having a political project 'nose' may help, particularly if the aroma is detected sufficiently early for pre-emptive avoidance action to be taken.

20.4.13   Opportunity Risk

Opportunity risk is the neglected child of project risk management. Understandably, protecting project objectives against the uncertainties of threat risks takes priority. For most risk management practitioners, the exigencies of dealing with threat risks leave little or no time to explore opportunities whereby project objectives may be exceeded, or potential benefit for future projects is gained.

Good opportunity risk management is deliberately proactive rather than reactively 'happenstance'. It looks for opportunities in project situations which can be exploited to improve project outcomes or organisational performance.

However, because of the intrapersonal cognitive, affective, and behavioural psycho-logical associations with risk, people who are good at threat risk management may not be best suited to deal with opportunity risks. To some extent, skills for the former are emphasised and inculcated in much professional education and training, thus also con-tributing to opportunity risk neglect. We advocate a much greater emphasis on creativ-ity in learning processes, but attitudinal shifts may be needed to change this skewed relationship between threats and opportunities.

20.4.14 Strategic Risk Management

The importance of strategic management is sometimes understated. It is always in the service of achieving objectives and relates to the 'how?' rather than the 'what?' of project processes. Factors that affect strategic risk management for projects include organisa-tional change, organisational culture, and technology change. Its intent is to promote risk awareness in the organisation and effective project risk management practice.

Strategic issues may concern the whole gamut of implementing and operating PRMSs, and of gathering and managing risk knowledge.

PRMS strategies must be determined at a high level in an organisation and then suc-cessfully diffused across it. However, strategic thinking should not be the sole preserve of a few individuals: contributions should be 'bottom-up' as well as 'top-down'.

20.4.15 Building and Maturing a PRMS

A PRMS has to reflect the strategic risk management intentions of an organisation, and appropriate implementation strategies must therefore be decided. The organisation must determine what system will be most appropriate and where it will be located in terms of responsibility (e.g. as a stand-alone system focusing entirely on projects, inte-grated with project management, or subsumed into another risk-related area such as safety management). While prescriptive legislation has directed much attention to safety requirements on projects, putting the whole of project risk management into this area may be counterproductive in the longer term.

A risk management 'champion' should drive PRMS implementation. As this is a pro-ject in itself, clear objectives and appropriate system design and planning are needed, together with sufficient resources to complete it. System start-up and roll-out proce-dures must be carefully considered, particularly with respect to operational staffing.

Issues of system decay must be dealt with, and performance review and benchmark-ing should be undertaken as part of a cycle of continuous improvement.

Since PRMS implementation may constitute radical change in an organisation, it should be undertaken with due attention to all the implications of change management.

20.4.16 Computer Applications

Any computerised PRMS application has to be judged against three requirements: appropriate information input capacity, sufficient information processing capability, and meaningful output deliverables. An overarching criterion is user-friendliness.

For commercial applications, it is a 'let the buyer beware' market environment, such is the plethora of system products available.

The ubiquitous spreadsheet probably offers the easiest introduction and route to PRMS computerisation. Despite its operational limitations, it provides a relatively easy and stable platform upon which to design and incorporate the stage requirements of the project risk management cycle. Spreadsheets are familiar to most people in many environments. They can accommodate quite complex modelling routines and deliver outputs in a wide range of format options.

More sophisticated project management systems do not necessarily include risk management as a primary focus, but add-on modules are usually available and add value to the application if users become adept in applying them. A caveat here is that, unless carefully managed, such systems do little for raising risk awareness in an organisation. They tend to 'distance' participants from the psychosocial construct of risk and promote a more 'mechanistic' view.

Project situation simulations, smart sensors, and aerial drone technologies probably constitute the next major advances in computerised information technology for project risk management, particularly where they are harnessed to decision support mechanisms such as AI and machine learning.

20.4.17 Communicating Risk

Greater risk awareness should be an objective for every project-driven organisation. This can only be achieved through better risk knowledge management and better risk communication. The importance of effective communication is paramount. If risk communication fails, so does risk management.

20.4.18 Case Studies

While we are obviously closer to the case studies in this book than our readers, we hope you will gain substantial value from them.

In **Case Study A** (a Public-Private-Partnership [PPP] correctional facility), the complicated procurement arrangements for this type of major long-term project are illustrated. The contractor in the Special Purpose Vehicle (SPV = private partner) group demonstrates a highly methodical approach to project risk management. In this case, an opportunity risk is exploited to launch a new business venture. Good risk knowledge management is shown.

Case Study B (a rail improvement programme) demonstrates project complexity at various levels. The whole programme is administered under a rail crossing removal authority, a temporary administrative organisation established by the state government and reporting to it. Under a chief risk officer, the removal authority operates a highly systematised risk management process compliant with ISO 31000. As a temporary organisation, the removal authority itself has no investment in the accrued risk knowledge beyond the completion of the final package in the project. The extension of the current programme with additional rail crossing removals (planned to continue until at least 2025) prolongs that investment, of course, but when the rail crossing removal authority is eventually disbanded its accumulated risk knowledge will largely be tacit rather than explicit. In the longer term, however, the benefits of the tacit risk knowledge should be felt across the relevant industry sectors involved, and in public sector administration and governance.

Case Study C (an aid-funded project) looks first at risk management from the perspective of a professional consultant specialising in offering project management services for aid-funded projects around the Pacific Rim. Operating a small practice where the costs of professional indemnity insurance are considered exorbitant, the consultant instead pays careful attention to raising his personal level of project risk awareness. He deliberately minimises growth in the consultancy that could dilute his personal control and is highly selective about the project commissions he accepts. This approach sharpens his 'risk wisdom'. The second part of this case study explores risk management for the aid-funded project to build a new Parliament building for an island government. For the most part, threat risk management was seen to be informal and ad hoc, other than that implemented by the funding government. Some evidence of reactive opportunity risk management was found. This case also confirmed the need for better capture and use of project risk knowledge.

Case Study D (a train mock-up project) revealed little evidence of formal risk management procedures in place in the subcontractor organisation, other than those required under industrial safety legislation and any needed for environmental management and quality assurance. Threat risk uncertainties were generally managed on a reactive basis. Risk knowledge management was fragmented, and not helped by the existence of a 'silo' inter-organisational knowledge culture. High reliance was placed upon engineering expertise (for technical risks), but the company was less well-equipped to deal with other types of risk.

For **Case Study E** (a hot-rod car project), no formal PRMS was adopted. However, as a private 'passion' project, it fell entirely within the interests and capabilities of the mechanical engineer. His confidence in his engineering knowledge, skills, and experience enabled him to deal with the uncertainties that arose on the project, helped also by the knowledge that time was not a severe constraint.

Case Study F (an aquatic theme park) concerned a civil engineering contractor's involvement in an unusual project. Risk and other management systems complied with relevant standards, and portable wi-fi enabled devices added administrative convenience and flexibility. However, fully integrated risk knowledge management is still some distance in the future for the contractor.

Greater value can be extracted from the case studies if readers use them as practice vehicles for project risk management. Each context could be changed to make it more relevant, and each stage of the project risk management cycle then pursued, with the possible exception of the risk knowledge capture process. A team approach would be ideal.

20.5 Closing Remarks

In retrospect, we have to acknowledge that effective project risk management is built upon a remarkably comprehensive framework of theory. Among the more easily detectable theory fields are those relating to:

- Cognitive, affective, and behavioural psychology
- Administration
- Human resource management

- Communication
- Decision making
- Gaming
- Organisational structure
- Systems
- Complexity
- Culture
- Politics.

Doubtless there will be others that we have overlooked. Dealing fully with all of them in a book of 20 chapters is impossible, of course, and 20 books would probably not be sufficient.

In several fields, competing theories exist, but we have not attempted to present all of them. Instead, we have chosen theoretical approaches that enjoy wide acceptance and best contribute to our views about how better project risk management can be achieved in practice. We make no apology for this limitation. Our 'theory deficiencies' can only be addressed by readers' willingness to engage with them individually in further private reading.

We have also recognised a distinct emphasis on *cycles* in our book. Projects have phase cycles (Chapter 3, Figures 3.2 and 3.3). Project risk management has a staged cycle (Chapter 4, Figure 4.2). Risk knowledge creation has an activity cycle, and risk knowledge application has an output–input cycle (Chapter 11, Figures 11.2 and 11.3). Clearly, we think cycles are good!

Uncertainty is the almost imperceptible thread that stitches all these theories and cycles together. It permeates the whole of this book, especially in terms of the systematic cycle of project risk management.

If we can understand uncertainty more fully, and develop greater confidence in dealing with it, we can manage project risks more successfully. To this end, in several chapters we have suggested ways of exploring project uncertainties, and have offered simple tools and techniques – many adapted from other project management purposes – for doing this. The power of matrices and organograms to reveal patterns, clusters, and complexities, and to show relationships, should not be underrated. We hope that these tools will also be valuable 'take-aways' for readers in their quest for improving project risk management practice.

Finally, good risk management entails a willingness to interrogate projects at many levels. The more questions you ask, and the more appropriately they are framed, the more effective will be your ability to manage project risks.

Case Study A

Public–Private Partnership (PPP) Correctional Facilities Project

Correctional facilities (prisons and youth detention centres) are a state, as distinct from federal, responsibility in Australia. Late in the first decade of the twenty-first century, an Australian state government acknowledged the need for additional prison accommodation. The rate of incarceration was increasing, paralleled by the incidence of recidivism. Harsher sentencing policies introduced by the government, and more stringent parole parameters, were also putting strain on the existing prison system.

Planning commenced for the development of a new correctional facility to be built and operated under a 25-year concession period (ending in 2042) using public–private partnership (PPP) procurement arrangements. Originally intended to accommodate 500 inmates, the project scope was subsequently increased to 1000, and then again to 1300 sentenced adult male prisoners. The design of the facilities was also influenced by a policy shift to programmes-based imprisonment.

The PPP uses an availability, fitness-for-purpose, and full-service provision model on a DBFOM (design, build, finance, operate, and maintain) basis. Besides cellular prisoner accommodation, the new prison includes facilities for medical and mental health care, pre- and post-release services, and 90 staff. State land was made available from a site formerly used for munitions manufacture and close to an existing youth detention centre. Besides the traditional time, cost, quality, and fitness-for-purpose objectives associated with all projects, the state's Department of Justice and Regulation has established operational objectives for such facilities, some of which are quite radical, and all affect the PPP. These include:

- New approaches to reduce the risks of reoffending
- An integrated and holistic model of care for prisoners experiencing mental illness
- A targeted approach for dealing with prisoners with challenging behaviours
- Recognition of the special needs of indigenous prisoners
- Improved responsiveness to the complexities of dealing with younger prisoners
- Programmes and services for prisoners serving shorter sentences
- Pre- and post-release services for prisoners.

The case study adopts the perspective of the successful construction contractor – a major international contractor specialising in large building and infrastructure projects. It spans the period from the call for expressions of interest (EOIs) to the financial closure of the PPP agreement. The subsequent construction activity and commencement of facility operations are not dealt with here.

Managing Project Risks, First Edition. Peter J Edwards, Paulo Vaz Serra and Michael Edwards.
© 2020 John Wiley & Sons Ltd. Published 2020 by John Wiley & Sons Ltd.

The EOI was issued in June 2013, and interview workshops with interested parties were held between July and September 2013. Briefing provided to parties included information about risk allocation and shared risks. In October 2013, the first accommodation scope increase (from 500 to 1000 prisoners) was notified. Towards the end of the year, the EOI response list was reduced to the two best qualified bidders, and the scope increased again to 1300 prisoners under the state government's 'scope ladder' provisions for PPP projects. A five-month tender preparation period, including workshops and site visits, ensued, culminating in tender presentations by the two bidders in April 2014. This was followed by an exclusive period of negotiation by the state government, and the announcement of the preferred bidder was made in August 2014. Financial closure for the PPP contract agreement between the public and private partners was achieved in early September 2014.

Figure A.1 presents a simplified organisational structure for the PPP correctional facility project.

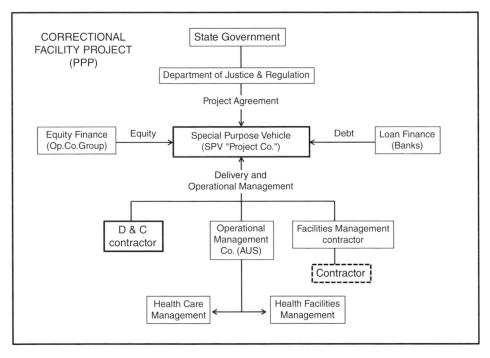

Figure A.1 State correctional facility project structure.

A typical PPP special purpose vehicle (SPV) company was formed, with the state government represented through its Department of Justice and Regulation. Private partners in the SPV included the construction contractor (appointed on a design and construct [D & C] basis), an operational management company (the Australian arm of an international group), and a facilities management (FM) company (again, an Australian subsidiary of an international company). Ongoing prisoner healthcare management and health facilities management were each made the responsibility of separate specialist companies under the operational management company. Debt finance to the SPV is provided by a consortium of banks. In this project, equity finance is also provided by the international parent company of the Australian operational management organisation.

After the completion of construction, the D & C contractor company created a new company to provide services to the FM contractor. Construction of the new prison lasted three years and was completed on time and to budget (AU $650 million) in September 2017. This was followed by three months of systems testing and commissioning, and a four-month ramp-up period to full occupation.

A project summary was published by the state government in February 2015. This contained a comprehensive risk allocation schedule between the state and the SPV.

The D & C contractor's project risk management approach is based on the framework established in ISO 31000 (see Chapter 4). It is also integrated with similar standards for Quality Assurance (ISO 9000), Environmental Management (ISO 14000), and Safety Management (ISO 45001). The procedures and processes of project risk management adopted by this contractor are shown in Table A.1. This approach is a policy requirement of the construction company and is used regardless of the type, size, or scope of projects or the procurement systems used.

Risk management responsibility for a project is generally devolved to an appropriately experienced member of the company's Enterprise Risk Management team, and the project development manager is ultimately accountable for all risk management activities during the tender process.

Project risk management activities are determined by the particular requirements of each of the tender preparation, construction, and ongoing operational stages for a project. The matrix format shown in this approach (Table A.1) is deliberate, since it acts as a risk management process checklist for each of the vital stages.

In addition to the process and procedural guide shown in Table A.1, the company also uses a risk management process flowchart which sets out the required inputs and outputs for each process. Inputs address concerns such as when an input is needed, who is involved in the process, and what informational references and resources will be available. Key outputs are described in terms of requirements for plans, permits, and risk treatment, monitoring, and control procedures.

The context establishment stage of project risk management (see Chapter 2, Figure 4.2, Stage A) is dealt with by this contractor in the prospect/EOI stage as a prospect risk assessment (PRA) process. It is conducted on a workshop basis and attended by up to 10 key project staff. The number of attendees is not stipulated, but workshops with more than 10 participants are avoided wherever possible.

The PRA workshop deals with a number of key issues. These are shown in Table A.2. They tend to be project-typical, and the workshop participants then brainstorm those that are thought to involve specific and potential risks for the project under review. Most of the issues relate to contract terms, financial matters, and precise definitions of terms and requirements. Participants do not arrive 'blind' at the workshop, but are already prepared through briefing information and draft contract documents.

The importance of the PRA is stressed as it not only informs the tender preparation process but also can act as an early 'gateway' for deciding whether or not to proceed further with an EOI. It is therefore mandatory in terms of company policy and requires completion of a template form. This ensures consistency in the process and provides an auditable decision trail.

Table A.1 D&C Contractor's project risk management processes and procedures.

No.	A	B	C	D	E	F
	Process	Procedure	RM Tools	Prospect/expression of interest (EOI) stage	Tendering	Delivery and operational
1	Prospect risk assessment (PRA)	Tendering	PRA template	Mandatory	Mandatory	
2	Contract risk assessment (CRA)	Tendering	CRA form		Mandatory	Becomes contract rights and obligations document
3	Qualitative risk assessment	Pre-contract risk assessment	Risk register	Optional unless required by client or joint venture (JV) partners	Optional unless required by client or JV partners	Optional unless required by client or JV partners
4	Traditional risk and opportunity (R & O) assessment	Risk management (commercial)	Risk register		Mandatory unless quantitative risk assessment required	Mandatory
5	Design development scope growth		Risk register Uncertainty calculation sheet		Mandatory if D & C contract	
6	Probabilistic R & O assessment	Probabilistic modelling	Risk register (probabilistic)		Mandatory if triggered for approval or if required by client or JV partners	
7	Estimate uncertainty		Uncertainty calculation sheet		Mandatory if probabilistic modelling used	
8	Schedule risk analysis (SRA)	SRA guidelines	Planning & scheduling software application PRA tool		Mandatory if triggered for approval or if required by client, JV partners, or management group	Mandatory if triggered for approval or if required by client, JV partners, or management group

Table A.1 D&C Contractor's project risk management processes and procedures (Continued)

No.	A Process	B Procedure	C RM Tools	D Prospect/expression of interest (EOI) stage	E Tendering	F Delivery and operational
9	Safety, quality, and environmental (SQE) risk management (RM) review	SQE RM	SQE RM Checklist		Mandatory	
10	Workplace risk assessment (WRA)		WRA form		Optional	Mandatory
11	Activity method statement (AMS)		AMS form			Mandatory when triggered
12	Task risk assessment (TRA)		TRA form			Mandatory when triggered
13	Start card (SC)		SC form			Mandatory
14	Project financial report (PFR) (R & O)		PFR			Mandatory

Table A.2 Contextual issues explored by D & C contractor.

Issue
Payment terms
Security of payment
Bonding requirements
Fitness-for-purpose and availability requirements
Insurances
Completion definition
Currency exchange
Caps/exclusions on liability (including abatement risk, liquidated damages, and delay caps)
Latent conditions and contamination
Extension of time and delay costs entitlement
Native title and cultural heritage requirements
Defect rectification requirements
Consequential loss
Look-forward tests
Changes in law
Termination of agreement
Indemnities (required by/from contractor)

Contract risk assessment (CRA) also uses a template form and is a required part of tender preparation, since risks arising from the terms of contract agreements shape much of the risk profile of projects. If the bid is successful, the CRA form becomes part of the document dealing with the contractor's rights and obligations. The assessment is carried out mainly by legal staff in the company, but is informed by technical staff where necessary. It addresses the pertinent legal issues raised in Table A.2.

For the PPP correctional facility project, the PRA identified a particular risk associated with latent conditions and contamination. The state government had carried out decontamination work with respect to dangerous munitions and toxic materials left on the site after the demolition works. A clearance certificate had been provided by the government's Environmental Protection Agency, but it became clear to the contractor that a residual site contamination risk remained, which the government expected the contractor to bear. The contractor regarded this as a major risk and subsequently negotiated a contingency arrangement with the government whereby any additional remediation required would be reimbursed.

In the company's approach to risk analysis (see Chapter 4, Figure 4.2, Stage C_1), qualitative subjective analysis is used first. This is done by the project risk manager in collaboration with staff associated with particular risk categories. Linguistic ordinal measures are used to assess likelihood of occurrence, impact, and severity (Chapter 2, Sections 8.2–8.5). These are not predetermined as part of company policy but are left to the discretion of each project risk manager in a 'traditional risk and opportunity' manner of engagement. The staff churn rate in the organisation is such that the adoption of uniform assessment descriptors across the entire company is considered impractical at this stage in its risk management maturity, although the achievement of this step is a longer term company aim. Sometimes, risk analysis is not conducted if it is patently obvious that a particular risk must be dealt with regardless of its severity level (i.e. where it would constitute a breach of compliance with regulations).

Generally, qualitative risk assessment (QRA) is deemed sufficient to inform the risk response process unless the risk parameters are such that quantitative analysis is deemed desirable (e.g. high uncertainty in probability or impact, or high level of severity). As a general rule, the 10 highest priority risks from the PRA are subjected to quantitative analysis.

The QRA of the correctional facility project revealed several matters of concern to the contractor. These included:

- Latent conditions – unexploded ordinance (UXO), asbestos, rock, contaminated fill, water table levels
- Environment – pollution, water discharge, and flora and fauna (e.g. golden sun moth and legless lizard breeding seasons; translocation considerations)
- Cultural heritage – impacts, artefacts, three indigenous groups with an interest in the site
- Stakeholders and community – disruption, protests, complaint handling, construction impacts
- Human resources – resource availability and industrial relations
- Design – fee growth, design scope definition and growth, fitness for purpose, abatement risk, warranties, quality and technical standards, life-cycle costing
- Programme – time delay, approval processes, design programme, schedule risk analysis (SRA) delay allowance
- Estimate uncertainty.

However, before probabilistic (quantitative) risk analysis is undertaken, an important task is to assess project design development in order to capture any significant changes to design or scope. This exposes any additional uncertainty, and the process is mandatory for all D & C projects undertaken by the company. Uncertainties are recorded via uncertainty calculation sheets. Data from these sheets provide the parameters for quantitative risk modelling using techniques such as Monte Carlo simulation.

While the need for probabilistic risk analysis is heavily influenced by the extent of project estimate uncertainties, the quantitative assessment process becomes mandatory if it is a prerequisite for project approval or if it is required by a client or a joint venture partner.

SRA, using analytical components incorporated into a commercial project management programming and scheduling software application, is conducted if needed or if mandated in the same way as probabilistic analysis.

A review of safety, quality, and environmental (SQE) risks is a mandatory requirement for tender preparation, and this review uses an SQE risk management checklist.

Workplace risk assessment (WRA) – a more detailed progression of the safety risk review – is carried out as an optional process during tender preparation. Where such assessment is deemed necessary, the WRA form captures the necessary information and decisions.

Workshops for SRA, SQE risks, and WRA are arranged on an as-needs basis. For some projects, the outcomes of these reviews trigger further risk reviews covering activity methods statements (using activity method statement [AMS] forms) and task risk assessment (using TRA forms).

On all projects, a start card form (SCF) must be completed and approved for each worker involved in construction activities on site after successful tender award, or in the ongoing operational activities of the company. The form includes evidence of attendance at an obligatory induction training programme.

A project financial report (PFR) provides the necessary vehicle for monitoring and controlling project risks. Its use is mandatory from the commencement of construction, and it should flow continuously and seamlessly through to operational management if necessary. The PFR automatically surfaces the 'top 10' most severe risks for the company. The ranking, as far as possible, is based upon financial measures. Those for the correctional facility project are listed in Table A.3.

Delays seriously affect the company's bottom line through their knock-on effects and impact on project overhead costs.

Latent conditions usually relate to geological conditions present on the site, particularly where soil tests still leave uncertainties about the underlying strata. Latent conditions could also relate to uncertainties about the condition of existing structures and buried services, particularly where expensive refurbishment projects are involved.

SRA invariably has to deal with a large number of uncertainties about project task durations and interdependencies. These can be exacerbated by project scope or design changes and will impact heavily on design costs, especially where professional design consultants are involved.

The quality of the delivered project is always a difficult problem to deal with, particularly with so much uncertainty surrounding quality expectations of clients and other stakeholders, and the quality performance of suppliers and subcontractors.

Table A.3 'Top 10' contractor risks identified for correctional facility project.

No.	Project risk issue
1	Delays
2	Latent conditions – rock
3	Latent defects
4	Design fee for any SRA amendments
5	Defect-free condition allowance at technical completion and commercial acceptance stage
6	Currency exchange rate variations – chillers, generators, etc.
7	Insurance coverage, exposure, claims, and excesses
8	Requirements set out in the Building Code of Australia (BCA) and the relevant Australian/New Zealand Standards
9	Abatement risk
10	Design fee growth costs

Currency exchange rate risks arise where expensive machinery and equipment have to be imported and quotations are not based upon landed prices in AU$ dollars. Besides the cost of the components themselves, transport costs are also likely to be affected.

Insurances costs are difficult to estimate with reliable certainty in advance, as premiums and terms relate wholly to the state of the insurance market at any given time. Insurance of imported materials and components is particularly affected in this way.

Given the plethora of regulations about buildings and construction processes, ensuring compliance – in every jurisdiction in which the contractor operates – is a costly and time-consuming process easily aggravated by changes to laws and regulations.

Abatement risk relates to any financial withholding (or penalties) for nonperformance attaching to the terms of the SPV agreement. This is a particularly difficult risk in PPP procurement, since the partnership is a long-term one and the public partner may elect not to pursue abatement in the early years in order to encourage better performance and to build a more positive relationship with the private partner. Additionally, some key performance indicators (KPIs) and criteria may have to be changed in order to better measure performance. Clearly, what was originally thought to be certain is often found to be uncertain in practice. Some public partner contract administrators insist on a strict 'black letter' interpretation of contract terms. Others are more flexible.

Where a D & C contractor employs professional design consultants on a project, additional fees may be charged for relatively small design matters. Given the complexity of most major PPP projects, this fee 'growth' can be substantial. Interdependencies between different design elements can exacerbate the growth.

Much of the company's project risk management focuses upon threat risks as a priority, since in the early stages of the project they will influence the tender price. Less attention is given to opportunity risks and the resources needed to exploit them.

Where the probability of occurrence of a particular threat risk is deemed too high, the organisation's strategy is to cost the *whole* risk impact into the tender price. While this could render the bid less competitive, the company's aversion to such risks makes this a wise risk avoidance response.

For projects where the company has a D & C role, plus an ongoing FM involvement, considerable attention is given to technical design risks that will also impact FM. For example, this could include design modifications to create better access for future repair or replacement of large items of equipment or services engineering components.

'Lessons learned' is considered an important project debriefing process within the organisation. At the end of each project, a meeting is organised to share the knowledge acquired in each project by each department. The way the company has decided to organise its knowledge is by the main activities involved:

- Commercial and strategic
- Tendering
- Contract
- Design (buildability, constructability, durability, operations, and maintenance)
- Planning and scheduling
- Financial
- Safety, Quality and Environment (SQE).

Each coordinator for these main activities is therefore responsible at the end of a project for producing a report with the lessons learned from that perspective.

The report is presented at a meeting of all coordinators and then uploaded to a common platform. For management and logistical reasons, more than one meeting may be held to share the knowledge created by the project, but good practice assumes that at least the bid manager and the project manager of the project should attend all such meetings for each main activity.

The online knowledge platform is called the 'Project Lounge', denoting its intended informal knowledge accessibility and sharing purpose, but it is also a contributory work platform. Each time the organisation considers a new project prospect, consultation with the knowledge management system (KMS) is mandatory in order to evaluate the risks associated with any decision to tender, the tendering process, and the project delivery and operational involvement. The whole array of organisational knowledge and experience is made possible and accessible through the debriefing information captured from previous projects.

In order to avoid communication problems in collecting and organising knowledge, the organisation has developed templates for each responsible coordinator or manager to fill in as the basis for reporting. Using templates in this way ensures that access to the right information is straightforward and consistent. The risk of losing important information or failing to identify risks is also reduced.

Similar information on each project can be compared and reused on future projects in the right place at the right time, because the debriefing reports become fully accessible as soon as the relevant stages of the project are completed.

Due to the uniqueness of each new project, referring to the knowledge base is mandatory, but this alone is not sufficient to make go/no-go decisions for pursuing an EOI. The team tasked with the responsibility to prepare a report for the board of directors needs to use information from past projects but is required to undertake its own research and analysis. In the case of the correctional facility project, the company not only used their past experience with other PPP projects but also visited other correctional facilities in the state and around the world.

Its procedure of project risk knowledge retrieval allows the company to create new knowledge based on its experiences. This approach helps the organisation to maximise the knowledge benefits. The organisation's project managers thereby create effective communities of practice, sharing experiences and different points of view to add specific knowledge. The knowledge of the organisation resides in its people and teams as together they develop projects, but it is backed up by its KMS.

Case Study B

Rail Improvement Project

After coming into power, an Australian state government implemented one of its election manifesto policies. The policy objective was to improve rail and road traffic systems, and increase their safety performance, through the programmed removal of the 50 most dangerous level crossing intersections in the rail transport system across the greater metropolitan area of the state's capital city.

The policy was developed according to accepted contemporary urban planning and design principles that seek to integrate transport, planning, architecture, landscape architecture, engineering, and finance expertise. The rail crossing removals were selected on the basis of the extent of road traffic congestion associated with them, the degree of danger they presented to the public (particularly pedestrians and cyclists), and the potential that their removal would release in terms of improving train service capacity and performance.

The removal programme constituted the largest rail infrastructure project in the state's contemporary history, valued at the end of 2017 at more than AU $8 billion. The programme can therefore be described as a mega-project. It is carried out under the auspices of a level crossing removal authority established as a temporary administrative organisation by the state government.

Costs have escalated from the original AU$6 billion estimate in 2014 and from the subsequent increase to AU$7.6 billion in the removal authority's 2016 business case. The estimated cost is now more than AU$8 billion. This is largely due to increased scope, upgraded technical requirements, and decisions to 'future-proof' some locations by carrying out works in the programme that would facilitate future planned improvements.

Besides removing the level crossings, the programme includes 20 stations to be built or upgraded, many kilometres of new track and signalling, and replacement or upgrading of other technical systems. The programme is scheduled to be completed by 2022, with half of the crossing works completed by the end of 2018. Crossing removal and associated construction works were completed at 11 locations by the end of 2017. Boom gates were removed (an important indicator of progress) at a further 14 crossings by mid-2018, and at that time advanced planning was underway for 20 more. In the lead-up to the November 2018 state election, the incumbent government announced that, if it was returned to power, the programme would be extended by a further three years and five more crossing sites. After winning the election, the programme extension has been confirmed.

Managing Project Risks, First Edition. Peter J Edwards, Paulo Vaz Serra and Michael Edwards.
© 2020 John Wiley & Sons Ltd. Published 2020 by John Wiley & Sons Ltd.

The case study is developed through the perspective of the level crossing removal authority tasked with the responsibility of overall project management for the crossing removal programme. This organisation acts as an overarching or 'umbrella' public administrative authority for all the separate rail crossing removal projects. Figure B.1 illustrates the organisational structure for the programme.

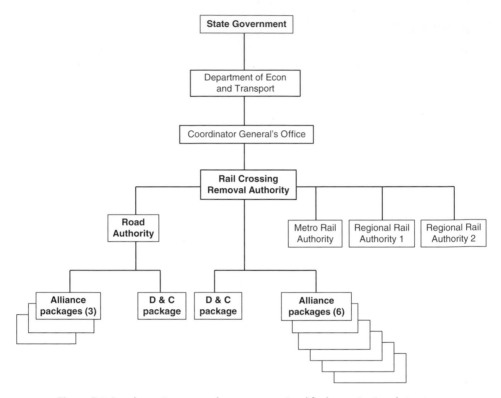

Figure B.1 Level crossing removal programme: simplified organisational structure.

The rail crossing removal authority reports to a Co-ordinator General's Office under the aegis of the appropriate department and minister in the state government. Other rail authorities sit alongside the rail crossing removal authority, representing the various rail system areas in which crossing removals are planned. The state roads department is also involved at this level. These agencies provide information, advice, and design services to the removal authority about matters such as rail and road engineering, land management, safety, compliance, and other relevant issues.

The rail crossing removal authority itself has a divisional structure based on function, and each divisional head is required to report regularly to its representative authority board member about progress and issues arising in its domain on each removal project.

The whole removal programme is undertaken on a phased basis, and the authority uses two different procurement systems for the crossing removals and associated work: design and construct (D & C) and alliances.

The main criterion for choosing one system over the other is complexity. Crossing removals that are considered straightforward in terms of design and construction, and which involve relatively few stakeholders and less uncertainty, are offered as single D & C contracts. There are two such contracts in the current (2022 completion) removal programme.

More complicated crossing removals (in terms of physical features or stakeholder management requirements) are dealt with as alliance packages and involve more nuanced risk sharing arrangements to accommodate their greater uncertainty. Within each alliance procurement package system, several removal site projects may be 'bundled' together and offered to qualified bidders. Alliances each comprise a main contractor, design consultants, and specialist subcontractors.

For the two D & C packages, it is considered that there is less risk uncertainty, and thus more risk transfer to each contractor is possible. The nine alliances reflect substantially greater risk uncertainty through more 'risk pain/risk gain' sharing.

Responsibility for project oversight falls upon either the removal authority (one D & C and six alliance packages) or the state roads authority (one D & C and three alliance packages), depending upon the predominant nature of the work involved.

As a further complication, some alliance packages are bundled and competitively tendered for all the sites in the package. Others are bundled and competitively tendered for the first one or two sites in the package, and then partially competitive prices are negotiated for the remaining sites. One package adopts a partially competitive price approach which offers all sites in the package to a preferred alliance subject to price negotiation. Six sites are not packaged but are awarded by the authority under partially competitive price negotiation to existing alliances on the basis of their performance with previous packages. Table B.1 presents the types and distributions of the level crossing removal packages.

Table B.1 Level crossing site packages and procurement systems.

Package	Sites	Alliance procurement	D & C procurement	Authority
1	4	Competitive (all sites)		State roads
2	4	Competitive (all sites)		State roads
3	9	Competitive (all sites)		Removal authority
4	2	Competitive (all sites)		State roads
5	1		Competitive	Removal authority
6	1		Competitive	State roads
7	4	Partial price competition (all sites)		Removal authority
8	6	Competitive (2 sites) Partial price competition (4 sites)		Removal authority
9	6	Competitive (1 site) Partial price competition (5 sites)		Removal authority
10	9	Competitive (4 sites) Partial price competition (5 sites)		Removal authority
11	6	Performance-based partial price competition (6 sites)		Removal authority
Total	52[a]	9 packages	2 packages	

a) Two sites added to original programme.

Although the original election manifesto promise (and the programme business case) refers to 50 rail crossing removals, two more were added later following a rationale to include them as they were part of a relatively short stretch of one rail corridor.

The state auditor-general's report into the removal programme and its business case questioned the criteria used for packaging decisions, and noted that the partially price-competitive options exposed the state government to the risk of bidders pricing low on competitively priced packages and then subsequently recouping any losses on the partially competitive bundles. The report also cited instances where the government might be the loser in terms of shared pain/gain risk arrangements in alliance packages (which constitute by far the largest proportion of the procurement options used).

Using the recommended 7% discounting rate, the benefit–cost ratio for the removal programme was calculated as 0.78 (i.e. a dis-benefit). With a 4% discounting rate, the ratio improves to 1.34. However, should costs increase by 20% over the AU\$7.6 business case estimate, the 0.78 ratio (with the 7% rate) would decrease to 0.65.

The rail crossing removal authority operates a risk management system (RMS) in accordance with ISO 31000. The system is objectives-driven. The objectives include:

- Use a simple process that meets compliance requirements.
- Implement a RMS that is relevant to the phases of the work package (e.g. development, delivery).
- Achieve consistency and alignment across programme projects.
- Ensure a defendable position for the authority and government.
- Promote a good risk culture through a top-down approach.

The RMS also has to comply with the state government's well-established risk management framework and attestation requirements.

The removal programme authority maintains an overall risk register, and this is informed by the project-based risk registers established under each of its functional divisions; these are, in turn, informed by the project teams and by information supplied by each package's alliance or D & C contractor. Periodic reporting about risk therefore escalates through the rail crossing removal authority to board level and then to the Co-ordinator's General's Office and the responsible government minister.

Under the rail crossing removal authority's RMS, context establishment is undertaken at two levels: one for the overall removal programme, and again for each alliance or D & C package. In each instance, workshops are arranged in which consideration is given to the nature of the work, its timing, and matters relating to the different stakeholders likely to be involved. For the initial overall programme workshop, attendance was determined by the number of divisions in the authority who needed to be represented. In essence, there are closed risk management loops for the key areas of engineering, land management and the environment, and safety.

Transfer of risks to package contractors is done through the contract awards and the terms of the contract agreements. Risks that are fully or partially retained by the rail crossing removal authority include reputational risk impacts, and risk impacts arising from 'risk pain' sharing arrangements.

Risk monitoring and control are achieved through regular review of the alignment of risks between the package contractors and the removal authority. The process also watches for emerging risks.

No evidence was found of any determined search for opportunities associated with project risks, although the 'future-proofing' rationale for the programme scope increase could be construed as exploitation of opportunity risk.

Risk knowledge is captured primarily at the project level. Project team and divisional area forums (i.e. 'communities of practice') discuss and disseminate risk knowledge timeously across the package contractors and across the removal authority.

The removal authority is a temporary organisation that will be disbanded upon completion of the rail crossing removal programme. Staff are recruited or seconded to it under short-term contract arrangements. Following a recent state election in 2018, the re-elected government has proposed to extend the programme by a further five years to deal with at least five further rail crossing removals. The cost of this extension has not yet been announced.

Note: Information for this case study was sourced from the public domain and is presented without warranty as to its accuracy or sufficiency. Referenced comments offered in the chapters of this book are the interpretive opinions of the authors and do not represent the views of the rail crossing removal authority.

Case Study C

PM Consultant and a Government Aid–Funded Pacific Rim Project

This is a two-part case study for an international aid-funded project carried out for one of the island nations of the Pacific Rim.

Part 1 deals with a sole practitioner consultant engaged to provide professional services on the project. Part 2 explores risk management for the project itself.

Part 1

Consultant

The interviewee is a professional consultant commissioned by the aid funding government to act on its behalf and provide advisory services to a Pacific Rim island government which is building a new House of Parliament facility.

The consultant currently operates as a sole practitioner and specialises in providing professional services for international development and disaster recovery projects, particularly in the Pacific Rim area. The consultant's practice objectives are:

- To value-add to development (indigenous and international) and to disaster recovery projects
- To engage in a variety of different project work
- To gain personal satisfaction from the project experiences
- To maintain a high reputation for the quality of professional services delivered.

Substantial international travel forms an essential part of the consultant's work, but is not a major attraction.

While he is familiar with project risk management, the consultant does not use a formal risk management system (RMS) for his professional practice, despite not holding professional indemnity (PI) insurance for his work. In his opinion, the premium costs of PI insurance are too expensive for project management practices, especially for a sole practitioner. Rather than expanding the practice to take on a greater workload and engage associates to spread the project responsibilities, he prefers to restrict the workload to a level that he feels comfortable with, and to projects that he believes will fulfil his objectives. Projects that do not show these prospects are declined. In most instances, clients who would normally stipulate PI insurance for consultants will usually waive this requirement for the benefit gained through the specialist skills and experience of his engagement.

Managing Project Risks, First Edition. Peter J Edwards, Paulo Vaz Serra and Michael Edwards.
© 2020 John Wiley & Sons Ltd. Published 2020 by John Wiley & Sons Ltd.

Where the consultant thinks that his project engagement would be enhanced by the contribution of other specialists, negotiation with the client usually ensures their engagement and payment by the client but under the consultant's control.

The consultant's practice risk management is therefore informal and largely intuitive, but does relate to the staged approach of systematic risk management.

Any invitation to respond to a call for an expression of interest is subjected to a thorough process of contextualisation. The primary considerations are the identity and nature of the project client, followed by the location of the project. Where both are familiar to the consultant from previous experience, this context establishment is straightforward and the knowledge requires only updating refreshment through the internet or personal resources. New client enquiries are explored more comprehensively and rigorously, particularly with respect to client-related factors, such as capacity to undertake the project, commitment to the project, reputation, and fee payment security.

The nature of the project is then investigated in terms of the appointment brief and any additional information that can be sourced. Since the consultant is sometimes engaged to deliver value management (VM) services as part of his appointment, the context establishment process incorporates preliminary consideration of what parts of the project might benefit from a VM intervention.

From this point, project risk identification, for the consultancy practice and for the project client, proceeds almost simultaneously in parallel. The basis for identification is invariably the scope of the project and whatever information is available about the planned work breakdown or staged activities. The risk identification questions thus explore what events might happen and what their consequences might be in terms of creating uncertainty in achieving client objectives and also uncertainty in the fulfilment of consultant objectives.

Assessment of project risks to the consultancy is largely qualitative, although quantitative assessment is regularly carried out later for the client's project risks (usually in terms of prescribed client risk management frameworks). For the consultant, the possibility of reputation damage is a primary concern in terms of the consequences of any risk event.

The consultant's practice risk management strategy is to avoid threat risks wherever possible, but it is often worthwhile to explore mitigation measures such as risk sharing (with the client, another consultant, or both), usually by means of flexible negotiation.

Monitoring and controlling consultancy risks on projects occur in parallel with the project milestone client reporting requirements and processes.

Risk knowledge capture for consultancy risks is an intuitive process which reinforces and augments the mental checklist maintained by the consultant for application to future projects. Project completion review reports for clients also contribute to knowledge capture. In his view, knowledge acquired through a 'lessons learned' project debriefing is seriously neglected in the project areas with which he is most often associated. He believes it is a process that should be more consistently adopted, and the findings applied through a systematic framework.

For the consultant, while project risk management tends to focus upon threat risks, the VM process (usually mandated for aid-funded projects) often provides the best medium for identifying and exploiting project opportunity risks.

Part 2

Project

The project with which the consultant is currently involved is the construction of a new facility to house the parliament of an island nation on the Pacific Rim. A funding contribution to the project (about US$10 million) comes from the foreign aid programme of the Foreign Affairs Department of the donor government which has appointed him as a senior advisor. His responsibility also extends to supporting the recipient government of the island nation, which is directly responsible for project implementation. In this sense, the consultant wears three 'risk management hats': one for the donor government, one for the recipient government, and one for his own practice. His role is advisory to both the donor government and the recipient government.

The consultant reports formally to the donor government every three months in terms of project progress and continuing compliance with the funding agreement. He also has advisory approval authority over project funding payments from the donor government to the recipient government, and over monthly progress payments from the recipient government to the project consultant, main contractor, and so on.

The funding agreement provides for payments in four tranches from the donor government to the recipient government. The first payment is made at completion of the funding agreement itself, the second at the completion of project design, the third at the appointment of the main contractor, and the final tranche upon completion of 40% of the construction value of the project.

Figure C.1 shows a simplified organisational structure for the project.

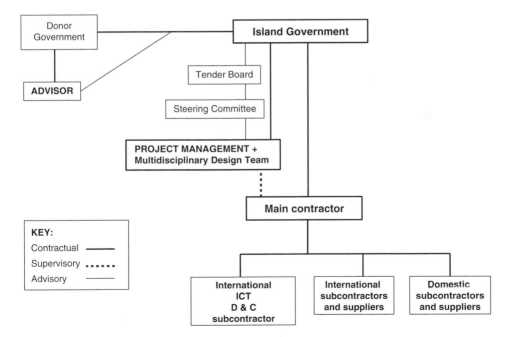

Figure C.1 Simplified organisational structure for aid-funded project.

The project procurement system is 'traditional' in that design and construction are separated for the construction work.

The project management and multidisciplinary design team is an integrated consortium of international consultants. This consortium was appointed after successfully responding to an international call from the island government for expressions of interest. The call attracted local and international interest, and bids were evaluated on the basis of technical ability and price. The senior advisor participated in the evaluation process.

In addition to providing design services, the project management team also prepared bills of quantities for tendering purposes. It acts in a supervisory capacity for the project on behalf of the island government, with a site superintendent providing direct oversight of all work.

A provisional sum was included in the contract bills of quantities for the information and computer technology (ICT) systems installation for the new parliament building. International bids were invited on a design & construct basis for this work. Advice on this aspect of the project was also provided by specialists from the donor government's own parliamentary ICT support group. An international bidder was successful in winning the ICT installation subcontract. The main contractor then became responsible for the ICT subcontractor and its work.

Constraints in local capacity entailed similar arrangements for other elements of the project. The complex steel roof structure (circular on plan and domed) was fabricated off-shore, and components were shipped to the island for assembly on site. Some specialised finishes were sourced internationally.

Three VM workshops were held on the project, with the senior advisor participating in the first two of these and providing advice to the third. The first VM workshop addressed the issue of the concept cost plan for the project being over budget, and dealt with this by tightening the project scope. The second workshop considered matters relating to individual cost elements, buildability, and operational maintenance. The third workshop mainly addressed the ICT installation in terms of components, materials, and their fitness for purpose. As uncertainty was prominent in all the VM agenda items, risks to the project objectives were exposed.

Project Risk Management

Risk management for this project is a mixture of formal and informal approaches. The Foreign Affairs Department of the donor government has an RMS that complies with the administrative requirements of government. For the aid-funded project, this process was informed by the reports of the senior advisor. As part of these requirements, a project completion review will provide limited risk knowledge capture, including an evaluation of the performance of the Project Management Consortium.

For the recipient island government, risk management tends to be more informal and relies on advice provided by the Tender Board, Steering Committee, and Project Management Consortium. There is no formal requirement in place for project risk knowledge capture.

The Project Management Consortium was required to submit risk management reports throughout the design and construction stages of the project, but does not operate a formal project risk management system (PRMS). It deals with uncertainties on an ad hoc basis and tends to treat project threat risks reactively as 'issues arising'. Other than a 'lessons learned' debriefing, there is no formal process for capturing risk knowledge on the project.

No information is available about the risk management practices of the main contractor and subcontractors.

Project uncertainties have related mainly to the time–cost–scope triangle, with the extent of work being more uncertain than its quality. Schedule delay constitutes the biggest uncertainty. The construction period was originally set at 12 months and then extended to 18 months at the contract award stage. Currently (at approximately 95% completion), the contract period is more realistically estimated to be 24 months, but no further extensions of time will be granted. The additional delay is attributed generally to a lack of robust construction management on the part of the main contractor.

Uncertainty in cost management arises largely from the absence of a reliable local cost database. The high level of innovation for the project, together with capacity limitations on local labour and material resources, contribute to this uncertainty.

Necessary re-scoping of project elements, in response to budget overrun, added further uncertainty. Unforeseen geotechnical issues also led to redesign and tender renegotiation.

For the senior advisor, risk knowledge capture from this project is important. It will be used to inform the financial, procurement, and ICT decision making for a similar aid-funded project (but with two donor governments working in association) proposed for another Pacific Rim island government.

Case Study D

High-Capacity Metropolitan Train Mock-up Project

A Chinese engineering company has been awarded a contract to manufacture new high-capacity metropolitan trains (HCMTs) for the metropolitan railway system in an Australian state.

Part of the contract included the fabrication of a full-size mock-up of the train super-structure (i.e. without the underframe and wheels) to be made available to the state client for review and used for inspection by the public; rail personnel; disability groups; fire, police, and ambulance authorities; and other interested parties.

The completed mock-up was transported to a city location and placed on display to the public for two weeks. As part of the review process, a survey was conducted by the state government to gather opinions about the features intended for the new HCMT. After the public display ended, the mock-up was taken to a state railway engineering and maintenance facility close to the city and stored there for possible further use in maintenance and training.

The mock-up project was treated as a subcontract, and this was awarded to an Australian company which is a division of a Chinese multinational transport engineering company. The parent company has other branches in the USA, Europe, and Asia.

The train mock-up project was valued at about AU \$1 million. Subcontract conditions imposed by the state government included requirements to have at least 35% of this value undertaken within the state.

Essentially this was an assembly and fitting-out project undertaken over three months, using components manufactured in China and shipped progressively to Australia over that period in shipping containers.

Apart from time and quality objectives, the subcontractor's main objective is to build relationships with the key stakeholders, in the interests of obtaining future work. Profit was therefore not a primary aim for the project.

The Australian subcontractor does not operate a formal risk management system, but does comply with prescribed state and federal requirements for matters such as safety and the environment. It also has ISO 9001 compliance for quality assurance.

In undertaking this project, the subcontractor knew from previous experience that the quality of the imported components constituted a large area of risk uncertainty, with the main problem being quality inconsistency. Because of the project's size, multi-national project workload, budget, and limited time frame, fine detail could have been overlooked by the main contractor who might also have been unwilling to question specifications to improve component quality for what it probably considered a relatively minor project.

Managing Project Risks, First Edition. Peter J Edwards, Paulo Vaz Serra and Michael Edwards.
© 2020 John Wiley & Sons Ltd. Published 2020 by John Wiley & Sons Ltd.

Also, while modifications could be proposed, reviewed, and approved (as change requests), the circularity of this process around the various stakeholders was quite protracted. Transport logistics also meant that, by the time manufacturing specifications were modified and approved, affected but unmodified components might already be at various points in the shipping chain (e.g. crated at the factory awaiting delivery to the Chinese port, at the docks for loading, on the high seas, being unloaded in Australia, on road transport to the subcontractor, or being unpacked at the company's premises). The subcontracting company had no clear strategy to mitigate this risk proactively, and reactive measures such as site modification often became necessary. The prospect of incurring liquidated damages for any project delay exacerbated this risk, and the subcontractor was very much aware of its relationship building objective.

Risk knowledge capture, in terms of lessons learned, is fragmented in the Australian subcontractor's organisation, and is not helped by a silo knowledge culture existing among the divisions of the parent company, in that communication does not always work well between them. Information and engineering knowledge are not fully shared. Despite the development of business relationships being a common objective for all divisions, senior management fear the potential loss of valuable intellectual property (IP) through becoming indirectly involved with the Chinese main contractor through the Australian partner.

Case Study E

Hot-Rod Car Project

Gary, an Australian mechanical engineer, has a passion for cars, including restoring and 'creating' them. His automotive passion stems from the mechanical abilities of his fore-bears, as both his father and his grandfather were skilled fitters and turners.

With a view to stimulating a similar interest in his son, and engaging in a joint project together, Gary proposed that they should build a 'hot-rod' car.

Overall, the project took over three years to bring to the point of roadworthy comple-tion, but in some respects it is still a 'work in progress'. The son's interest was very active at the beginning, but waned towards the end as his own career in the automotive indus-try began to develop and take priority.

The project was mostly undertaken in Queensland, Australia, but the car was subse-quently transported to another south-eastern state, Victoria, following Gary's own work relocation. Still not totally finished (at least in Gary's mind), it is planned that the car will be taken back to its home state at some time in the near future.

The hot-rod is based upon a 1942 Chevrolet 2-ton pickup truck purchased from a Queensland sugar mill. Figure E.1 shows the hot-rod in its current state, which has been deemed sufficient for exhibition in several custom-built and hot-rod car shows.

As with all home- or self-built vehicles in Australia, one large area of uncertainty was achieving registration of the vehicle for use on public roads.

Before commencing the project work, Gary had to obtain approval for the proposal under the 'hot-rod' classification system, one of two systems used in Australia to sanc-tion such automotive projects. The proposal submission requires information about what is intended to be built and how much modification will be carried out on the origi-nal vehicle. Initial approval is then followed by three on-site inspections carried out by licensed certifiers, with all inspection costs borne by the applicant.

The first inspection deals with the chassis frame; the second with the engine, gearbox, transmission, suspension, and running gear in place (as an assembly); and the third inspection covers the completed vehicle. The federal authority also issues a detailed guide about how such builds should be carried out, accompanied by minimum specifi-cations that will satisfy Australian Automotive Design Rules with only limited areas of exception available.

The chassis of Gary's car (modified and welded from the original) passed the first inspection. However, significant uncertainty emerged with the second inspection where the assembly was deemed non-compliant in two respects: the braking system and the road wheels.

Managing Project Risks, First Edition. Peter J Edwards, Paulo Vaz Serra and Michael Edwards.
© 2020 John Wiley & Sons Ltd. Published 2020 by John Wiley & Sons Ltd.

Figure E.1 The 'finished' hot-rod car. *Source:* Photograph used with kind permission of the owner.

The inspection authority wanted disc brakes fitted, at least to the front wheels. Gary believed that, despite the acknowledged performance superiority of disc brakes, the massive drum brakes from the original truck would be sufficient, especially given the substantially lighter mass of the hot-rod. His arguments were eventually accepted.

The road wheels proved to be a more difficult issue. For the second inspection, the original road wheels were simply placed alongside where they would be fitted. These wheels were originally designed as conventional two-piece split pressed steel units to accommodate *tubed* tyres (i.e. a thin rubber inflatable inner tube is placed on the metal rim, and the outer heavier rubber tyre casing stretched over it). This construction left the existing wheels with no mechanical means for preventing a *tubeless* tyre from rolling off should it become deflated on the road. On modern vehicles, which are now universally fitted with tubeless tyres, the wheel rims have a metal 'J'-section bead around the circumference on each side to prevent deflated tyres leaving the rim at speed on the road, and are thereby stamped as 'JJ' rims. The inspectors required JJ-rimmed wheels to be fitted on the finished car.

For all practical purposes, this meant new wheels were needed, but there was no obvious 'off the shelf' solution as the existing hubs would not accommodate the fixing stud configuration of most modern wheels. Figure E.1 shows the added complication of different front and rear wheel sizes, and the double rim configuration for the rear wheels.

The road wheel problem was eventually solved by a combination of luck, engineering skill, and business knowledge. Front wheels with JJ rims were sourced from an American version of a popular Japanese sports utility vehicle (SUV), and Gary was able to re-engineer the locating stud configuration to suit them. His long career involvement with engineering companies in China then enabled him to find one that was manufacturing double-rimmed JJ aluminium wheels for a European truck company which were almost identical to his vehicle's original steel wheels. The Chinese company was willing to supply him (at a reasonable price) with polished stamped aluminium wheels in the configuration that he needed.

Other build issues for the hot-rod were comparatively minor. Lowering the original cab floor entailed modifications to the front firewall. The radiator placement had to be changed from vertical to horizontal as the engine was dropped lower in the chassis, with some resulting uncertainty about its cooling effectiveness. The exhaust system required extensive modification.

All these issues, essentially a matter of re-engineering from the original, were resolved through knowledge and ability, patience and determination.

The original 70-year-old engine also had to be replaced. This six-cylinder in-line unit was a stalwart for General Motors for nearly three decades from the late 1930s to the early 1960s. While finding a suitable replacement engine was not easy, it was not impossible.

The hot-rod car successfully passed the third inspection, but the now fully registered roadworthy vehicle is not 'totally' finished. Nor is it perfect. The ratios of the original gearbox for the vehicle are not entirely suitable for its new purpose. The car's road manners are not always forgiving (especially in modern traffic conditions), and, as with many vehicles surviving in some form or other from former eras, it can be cranky and somewhat unreliable. Passenger comfort is not a primary consideration. Instrumentation is sparse.

For Gary, much of his satisfaction lies in having undertaken this project with his son. It perhaps also reassures him that he can take a well-deserved place in the family engineering tradition. Hence the commemorative decal on the passenger side door in Figure E.1.

Given the 'passion-driven' nature of this project, it was not unexpected to find that no formal risk management was undertaken. While uncertainties were anticipated, few were addressed proactively. Instead, reliance was placed upon engineering knowledge and capability, harnessed to a substantial measure of ingenuity. The outcome suggests that this trust was not misplaced.

Case Study F

Aquatic Theme Park Project

This case study takes the perspective of the civil engineering main contractor associated with a new aquatic theme park project being built on a development site located at the perimeter of a major city.

The interviewee is a civil engineer with more than 30 years of project management experience gained in several countries. He is one of two senior project managers in a regional branch of a national civil engineering company.

An organogram for the branch company is shown in Figure F.1.

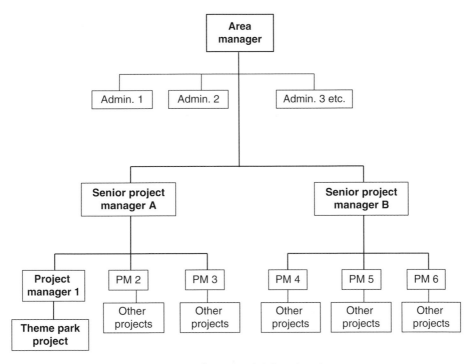

Figure F.1 Organogram for regional civil engineering contractor.

Managing Project Risks, First Edition. Peter J Edwards, Paulo Vaz Serra and Michael Edwards.
© 2020 John Wiley & Sons Ltd. Published 2020 by John Wiley & Sons Ltd.

The company's typical civil engineering project activities include landfill containment, minor roads and bridges, car parks, parks and landscaping, golf courses, land subdivisions, drainage, and pipeline reticulation. Contract values for projects are generally in the range of AU $1–10 million.

The regional company operates with a core complement of administrative staff and tradespeople and has basic plant and equipment resources. Subcontractors and specialised plant and equipment are engaged or hired on an as-needs basis.

The aquatic theme park project is a new venture for the civil engineering contractor, and it has taken a small equity share in the project, which has an estimated capital cost of just under AU $40 million. The company is thus involved in operational as well as construction risks for the project. Under the oversight of the senior project manager, a project manager, four site engineers, and three site supervisors are employed. The functional management matrix is shown in Table F.1.

Table F.1 Functional management for aquatic theme park project.

Senior project manager A	Project manager 1	Site engineer	Quality assurance
		Site engineer	Purchasing and contract administration
		Site engineer	Subcontractors
		Site engineer	Assistant/trainee
		Site supervisor	Structural and services works
		Site supervisor	Civil and building works
		Site supervisor	Pool area floor construction

Theme park projects are inevitably commercial ventures, with investment returns relying on the availability of sufficiently cheap land and reliable demand level and growth forecasts. The latter constitute the major operational risk and drive the business case through the ramping up of cash inflows to a stable and profitable level as rapidly as possible.

The major element of the aquatic theme park is a simulated surfing environment created by the hydraulically engineered generation of artificial 'waves'. Wave conditions are therefore a function of generation power and pool area configuration.

Although wave generation technology is not entirely new (it has probably been developed over more than 30 years), the project incorporates a unique wave generation system used only once before.

The wave pool itself covers nearly 2 hectares. It has a maximum depth of 4 m ramping up to 0 m at the far end.

For the civil engineering company, the construction work (including the structural components and ancillary buildings) is estimated at about one-third of the total capital cost. The contract agreement is based upon a negotiated construction management fee plus actual construction costs. The equity investment ensures the contractor's interest in constructing efficiently.

Structural design and other engineering services design are undertaken by professional consultants managed by the contractor, but the wave generator installation is carried out on a design and construct (D & C) basis by a specialist subcontractor.

This is a critical part of the project, since the configuration of the affected water retention area for the pool is determined by this technology.

The solution adopted for the wave pool floor is a proprietary system which combines the strength of concrete with the plasticity of polymer-based admixtures to provide sufficient flexibility. The site supervisor managing this construction element is not an expert in this form of construction, but has considerable experience with complex floor systems and is known for his ability to deal rapidly and effectively with on-site construction processes and difficulties.

The delivery schedule for the project has been tight. While the wave generation technology is unique, its installation does not present major difficulty. The remaining civil engineering and building work, other than the pool floor, is relatively straightforward. A 12-month construction period was originally planned. However, while certain elements of the design solution have taken time to resolve, it has not seriously impacted the progress of other work. Completion delay is estimated to be not longer than two months.

Management Systems

The civil engineering company is extensively systems-driven in terms of management and administration, with a genuine commitment to avoiding unnecessary 'paper trails' through computerisation. Portable wi-fi enabled information technology (IT) devices are used by relevant site personnel. While the risk management, quality assurance, environmental management, and safety management systems each comply with relevant Australian and international standards (adhering to an established strategic management policy), the systems tend to be configured on a separated 'silo' basis in the organisation, rather than through a fully integrated knowledge management approach. The systems are appropriately process-focused, but systems learning is not accorded a high priority, and staff induction training does not currently include any wider aspects and implications of knowledge management.

Project Risk Management

Project risk management in the organisation is based upon ISO 31000. Processes relate primarily to project stages.

Following an expression of interest (EOI) call, a preliminary workshop establishes the project context, particularly in relation to the company's standard business policy for taking on new projects. This is essentially a checklist that seeks to assess alignment of the potential project with the company's business case requirements. Some of the relevant checklist requirements are listed in Table F.2.

Context establishment also explores the company's current resource capability in terms of the project and its capacity to take on additional work. Compatibility of the project with current organisational systems is also considered, and any opportunities to negotiate a client agreement are always regarded favourably.

For tender preparation, a procedural 'tender book' is followed and a project risk register (PRR) is commenced. Risk identification largely uses a checklist approach developed from previous experience, but workshops are used to identify and explore new risks. While the PRR is capable of incorporating opportunity risks, the main focus is always upon threat risks.

Table F.2 Company–project alignment policy.

No.	Project–business case alignment parameters
a	Seek opportunities that will benefit from early involvement, innovative solutions, and alternative tender arrangements.
b	Avoid overcrowded tender markets, especially for simple projects where there are more than six tenderers.
c	Ensure that the proposed project satisfies the majority of strategic criteria including: • We have worked successfully with this client before. • The work is sufficiently within our areas of expertise. • The project location/region is familiar to us. • We have undertaken similar work in the past without experiencing any major difficulty. • The proposed timeline is not impossible or over-rigid. • Payment security is acceptable.
d	Avoid subcontracting to another organisation where the head contract for the project is rigidly tied to a fixed lump sum.
e	Know the contract conditions, and understand all their implications.
f	Conduct appropriate due diligence processes for new clients.

Risk analysis and assessment are generally qualitative, but quantitative analysis is used where it is practicable (e.g. for weather risks if reliable historical and seasonal data are available).

Avoidance or mitigation are preferred as risk response and treatment options, but no strategic 'as low as reasonably practical' (ALARP) principle is followed. For risks with very high probability and very high impact, the whole financial impact cost may be incorporated into the bid. The company is aware that this risk decision may affect its ability to compete successfully in the tender market. Potential impacts such as this drive the company's desire to negotiate wherever possible, or inform a decision not to tender.

Levels of risk monitoring and control are determined by the nature and perceived severity of each risk.

In terms of project risk knowledge capture, a 'lessons learned' debriefing is undertaken after project completion. The debriefing process is rarely formal and sometimes desultory, mainly because of staff preoccupation with the next project. Beyond the PRR, no wider organisational risk knowledge management system is maintained.

Index

Managing Project Risks, First Edition. Peter J Edwards, Paulo Vaz Serra and Michael Edwards.
© 2020 John Wiley & Sons Ltd. Published 2020 by John Wiley & Sons Ltd.